Regeneration

Ending the Climate Crisis in One Generation

Paul Hawken

PENGUIN BOOKS

PENGUIN BOOKS

An imprint of Penguin Random House LLC
penguinrandomhouse.com

Excerpt adapted from *The Overstory* by Richard Powers, copyright © 2018 by Richard Powers.
Used by permission of W. W. Norton & Company, Inc.

Excerpt from *Wilding: Return of Nature to a British Farm* by Isabella Tree, copyright © 2018 by Isabella Tree.
Used by permission of Picador Books.

Excerpt from *Psychic Numbing: Keeping Hope Alive in a World of Extinctions* by Carl Safina, copyright © 2020 by Carl Safina.

Excerpt from *Call of the Reed Warbler* by Charles Massy, copyright © 2017 by Charles Massy.
Used by permission of the University of Queensland Press.

Excerpt from *Letter to Nine Leaders by Nemonte Nenquimo*, copyright © Nemonte Nenquimo.
Published by arrangement with the author.

The Forest as a Farm by Lyla June Johnston, copyright © 2019 by Lyla June Johnston.
Published by arrangement with the author.

Excerpt from *Soul Fire Farm* by Leah Penniman, copyright © Leah Penniman.
Used with permission of Chelsea Green Publishing.

Acts of Restorative Kindness by Mary Reynolds, copyright © 2020 by Mary Reynolds.
Published by arrangement with the author.

Who's Really Trampling Out the Vintage? by Mimi Casteel, copyright © 2020 by Mimi Casteel.
Published by arrangement with the author.

Philanthropy Must Declare a Climate Emergency by Ellen Dorsey, copyright © 2020 by Ellen Dorsey.
Published by arrangement with the author.

Excerpt from *We Are the Weather: Saving the Planet Begins at Breakfast* by Jonathan Safran Foer, copyright © 2019 by Jonathan Safran Foer.
Used by permission of Farrar, Strauss and Giroux.

Credits for images appear on page 256.

LIBRARY OF CONGRESS CATALOGING-IN-PUBLICATION DATA
Names: Hawken, Paul.
Title: Regeneration : ending the climate crisis in one generation / Paul Hawken, author.
Description: First. | New York : Penguin Books, 2021.
Identifiers: LCCN 2021009872 (print) | LCCN 2021009873 (ebook) | ISBN 9780143136972 (paperback) | ISBN 9780525508496 (ebook)
Subjects: LCSH: Global warming. | Climatic changes.
Classification: LCC QC981.8.G56 R427 2021 (print) | LCC QC981.8.G56 (ebook) | DDC 363.738/74—dc23
LC record available at https://lccn.loc.gov/2021009872
LC ebook record available at https://lccn.loc.gov/2021009873

Printed in the United States of America
1st Printing

Thakgil Canyon near Vik in the south of Iceland is surrounded by glaciers, rivers, ice caves, black sand beaches, and some of the country's best hiking trails.

Contents

We accumulated thousands of references, citations, and sources in the process of researching and writing Regeneration. Although they are too numerous to be published in the book, they may be found at www.regeneration.org/references.

Foreword
Jane Goodall

When I was studying chimpanzees, I learned about the interconnectedness of all life in the rainforest. How every species of plant and animal has a role to play in the tapestry of life. When a species becomes extinct, a hole is made in that tapestry. When it becomes too torn and tattered, an entire ecosystem may collapse. The thing is, we too are part of the natural world. We depend on it for oxygen, food, water, clothing—everything. And it too has become torn and tattered.

What separates us most from our closest living relatives, the chimpanzees—and from all other animals—is the explosive development of our intellect. Animals are far more intelligent than was once thought, but no animal could come up with a theory of relativity or land on the moon. How bizarre that we, the most intellectual of all species, should be destroying our only home. There seems to have been a disconnect between our clever brains and the love and compassion that, poetically, we seat in the human heart. Only when head and heart work in harmony, I think, can we attain our true human potential.

We have many self-inflicted problems to work out and, as Paul Hawken emphasizes so convincingly here, they are all interconnected, and we need to understand and solve them in an integrated way. We must alleviate poverty, address unsustainable lifestyles in higher-income countries, achieve social justice, provide universal healthcare, and make available education for all. Fortunately, people are finding innovative solutions to these problems, and it is these that Paul discusses in this book.

I have had firsthand experience with much that has gone wrong. When I began working with the chimpanzees in 1960, the Gombe National Park was part of a forest that stretched across equatorial Africa. By the mid-1980s, it was a small island of forest surrounded by bare hills. People were living unsustainably, farmland was overused and spent, and trees were cut down to make room for farming and charcoal production. People were struggling to survive. That's when I realized that if we can't help these communities find ways of making a living without destroying their environment, we could not protect the chimps. You cannot save an animal species unless you protect its environment, which is not possible without the participation of local communities, and that will not happen if they are living in poverty.

The Jane Goodall Institute began TACARE, Lake Tanganyika Catchment Reforestation and Education, a holistic method of community-based conservation. We chose local Tanzanians to go to the villages around Gombe and ask the people how we could help. Their response and needs were clear. They wanted to grow more food and have better health and education. With a small grant from the European Union, we helped them restore fertility to the land without the use of chemicals. We worked with the local Tanzanian

government to improve existing schools and improve or create village clinics. We introduced water-management programs, agroforestry, and permaculture. Utilizing geographic information systems and satellite imagery, we enabled villagers to make land use management plans. Volunteers learned to use smartphones to record the health of their village forest reserves. Scholarships enabled girls to go into secondary education, and microcredit programs enabled villagers—especially women—to start their own sustainable businesses. Family planning information is eagerly received as parents want to educate their children, but schools are expensive. In this way, TACARE advances both environmental and social well-being.

There are close to 7.9 billion of us on the planet now. In many parts of the world our finite natural resources are diminishing faster than nature can replenish them. By 2050, it is estimated there may be as many as 10 billion of us. Livestock numbers are also increasing, using up ever more land and water and producing vast amounts of methane gas. Moreover, as people rise out of poverty, they understandably seek to emulate the unsustainable standard of living that we know we must change. If we carry on with business as usual, the future looks, well . . . grim is hardly a strong enough word.

We must develop a new relationship with nature and ensure that our children are equipped to deal with the problems we have created. There are many programs that promote environmental education and others that discuss social justice. I began Roots & Shoots, an environmental and humanitarian movement for youth, in 1991. It now has thousands of groups, preschool through university, in sixty-eight countries. To help them understand that social and environmental problems are connected, each group is required to choose and participate in projects in three areas—people, animals, and the environment. By taking action, the members realize their ability to effect change. It is empowering and is giving hope to thousands of young people who are working together on issues as diverse as wildlife trafficking, homelessness, women's rights, animal rights, and discrimination.

I have three reasons for hope: the energy and commitment of youth; the resilience of nature (the forest has returned around Gombe) and the way animal and plant species can be rescued from extinction; and the human intellect, which is focusing on how we can live in greater harmony with nature.

Paul, in his usual inimitable way, describes the most important solutions to the environmental and social problems we have brought upon ourselves, and shows how they are inseparably linked. *Regeneration* is honest and informative, a rebuttal to doomsayers who believe it is too late. He echoes my sincere belief that we have a window of time, that there are practical solutions, and that we and all our institutions can initiate and implement them in order to restore life climatic stability and on Earth. Let us work to live up to our scientific name: *Homo sapiens*, the wise ape. ●

Orphans Kudia and Ultimo hug each other at the JGI Tchimpounga Chimpanzee Rehabilitation Center in the Republic of Congo.
One hundred sixty chimpanzees are cared for, for life, at the sanctuary.

Aerial view of a small forest lake in Karula National Park,
Valgamaa County, Southern Estonia.

Regeneration

Regeneration means putting life at the center of every action and decision. It applies to all of creation—grasslands, farms, people, forests, fish, wetlands, coastlands, and oceans—and it applies equally to families, communities, cities, schools, religions, cultures, commerce, and governments. Nature and humanity are composed of exquisitely complex networks of relationships, without which forests, lands, oceans, peoples, countries, and cultures perish.

Our planet and youth are telling us the same story. Vital connections have been severed between human beings and nature; within nature itself; and between people, religions, governments, and commerce. This disconnection is the origin of the climate crisis, it is the very root—and it is where we discover solutions and actions that can engage all people, regardless of income, race, gender, or belief. We live on a dying planet—a phrase that may have sounded inflated or over the top not long ago. The earth's biological decline is how it adapts to what we are doing. Nature never makes a mistake. We do. The Earth will come back to life no matter what. Nations, peoples, and cultures may not. If putting the future of life at the heart of everything we do is not central to our purpose and destiny, why are we here?

The proximate causes of the climate crisis are cars, buildings, wars, deforestation, poverty, oil, corruption, coal, industrial agriculture, overconsumption, and fracking, among others. All have the same origin and impact: the economic structures created to support human well-being degenerate life on earth, creating loss, suffering, and a heating planet. The financial system is abetting and investing in planetary liquidation—a short-term source of monetary wealth and a near-term cause of biological depletion, poverty, and inequality.

For the past forty years, the most powerful way to reverse global warming was largely overlooked. Fossil-fuel combustion is the primary cause of warming and must cease rapidly; without this, there is no cure. However, in order to stabilize the climate, we need to draw down carbon dioxide and bring it back home. The only effective and timely way to reverse the climate crisis is the regeneration of life in all its manifestations, human and biological. It is also the most compelling, prosperous, and inclusive way. Biological degeneration has brought us to the brink of an unimaginable crisis. To reverse global warming, we need to reverse global degeneration.

Our economic systems, investments, and policies can bring about the degeneration of the world or its regeneration. We are either stealing the future or healing the future. One description of the current economic system is *extractive*. We take, we dam, we enslave, we exploit, we frack, we drill, we poison, we burn, we cut, we kill. The economy exploits people and the environment. The ongoing cause of degeneration is inattention, apathy, greed, and ignorance. Climate change may leave people feeling as if they have to make a choice between "saving the planet" and their own happiness, well-being, and prosperity. Not at all. Regeneration is not only about bringing the world back to life; it is about bringing each of us back to life. It has meaning and scope; it expresses faith and kindness; it involves imagination and creativity. It is inclusive, engaging, and generous. And everyone can do it. It restores forests, lands, farms, and oceans. It transforms cities, builds green affordable housing, reverses soil erosion, rejuvenates degraded lands, and powers rural communities. Planetary regeneration creates livelihoods—occupations that bring life to people and people to life, work that links us to one another's well-being. It offers path out of poverty that provides people with meaning, worthy involvement with their community, a living wage, and a future of dignity and respect.

In December 2020, Dr. Joeri Rogelj of the Grantham Institute in London and a lead author of the Sixth Assessment of the Intergovernmental Panel on Climate Change made a remarkable statement: "It is our best understanding that, if we bring carbon dioxide [emissions] down to net zero, the warming will level off. The climate will stabilize within a decade or two. There will be very little to no additional warming. Our best estimate is zero." This was a remarkable change in scientific consensus. For decades it was assumed that if we were able to stop our carbon emissions, the momentum of warming would continue for centuries. That was mistaken. Climate science now indicates that global warming would begin to recede after we achieve zero carbon emissions.

This is a watershed moment in history. The heating planet is our commons. It holds us all. To address and reverse the climate crisis requires connection and reciprocity. It calls for moving out of our comfort zones to find a depth of courage we may never have known. It doesn't mean being right in a way that makes others wrong; it means listening intently and respectfully, stitching together the broken strands that separate us from life and one another. It means neither hope nor despair; it is action that is courageous and fearless. We have created an astonishing moment of truth. The climate crisis is not a science problem. It is a human problem. The ultimate power to change the world does not reside in technologies. It relies on reverence, respect, and compassion—for ourselves, for all people, and for all life. This is regeneration.

Agency

The climate crisis is not the warming of the planet. What unnerves scientists is what warming will do to life *on* the planet. Changes in atmospheric temperature, ocean currents, and melting polar ice could trigger runaway disruption on multiple fronts in rapidly succeeding tipping points. Losses could include more frequent droughts in the tropics that would convert the world's rainforests into fire-prone savannas. Changes in ocean circulation would dramatically alter worldwide weather and agriculture. The rapid increase in fires and pests could lead to the collapse of our northern forests. Ocean heating and acidification could cause the death of every coral reef in the world. The accelerating melting of the Thwaites Glacier in Antarctica would cause a three-foot rise in sea levels. The melting of Arctic permafrost would release massive amounts of ancient stored carbon dioxide and methane. It is difficult, if not incomprehensible, to imagine how these events would affect families, cities, economies, companies, food, politics, and children in more temperate climates. However, it is not difficult for more than two dozen Arctic cultures who are experiencing the impact of the melting Arctic directly and quickly: the Inuit, Yupik, Chukchi, Aleuts, Saami, Nenets, Athabaskan, Gwich'in, and Kalaallit, cultures that have occupied their lands for up to ten thousand years.

Accurate as they may be, climate predictions can obscure another set of tipping points, the numerous small changes and crucially important outcomes and that lead to people's involvement and participation, rather than passivity and fear. These are actions that slow, forestall, and transform the climate crisis. Ending the climate crisis means creating a society that is going in the right direction at the right speed by 2030, a rate of change that will lead to zero net emissions before 2050. That means halving emissions by 2030 and then halving again by 2040. Tens of thousands of organizations, teachers, companies, architects, farmers, Indigenous cultures, and native leaders know what to do and are active in implementation. The current growth of the climate movement is magnificent, but it remains a small fraction of the world. Hundreds of millions of people need to realize that they have agency, that they can take action, and that collectively it is possible to prevent runaway global warming.

The agent who can head off the climate crisis is reading this sentence. Logically, this seems like nonsense—surely individuals are powerless to counter the global drivers and momentum of global warming. That's a fair conclusion if we assume that yesterday's institutions should or will do it for us. There is a debate as to whether individual behavior or government policy is the key to solving the climate crisis. There shouldn't be. We need the involvement of every sector of society, top to bottom, and everything between.

It is engaging and fascinating to calculate one's own carbon footprint, but *Regeneration* takes a different and wider tack, because there is no such thing as a single individual. *Thinking* you are an individual is self-identity. *Being* an individual is an ongoing, functional, and intimate connection to the human and living world. When we look at our networks, each of us is multitudes. We have different skills and potential, including sharing, electing, demonstrating, teaching, conserving, and diverse means of helping leaders, cities, companies, neighbors, co-workers and governments become aware and able to act.

Worried that you are not an expert? Almost no one is. But we understand enough. We know how greenhouse gases function and warm the planet; we are seeing greater climate volatility and extreme weather; and we know the primary sources of carbon emissions. We want a stable climate, food security, pure water, clean air, and an enduring future that we can become ancestors to. Cultures, families, communities, lands, professions, and skills vary with every person. The situations we find ourselves in differ. Who better to know what to do at this time, in this place, with your knowledge, than you?

Nevertheless, solving the climate crisis is an unnatural act, one that human beings are ill-equipped to do. Our minds just don't work that way. The idea of a future existential threat is abstract and conceptual. War metaphors about fighting, battling, and combating climate change don't connect either. Who wakes up in the morning excited about mitigating or getting to "net zero" in thirty years? Most people ignore climate headlines, and for good reason. The overwhelming majority focus on current dilemmas, not distant ones, obstacles that impact one's life now, not in 2050. On the other hand, humans are notably brilliant at joining together to solve problems. Give us immediate threats like an impending cyclone, flood, or hurricane, and we are all over it. If we are going to engage the bulk of humanity to end the climate crisis, the way to do it is counterintuitive: to reverse global warming, we need to address current human needs, not an imagined dystopian future.

If we want to get the attention of humanity, humanity needs to feel it is getting attention. If we are going to save the world from the threat of global warming, we need to create a world worth saving. If we are not serving our children, the poor, and the excluded, we are not addressing the climate crisis. If fundamental human rights and material needs are not met, efforts to stem the crisis will fail. If there are not timely and cumulative benefits for an individual or family, they will focus elsewhere. The needs of people and living systems are often presented as conflicting priorities—biodiversity versus poverty, or forests versus hunger—when in fact the destinies of human society and the natural world are inseparably intertwined, if not identical. Social justice is not a sideshow to the emergency. Injustice is the cause. Giving every young child an education; providing renewable energy to all; erasing food waste and hunger; ensuring gender equity, economic justice, and shared opportunity; recognizing our responsibility and making amends to myriad communities of the world for past injustices—these and more are at the very heart of what can turn the tide for all of humanity, rich and poor, and everyone between. Reversing the climate crisis is an outcome. Regenerating human health, security and well-being, the living world, and justice is the purpose.

This requires a worldwide, collective, committed effort. Collectives do not emerge from the tops of institutions. They begin with one person and then another, the invisible social space where commitment and action join and come together to become a dyad, a group, a team, a movement. To put it simply: no one is coming to help. There is not a brain trust that is going to work out the problems while we ponder and wait. The most complex, radical climate technologies on earth are the human heart, head, and mind, not a solar panel. Just as we stand at the abyss of a climatic emergency, we stand at another remarkable threshold. The rate of understanding and awakening about climate change is increasing exponentially, even skyrocketing. Climate change is becoming experiential rather than conceptual. As weather becomes ever more disruptive, and awareness and concern increase, the movement to reverse the climate crisis will likely become the largest movement in the history of humankind. It took decades to create this moment.

It is natural to worry that it matters little if you are taking action if others are not. From the planet's point of view, there is no difference between a climate denier and someone who understands the problem but does nothing. The number one cause of human change is when people around us change. Research by Stanford neuroscientist Andrew Huberman upends the idea that beliefs determine what we do or what we can do. It is the opposite. Beliefs do not change our actions. Actions change our beliefs. Do you believe there is nothing you can do to make a difference? Logical. Do you fear the future? Understandable. Do you feel stressed about climate change? Sensible. However, stress is your brain telling you to act. Stress is a signal; it is urging you to do something. Not only do actions change your beliefs, your actions change other people's beliefs.

When honeybee scouts find a bounty of blooms and nectar, they return to the hive, where they do a symbolic waggle dance at the entrance of the hive. The dance signals the precise direction and distance to the flowering plants or trees. The more vigorous the waggle, the richer the source of nectar. Once worker bees have seen the dance, they have the necessary information and fly straight to the source. It is time for humanity to create waggle dances unique to their knowledge, place, and determination. Another way to look at this time in history is this: we are being homeschooled by the planet, our teacher. This book is an attempt to reflect those teachings.

—Paul Hawken

An advocate for combining traditional knowledge with science and technology, Hindou Oumarou Ibrahim is a leader in the movement to elevate the role of Indigenous women in shaping policy and practices that affect the future of the planet.

How to Use This Book

The purpose of *Regeneration* is to end the climate crisis in one generation. Ending the crisis does not complete the challenge of global warming. That is a century-long commitment. Ending the crisis means that by 2030, collective action by humanity will have reduced total greenhouse gas emissions by 45 to 50 percent. At this writing, we are going backward and increasing emissions.

The book and its companion website plot a pathway to achieve goals outlined by the special report, "Global Warming of 1.5°C," published by the Intergovernmental Panel on Climate Change (IPCC) in October 2018. The report calls for 45 to 50 percent reductions from 2010 levels in global greenhouse gas emissions in each of the next two decades in order to avoid exceeding a 1.5°C (2.7°F) rise in global temperatures. The most common question about the crisis is "What should I do?" How can a person or entity create the greatest impact on the climate emergency in the shortest time? Most people do not know what to do, or may believe the things they can do are insufficient. We think otherwise.

Our approach to reversing climate change differs from other proposals. It is based on the idea of regeneration. We do not oppose other strategies and plans. To the contrary, we praise and are grateful for all approaches. Our concern is simple: most people in the world remain disengaged, and we need a way forward that engages the majority of humanity. Regeneration is an inclusive and effective strategy compared to combating, fighting, or mitigating climate change. Regeneration creates, builds, and heals. Regeneration is what life has always done; we are life, and that is our focus. It includes how we live and what we do—everywhere.

We conclude the book with Action + Connection. There, we show that the solutions detailed here, scaling and growing, meet the goals established by climate scientists and the IPPC. All of the solutions described are doable and realistic. They require one thing: broad participation. You are welcome to read the end of the book first if that is helpful.

Frameworks. The following are six basic frameworks for action to solve the climate crisis. They overlap in many ways; however, each category holds multiple levels of discovery, innovation, and breakthrough. The Action + Connection section links to the imaginative and effective ways people, communities, companies, neighbors, counties, schools, corporations, and countries can make a difference. What is holding us back today is not lack of solutions. It is the lack of imagination of what is possible. If you are feeling pessimistic or defeatist, read some or all of the book, and then turn to the end. It may change your mind.

Equity. This comes first because it encompasses everything. All that needs to be done must be infused by equity. Fairness is about social systems—how we treat one another, how we treat ourselves, and how we treat the living world. The planet has been transformed in a blink of an eye. If we are to transform the climate crisis, we need to transform ourselves, and we had best not blink. Time is of the essence. Social systems require the same level of care, attention, and kindness as ecosystems. They are incomparable yet inseparable. The state of the environment accurately reflects the violence, injustice, disrespect, and harm we do to people of different cultures, beliefs, and skin color. As Jane Goodall points out in her foreword, you save forests and species by helping to create better lives for people.

Reduce. The primary method of reversing global greenhouse gas emissions is simple: stop putting them into the atmosphere. It is also the most difficult, while being the greatest economic opportunity. The amount of carbon-emitting fossil fuel consumed is astonishing. Every day, the world burns 100 million barrels of oil, 47 billion pounds of coal, and 10 billion cubic meters of natural gas, which together emit 34 billion tons of carbon dioxide every year. Replacing the coal, gas, and oil we currently depend on is a formidable undertaking. *Reduce* includes the carbon and methane emissions from agriculture, food systems, deforestation, desertification, and destruction of ecosystems. The implementation of renewable energy from wind, solar, energy storage, and microgrids are critical, and well on their way. Less discussed but equally important is the reduction of energy and material use. *Reduce* solutions include electric vehicles, micro-mobility, carbon-positive buildings, walkable cities, carbon architecture, electrified buildings, minimized food waste, and the next category: *Protect.*

Protect. This is synonymous with preserving, securing, and honoring. You will find essays about pollinators, wildlife corridors, beavers, habitats, bioregions, seagrasses, wildlife migration, and grazing ecology, subjects not normally associated with solving the climate crisis. How could these be some of the most important solutions to the climate crisis? Because they are essential and critical to the living systems we need to defend and strengthen. Terrestrial systems hold 3.3 trillion tons of carbon in and above ground. That is about four times more carbon than is in

phytoplankton take in carbon dioxide and convert it to oxygen and carbohydrates. Roughly 25 percent of our carbon emissions are absorbed by oceans and transformed into fish, kelp, whales, shells, seals, and bones, but most of it is converted to carbonic acid, which is slowly killing sea life and is leading to a dead ocean. The primary way human beings can sequester is through regenerative agriculture, managed grazing, proforestation, afforestation, degraded land restoration, replanting mangroves, bringing back wetlands, and protecting existing ecosystems. The oft-used term *net-zero emissions* is not the goal. It is the threshold where the world begins to reduce atmospheric carbon levels back to preindustrial levels.

Influence. This encompasses laws, regulations, subsidies, policies, and building codes. For example, it is one thing to cease using plastic bags. It is better to get single-use plastic banned. As each of us endeavors to examine and modify our own impact, we gain insight into the causes and sources of degenerative processes, products, and services. You can't fix pollution, degradation, or plastic downstream. The cause is upstream, and that is where influence needs to be directed. It can start with the purchasing policies and habits of one's school, city, or business. Your influence can be exerted in the form of letters, emails, or messages to corporations and trade associations. It can mean speaking with or writing to city councilmembers, provincial or state legislators, governors, presidents, and member of Congress or Parliament. It can take the form of boycotts and protests. Each of us has but one voice. When one voice becomes "we," change happens.

Support. In virtually every area of climate, social justice, and the environment, there are organizations that are highly competent at what they do, that are ahead of the curve and embody knowledge and networks that make them the most effective change agents. Links within the Action + Connection section offer lists of the organizations around the world that are true regenerators, leaders often working with very limited resources, people doing the extraordinary activities that governments and big business do not. The lists are specific to place, ecosystems, species, social justice, food, pollution, water, and more. You can find them quickly and easily to match the geographies and areas in which you want to help make change. ●

the atmosphere. The carbon is contained in forestlands, peatlands, wetlands, grasslands, mangroves, tidal salt marshes, farmland, and rangeland, and we need it to stay here on earth. Each year, some portion of each of these ecosystems is degraded, developed, converted, or lost. It is a relatively small percentage, but it adds up. When living systems break down or are destroyed, the plants and organisms below and above ground die, resulting in carbon dioxide emissions. If we lose 10 percent of the earth's terrestrial systems, those emissions could increase carbon dioxide in the atmosphere by as much as 100 parts per million. *Protect* maintains the healthy function of living systems, resulting in the sequestration and storage of more carbon rather than less. When we lose an ecosystem, the birds, reptiles, rodents, mammals, insects, and creatures that dwell there lose their homes, the primary cause of the extinction crisis. Conversely, if we lose the species that occupy our forests, wetlands, or grasslands, those systems fall apart. Hummingbirds, hawk moths, and sharks may seem irrelevant to climate change, but it is the other way around. Biodiversity, humankind, the land, cultures, the oceans, and climate are inseparable.

Sequester. There is a natural carbon cycle that has been functioning for hundreds of millions of years. Carbon moves in and out of the atmosphere. Forests, plants, and

A group of juvenile spotted owlets (*Athene brama*) in Tamil Nadu, India.

Reference Guide

Is there a difference between climate change and global warming?

Global warming refers directly to the accumulation of heat in the earth's atmosphere, land, and oceans resulting from increased greenhouse gases in the atmosphere. Climate change describes the broader set of changes, including shifting rainfall patterns, drought, glacial melting, and flooding caused, in part, by the increased levels of water vapor that can be held in a warmer atmosphere.

How much has the earth already warmed?

The 2020 average surface temperature was 0.98°C (1.76°F) above the preindustrial average temperature. Since the 1980s, the average temperature has been increasing by 0.18°C (0.32°F) every decade.

Are forecasts of global warming accurate?

The current increase in global warming matches scientific temperature forecasts made thirty years ago. However, science did not foresee the full impacts of warming. The rates of melting of polar ice, sea-level rise, and drought intensity are greater than had been predicted.

When did we discover the mechanism of global warming?

In 1824, Joseph Fourier, a French physicist and mathematician, showed how atmospheric gases trap heat and regulate the atmosphere. In 1856, American physicist Eunice Newton Foote determined that carbon dioxide had the greatest warming potential of atmospheric gases. Irish physicist John Tyndall's studies in 1859 are credited with establishing the greenhouse effect. In 1896, Swedish scientist Svante Arrhenius showed that increases in carbon dioxide were coming primarily from industry and that a 50 percent increase would raise global temperatures 5° to 6°C. Were it not for greenhouse gases, the earth would be a frozen, icy rock, and life as we know it would not exist. As carbon dioxide levels increase far beyond what human civilizations have ever experienced, they are in effect double-glazing the planet—more heat is trapped and less escapes into space.

How much carbon dioxide and other greenhouse gases are in the atmosphere?

The amount of carbon dioxide in the atmosphere is 419 parts per million, a 50 percent increase since the beginning of the Industrial Age. However, there are other greenhouse gases, including methane, nitrous oxide, and refrigerant gases—methane foremost due to its ubiquity and impact. They are measured according to their global-warming potential as compared with carbon dioxide. In this book, we describe these other greenhouse gases by their global-warming potential compared with carbon dioxide over a hundred years, a unit we call "carbon dioxide equivalent." If we include these gases, the equivalent level of carbon dioxide in the atmosphere is 500 ppm, the highest parts per million this planet has seen in more than 20 million years.

What is the difference between carbon and carbon dioxide?

Carbon is an element. It becomes a gas—carbon dioxide—when combined with two molecules of oxygen. In the atmosphere, carbon levels are measured as carbon dioxide. In soil and plants, it is measured as carbon only. One ton of carbon converts to 3.67 tons of carbon dioxide.

How much carbon exists on earth?

There are approximately 121 million gigatons of carbon on or near the surface. About two-thirds, 78 million gigatons, is in the form of limestone, sediments, and fossil fuels. Of the remaining carbon, 41 million gigatons is in the deep and near ocean, 3,300 gigatons is held on the land, and only 885 gigatons is in the atmosphere in gaseous form as carbon dioxide.

How big is a gigaton of carbon dioxide?

A gigaton is one billion metric tons. A one-gigaton ice cube would be about one kilometer high, long, and wide. The world burns 17 trillion pounds of coal per year, and each pound emits an average of 1.87 pounds of carbon dioxide, creating 14.5 gigatons of carbon dioxide.

What can be done to reverse warming?

There are three things we can do about planetary heating. We need to reduce and cease net carbon dioxide emissions over time. We need to protect and restore the enormous stores of carbon contained in our forests, wetlands, grasslands, salt marshes, oceans, and soils. And we need to bring carbon from the atmosphere back to earth by sequestering carbon dioxide.

What is sequestration?

Sequestration removes carbon dioxide from the atmosphere through photosynthesis, some of which is stored in soil, plants, or trees. When carbon dioxide is captured by a plant, it releases oxygen into the air and combines carbon and water into sugars that feed the plant, the roots, and underlying soil organisms. Virtually all ecosystems, including grasslands, ocean algae, mangroves, forests, and peatlands, are actively sequestering carbon. There are artificial methods being developed to sequester carbon, such as direct air capture, but it is still too early to determine whether these techniques will prove to be practical and affordable at scale.

What is the Paris Agreement?

Actions to prevent runaway global warming have been discussed at the UN-sponsored Conference of the Parties (COP), held annually in capital cities around the world. In 2016, a year after COP 21 in Paris, an accord was reached that committed 191 parties to reduce emissions in order to keep global warming under 1.5°C; this is called the Paris Agreement.

Where does the world stand right now with respect to the Paris Agreement?

Of the 191 signatory states, only eight of the country pledges were in alignment with the original 2°C goal and only two countries have targets consistent with the 1.5°C limit—Morocco and Gambia. No G7 country—the United States, Canada, France, Germany, Italy, Japan, or the United Kingdom—has come close to setting targets in alignment with the Paris Agreement.

What units are we using?

Every number is reported in the imperial system of units unless explicitly defined. The notable exception is that all tons are metric tons for consistency with common measurements (e.g., gigatons of carbon dioxide are always reported in metric tons).

Oceans

Oceans absorb the greatest impact of human activity on the planet and receive the least coverage. Ten percent of the population directly depend on fisheries and three billion more rely on oceans for at least 20 percent of their protein. Yet most people are not aware of how rapidly oceans are changing as a consequence of global warming and rampant pollution.

Oceans are beginning to fail under the effects of heating, acidification, predatory overfishing, and unchecked pollution, both chemical and plastic. Oceans are the largest carbon sink on earth. They contain twelve times more carbon than land and forty-five times more than the atmosphere. The oceans have absorbed 93 percent of increased atmospheric heating and 25 percent of carbon dioxide emissions, resulting in warming ocean temperatures and ocean acidification, the decrease in pH that occurs when carbon dioxide from the atmosphere dissolves in seawater. Acidification removes carbonate ions from the water, which many organisms rely on to build their shells, including some species of phytoplankton that are crucial to the ocean's ability to sequester carbon. The capacity of the oceans to continue as carbon and heat sinks is reaching its limit. Heat waves are rising in bodies of water around the world, leaving massive areas warmer than their historical temperatures. In 2020, a patch the size of Canada stretching west from the California coast was up to 7 degrees Fahrenheit warmer than usual. Ocean heating can significantly reduce forage fish populations and cause mass strandings of seabirds and marine mammals. It can also cause mass coral bleaching and shifts in phytoplankton distribution.

Oceans have become the largest "storehouse" of pollution. Eighty percent of marine plastic pollution comes from land-based sources. The rest arises from maritime use, including shipping, fishing, drilling, and direct dumping. Coastal waters contain thousands of different types of pollution, including industrial chemicals, petroleum and fracking waste, agricultural runoff, pesticides, pharmaceuticals, raw and treated sewage, heavy metals, and street waste from urban runoff. Up to 12 million tons of plastic waste enters the oceans annually, and the short- and long-term impacts are only beginning to be fully understood.

Oceans and the atmosphere are inseparable. In polar regions where the water is colder and can dissolve more carbon dioxide, it is also saltier and denser. Through a process called deepwater formation, the dense carbon dioxide–rich water can sink to the deepest parts of the ocean where currents then move that water, and with it the carbon dioxide, to all the corners of the oceans. The currents are brimming with interconnected life that consumes, cycles, and eventually sequesters the carbon dioxide. The first to consume the gas are the phytoplankton at the surface. Through the process of photosynthesis, these microscopic plants use sunlight to combine water and carbon dioxide, producing the foundation of the complex ocean food web. Phytoplankton are a food source for everything from microscopic zooplankton to shrimp and fish. Smaller animals are eaten by bigger ones, and the carbon is cycled throughout life in the oceans.

While all marine species contain carbon, phytoplankton are the standout. Globally, they hold 0.5–2.4 billion tons of carbon, almost equivalent to the carbon dioxide sequestered by all the trees, grasses, and other land-based plants combined. While most phytoplankton are consumed, a small yet important fraction dies, sinks, and becomes part of the long-term carbon sequestered in the sediments on the ocean floor. Phytoplankton and a few microscopic animals, while small in size, account for almost all the transfer of carbon to the deep ocean and hence for the long-term removal of carbon from the oceans and atmosphere.

Marine animals play an essential role in cycling carbon through the oceans by building up carbon in their bodies and releasing it when they breathe, defecate, and die. Some species, like whales, accumulate large amounts of carbon in their bodies that eventually sink to the ocean depths upon their deaths. Additionally, when these immense animals defecate, the nutrients and carbon feed phytoplankton and other small animals at the base of the food chain, encouraging further carbon removal from the water and atmosphere and expanding the cycle of life in the oceans.

Most of the attention being paid to oceans is about protection—how to prevent the degradation, pollution, and acidification that are increasing yearly. In this section, we explore the means of protecting and regenerating oceans that also meet human needs. Because oceans cover 70 percent of the planet, the possibilities are both extensive and global. An important step is to cease using the ocean as a dump.

The second step is the creation of marine protected areas—expanses of ocean that exclude fishing, mining, drilling, and other forms of exploitation. When key areas of the ocean are set aside, fisheries rebound, not only within the protected areas but in the waters surrounding and beyond them. By doing less, we ultimately enable the presence of more fish, kelp, phytoplankton, and shellfish, because the innate regenerative capacity of oceans is allowed to operate unhindered. There is also a burgeoning movement to return to the sea as cultivator, farmer, and steward, interacting with ocean systems regeneratively, methods that not only sequester carbon but that feed billions of people while restoring coastal waters. ●

Marine Protected Areas

For many thousands of years, Indigenous peoples lived with and relied on a thriving ocean. Cultures across the Pacific Islands depended on healthy reef fish populations, the Chumash people in the Channel Islands ate abundant abalone, and the Aleut people lived off marine mammals in the Bering Sea. When Spanish colonists first reached the Caribbean, so many swimming sea turtles hit the wooden hulls of their ships that the captains recorded the thunking as a danger in their logs. "It seemed the ships would run aground on them and were as if bathing in them," wrote Andrés Bernáldez of Columbus's second voyage in 1494. When the Venetian explorer John Cabot fished off what became the Grand Banks of Canada in 1497, sailors dropped wicker baskets weighted with stones overboard; when they pulled them up, they were wriggling with cod. Massive reefs of oysters protected New York when the Dutch first arrived. Until the early twentieth century, oysters filtered and cleaned all of the

Chesapeake Bay in a week. Now, with the oysters mostly gone and the waters polluted by fertilizer runoff and pig waste, the Chesapeake Bay is toxic. Salt marshes held back hurricanes south of New Orleans. Mangroves broke the rush of tsunamis throughout South Asia, and castles of alabaster coral created robust ecosystems of shallow cities from Lizard Island, off the coast of Queensland, to the isle of Bonaire, in the southern Caribbean. Today, the turtles, cod, and coral are gone, damaged, or disappearing. For an excited child on holiday with a mask and a snorkel, the underwater world she finds still seem marvelous, and it is, but it is nothing next to what it once was. This phenomenon has a name, "shifting baselines." What looks alive and wondrous today cannot compare to what colonizers witnessed in stunned shock and what Indigenous peoples knew, enjoyed, and sustainably harvested. This is applies to our perceptions of what a "normal" climate is as well.

Marine protected areas (MPAs) preserve glorious expanses of wilderness areas in the planet's oceans and coasts. They also regenerate degraded, overfished, and over-shelled areas, be they open ocean or coastal. MPAs enhance and recover biodiversity, that lush mix of species and ecosystems, from sharks to seagrass. When well designed and enforced, they serve to protect human cities from storm surge, hurricanes, and sea-level rise, and the ocean itself from further acidification. Perhaps most important in our current situation, the plants and animals within them serve to draw down carbon and bury it for hundreds of years.

Not all MPAs are created equal. Some are designed to increase fish populations outside their boundaries, what is called spillover, and some, usually coastal, serve to keep carbon bound up and buried. There is a difference between protected zones in nearshore coastal areas—what is known as green ocean—and those created in the open or blue ocean areas. Through extensive experimentation, scientists have concluded that if 30 percent of the planet's seas can be protected by 2030—what is called the 30 by 30 approach—there will be more fish, not fewer, carbon dioxide will be drawn down and sequestered, and oxygen levels from phytoplankton will increase for us terrestrial creatures. (One of every two breaths we take originated in our imperiled seas.)

The idea of parks in the ocean or coastal waterways got off to a slow start. Not until 1966 did a movement coalesce to protect marine waters. Indigenous peoples have been practicing community-based management of local marine resources for centuries, if not millennia. Restrictions, called *kapu* in Hawaii and *bul* in Palau for example, were placed on the harvest of certain fish during certain seasons to support healthy fish populations that would sustain the community. However, by the early twenty-first century, some 90 percent of large predatory fish had disappeared from the world's oceans, and 30 percent of fish stocks were overfished to biologically unsustainable levels. The idea that marine no-take zones—meaning no fishing, shelling, or industrial use, such as kelp cropping or sand mining, would be permitted—was simple, bold, and wildly controversial. But it worked. It didn't bring back abalone or California sardines, because ecosystem baselines had shifted out of reach. But for other species, the eggs, larvae, and fingerlings prospered in MPAs on both coasts, and to the wonder of fishermen, almost all of whom had opposed protected areas, fishing improved outside the reserves, often dramatically. Similar improvements came from addressing indiscriminate fishing techniques directly by outlawing practices such as gillnetting and longlining outside of reserves as well as inside a nation's territorial waters.

What makes a successful marine reserve? First, protection: absolute no-take zones. If fishing or extraction is allowed, the system weakens, even collapses. Protecting keystone species like sharks and otter helps to establish balance and order. Currently, the open ocean is a lawless commons fished by those who have the resources to do so, primarily Chinese industrial fishing fleets. At the same time, unless rigorously protected, nearshore reserves are fair game for poachers. The big fleets need to be reduced and watched, with best practices strictly enforced. Reserves in wealthier nations tend to be easier to protect due to access to funds and stronger governance. In nations where much of a village's or region's protein is overfished, developing alternate sources of income can help to restore

Hawaiian green sea turtles (*Chelonia mydas*) crowded into a small seaside cavern to bask at sunset. Resting on shore is a rare behavior for sea turtles, except in Hawaii and the Galápagos Islands, Ecuador.

sustainability. In the small Cabo Pulmo reserve, on Mexico's Pacific coast, for instance, local fishermen petitioned the government to create a no-take zone, which became spectacularly successful. The number of fish inside the reserve increased over four times by weight, fishing greatly improved outside the reserve, and many citizens found employment in patrolling and maintaining the reserve and the shore as well—not so different from park rangers in Africa. Tourism jumped. Such local buy-in and stakeholder engagement is crucial in regions from the Philippines to Gabon.

Second, size is crucial: a successful protected area is defined as one larger than forty square miles, since fish, fingerlings, and larvae tend to migrate too easily outside the kinds of tiny reserves often set up for show and politics. If surrounded by deep water or sand, all the better.

Third, recovery takes time. Ten years is the most cited repair period. Marine reserves help create and expand ocean alkalinity, which counters the effect of acidification, caused primarily by atmospheric carbon dioxide. The most abundant vertebrates on the planet happen to be the fish that live 600 to 3,300 feet below the surface in the middle depths of the ocean, called the mesopelagic zone. These fish hang at the bottom of this range during the day to avoid predators and then rise en masse to the surface at night to feed on plankton and other small organisms. The fish digest their food at depth in their alkaline-based stomachs but poop it out as calcite crystals of carbonate when they rise up to the surface, thereby countering the surface acidity of the oceans. Scientists call this an "alkalinity" or "biological pump." The industrial fleets plying the high seas are now turning to the vast reserves of these fish to replace the stocks of other species they have fished out. If marine reserves were created in the open ocean, these fish would continue to help reduce acidification. It is a complicated, fascinating, and almost magical concept, this alkalinity pump orchestrated by billions of little fish that most of us have no idea even exist, helping to regulate the pH of our planet's oceans.

Fourth, MPAs draw down carbon in significant amounts. Nothing sequesters carbon better than the giant bladder kelp (Macrocystis pyrifera). It is the towering brown algae we see on nature specials with seals cavorting among waving branches, chasing bright orange damselfish into stone caverns. Kelp live fast and die young: Macrocystis pyrifera can grow two feet a day in ideal conditions. MPAs also protect coastal seagrass meadows. Although seagrass beds account for less than 0.1 percent of the world's oceans, they sequester approximately 10 percent of the carbon buried in ocean sediment annually. Coastal mangroves sequester twice as much carbon per acre than tropical forests. We need to prevent sand dredging, development, oil and gas canals, shrimp farming, and mining, all of which destroy the mangroves, oysters beds, salt marshes, seagrass beds, and kelp that draw down and bury significant quantities of carbon.

The high seas yield less than 2.4 percent of total seafood production, including aquaculture, or only 4.2 percent of the total global fish catch in the oceans. These are the tunas, Patagonian toothfish, billfish, and others, virtually all destined for luxury markets in food-secure countries like Japan, the United States, and Europe. These fish are caught mostly by heavily subsidized fleets from China, Taiwan, Japan, South Korea, and Spain, fleets that burn extensive fossil fuel getting to their prey. There is good data showing that high-seas closures to fishing would be more than compensated for by increased catches closer to shore.

As the value of MPAs has become increasingly understood, the percentage of protected ocean has increased from 0.7 percent in 2000 to some 5 to 7 percent designated or proposed in 2020 (in contrast to 15 percent on land). That is a tenfold increase, a surface area as large as North America, in more than fifteen thousand MPAs. The largest U.S. reserve is the Papahānaumokuākea Marine National Monument, in the northwest Hawaiian Islands, encompassing 582,578 square miles—greater than all U.S. land national parks combined—and protecting pristine reefs and the deep water between islands. The Phoenix Islands Protected Area, one of the largest MPAs in the world, was implemented by the Republic of Kiribati. Every year another country steps up—Palau, France, Argentina, Chile, Peru, Gabon—often initiated by the Pristine Seas project, founded by oceanographer Enric Sala. When Sala was a professor at the Scripps Institute, he realized his scientific papers were essentially obituaries for the ocean. He quit and founded Pristine Seas, sponsored by the National Geographic Society, which has caused more than two million square miles of ocean to be set aside as wild and protected. Every MPA has unique and specific conservation goals. The biomass of fish and marine life inside a fully protected reserve is one of the measures used to evaluate effectiveness.

MPAs are a low-tech, cost-effective strategy that sequesters carbon and helps protect and enhance the coastlines of the world. The world's 372,000 miles of coastline are populated by one-third of humanity. The goal of protecting 30 percent of the oceans by 2030 is a triple win: more wild fish to feed a growing population; restored biodiversity, which brings resilience to climate change; and secured carbon. ●

Black-striped salema (Xenocys jessiae) group schooling, and two Galápagos sharks (Carcharhinus galapagensis) fifty feet deep, Galápagos Islands, Ecuador.

Seaforestation

Strap on some scuba gear and plunge into the rocky coastal waters of Monterey Bay. The light dims just a few feet down. You are already in the shade of frondlike seaweed blades that soften the ocean current with their languid undulations. These marine leaves branch off hundred-foot-long stalks, the stipes of giant kelp—the aqueous equivalent of tree trunks. Through your mask they look like frayed rope tying the seafloor to the sky.

The seas can turn carbon into forests at a rate exceeding that of the lushest parts of the Amazon. Giant kelp is a type of brown forest-forming "macroalgae," a designation that encompasses 14,000 species of brown algae but also red and green algae such as the nori of sushi fame and the wakame of seaweed salads. Every year, a single square foot of healthy macroalgae forest can draw down more than two pounds of carbon dioxide from the atmosphere. In a single day, the giant kelp of Monterey Bay can grow over two feet in length.

But there's more. There's a key difference between the carbon fixed by marine forests and the carbon stored in their terrestrial counterparts. On land, much of the carbon is eventually returned to the atmosphere via the decomposition of leaves and wood. Marine forests shed small particles of organic carbon and dissolved organic carbon just as humans shed skin. Consequently, marine forests can be likened to "carbon conveyor belts," exporting carbon that ends up plummeting down into the ocean depths, where it won't be able to contribute to the greenhouse effect for centuries or millennia (if not longer). The upshot is that efforts to increase the coverage of marine forests may restore this natural process regionally and have a larger potential to draw down carbon than growing plants on land.

The starting point is marine reforestation, bringing back kelp forests where they used to occur but are currently absent. In many parts of the West Coast of the United States, humanity has witnessed two sequential decimations of kelp forests over the past decades. The first occurred during the rollout of industrialized farming practices in the first half of the twentieth century. U.S. Geodetic Survey maps from the 1850s to the 1900s show the first decimation of kelp coincided with the rise of industrial agriculture increasing siltation and runoff to the sea, preventing juvenile kelps from growing in deeper waters. The second occurred just in the last decade of warmer oceans, as a big, warm blob of seawater from Alaska stifled the California Current upwelling, followed by the biggest El Niño on record in 2015–16. This stifling of upwelling deprives kelp of its nutrient supply. The warm waters also rev up the metabolism of the sea urchins. The combined effects have turned major portions of the kelp forests from Northern California to Santa Barbara into sea urchin barrens. The habitats earned their name because the spiny creatures act as marine lumberjacks, chewing away the holdfasts of kelp, un-anchoring them from the seafloor and setting them adrift, where they will ultimately die. Decades of ecological studies show that the disappearance of urchin-consuming sea otters also contributed to the loss of kelp.

Sea otters are the cutest member of the weasel family, but to their detriment, they have the thickest and silkiest pelt in the animal kingdom—nearly one million hairs per square inch. Those pelts, so valuable they were referred to as soft gold, were greedily sought in what is known as the Great Hunt, from 1741 until 1911. Hundreds of thousands of otters were killed and sold by Russian, Spanish, and Native American hunters. Hunting expeditions scoured the Pacific Coast from the Aleutian Islands, off Alaska, to the final southern stronghold, near Santa Barbara, California, reducing the sea otter population to less than two thousand globally. Absent otters, the magnificent kelp forests of the Pacific Northwest were decimated by the exploding sea urchin population whose appetites and metabolisms were upregulated by climate warming and marine heatwaves.

While ensuring the recovery of sea otters and restoring lost kelp forests should be a major conservation and climate change priority, there are only so many potential locations for marine reforestation. Kelp requires shallow, rocky bottoms and cool, nutrient-rich waters. The larger potential role for marine forests in our quest for a better climate future comes from seeding kelp in nearby waters where they could not grow on their own, no matter how few urchins. This process was recently termed seaforestation, a portmanteau of "sea afforestation." Just as afforestation on land means growing forests where they wouldn't normally exist without the help of humans, seaforestation involves finding ways to grow marine forests in nearby empty-ocean conditions where they would not normally occur.

Seaforestation does not demand advanced technology. Humans have farmed non-kelp macroalgae since at least the fifteenth century. There are reports that as far back as 1670, farmers in Tokyo Bay used bamboo stakes to give young seaweeds something to attach to in the muddy shallows. The bamboo were then moved into estuaries, to bathe the seaweed in nutrient-rich water. Though this method was simple, it captures each of the essential elements of seaforestation: provide a firm site of attachment; ensure ample sunlight; keep the water cool; provision with high levels of nutrients. Whereas the Tokyo Bay seaweed farmers were primarily interested in growing food, modern seaforesters, aided by new technologies and markets, have envisioned a far greater number of applications, including drawing down gigatons of carbon from the atmosphere.

There is one more surprising similarity between the early seaweed farmers and modern seaforesters. Both took a special interest in river outlets. While the former group were solely interested in speeding the growth of their seaweeds, the latter group have realized that by targeting nutrient-rich runoff, they can create a win-win solution for waterways that have become polluted with excess nitrogen and phosphorus, two key nutrients for both macro- and microalgae. The idea is called nutrient bioextraction.

Normally, when rain washes excess fertilizer from farms or lawns, or from point sources like unchecked sewage discharge, it loads waterways with so many nutrients that harmful microalgae or bacteria blooms can form. While these algae and bacteria are a normal part of marine and freshwater ecosystems, when they become overabundant, they can produce large amounts of toxins that harm fish and shellfish and sicken or even kill people. When the overabundant microalgae and bacteria start to die off, their decomposing cells can rob the water of oxygen, creating huge dead zones in some nearshore marine habitats.

Like the Tokyo Bay farmers using bamboo poles to give the seaweed an anchor, seaforesters believe that they can seed kelp forests on large submerged platforms. And just like in Tokyo Bay, these platforms could be moved into place near the mouths of polluted rivers. There they could reduce algal blooms by soaking up excess nitrogen and phosphorus and speeding up the growth of their kelp. These kelp forests would produce oxygen during the day to reduce dead zones and could be grown along with a variety of fish and bivalves like scallops, clams, oysters, and mussels, which all filter microalgae out of the water and turn it into delicious protein. It would be a win for drawing down carbon, a win for the marine environment, and a win for lovers of fresh seafood.

Surprisingly, nutrient bioextraction is not the only way seaforestation could help solve problems created by industrial agriculture on land. Using the same techniques for growing new kelp forests, seaforesters could grow more modestly sized red algae including *Asparagopsis*, one of many seaweeds with an unusual superpower: they can dramatically reduce the outsized climate-warming emissions of the livestock industry. Normally, the oxygen-free part of a cow or bull's gut is filled with special microbes called methanogens, which take up to 11 percent of the carbon in the cattle's forage and turn it into methane, a potent greenhouse gas. The gas can't build up inside a cattle indefinitely, however, so most of it ends up getting released as a burp, with a small percentage exiting in the other direction. While an individual cow burp might not contain much methane by itself, there are billions of cattle, goats, sheep, and other ruminant livestock on earth. In fact, by one estimate they outweigh all wild mammals on the planet by roughly fourteen to one. In aggregate, cattle burps are thus one of the biggest sources of agricultural methane we

---

The Great Barrier Reef, the only living structure visible from outer space. It is home to fifteen hundred species of fish, four thousand mollusks, and five hundred types of seaweed, making it one of the most biologically diverse environments on earth. It is dying due to acidification and warming. A massive infusion of marine kelp platforms could reverse its demise.

have. However, by replacing just .5 to 5 percent of a cow's diet with *Asparagopsis*, researchers have found methane reductions of 50 to 99 percent of the methane created by their digestive gases, while facilitating weight gain. Other seaweeds can help as well, supplementing the natural diet of wild deer in New Zealand, reindeer in Svalbard, and sheep in the Orkney Islands over the past 5,000 years.

Seaforestation includes the cultivation of macroalgae like *Asparagopsis*, nori (the seaweed used to wrap your sushi roll), and sea lettuce—these are miniature sea forests and are already becoming widespread. They have given purpose and employment to coastal people from Alaska to Tasmania. According to the World Bank, the seaweed oil industry could expand to support around 100 million jobs worldwide. These farms mostly grow edible macroalgae for food, but it may be possible for sea forests growing macroalgae for other uses to also be profitable. Indeed, investors have already begun to speculate that they could meet billions of dollars of need for a diverse array of natural compounds in the cosmetics, agricultural, and nutraceuticals industries. Many of the natural oils in kelp and other seaweeds already find their way into vitamins and skin creams, for example. Macroalgae foliar biostimulants can promote budding, higher yields, and stress resilience of most flowering crops. Seaweed soil amendments also promote root growth.

Huge, varied, and lucrative markets for seaweed derivatives have led marine afforestation aficionados at the Climate Foundation to adopt the unofficial motto "The First Gigaton's on Us." In essence, they are suggesting that they can cover the costs of drawing down an entire gigaton of carbon dioxide from the atmosphere simply through sales of kelp and red algae products, making early seaforestation efforts a cost-negative climate solution, even without subsidies, tax credits, or a price on carbon. But as the saying also implies, their vision does not end with just one gigaton of carbon drawdown. As independent validators confirm the ecological merits of seaforests, the potential scale of seaforestation grows far larger, even though there is only so much demand for skin creams and plant stimulants. With a modest price on carbon funding their seaforestation activities, some researchers believe that the Climate Foundation's seaforestation development could help draw down atmospheric carbon while feeding the planet and providing kelp forest and coral reef ecosystem life support.

Natural marine forests cast an estimated 11 percent of their carbon into cold storage in the ocean depths. But that is because only a small fraction of dead kelps end up washed off the edge of the shallow seas where they grow and into the abyssal depths of the ocean. Through seaforestation, up to 90 percent of the carbon from harvested kelps could be deliberately sunk into the deep ocean, where it would remain locked away from the atmosphere for centuries to millennia. Seaforestation is thus a climate solution as massive in potential as the ocean is vast. Just imagine how much carbon could be drawn down if oceans were available to grow new forests.

There is one major obstacle to sinking enough kelp or other seaweeds to restore our atmosphere's chemistry to preindustrial levels: kelp won't grow in a desert.

To the untrained eye, the oceans may look uniform. However, in terms of biological activity, sailing a hundred miles can be the equivalent of trekking from the Sahara to the Congo. Most of the ocean's surface is relatively empty. That's particularly true of the large subtropical areas hovering above the deep ocean where carbon can be stored in the form of sunken seaweed. For seaforestation to work on a grand scale, it needs to provide the seaforest with the four conditions they need to thrive: anchoring sites, sunlight, cool water, and lots of nutrients. The underwater platforms solve the first two problems, but how could cool, nutrient-rich water find its way to a marine desert without a massive, complex infrastructure?

According to the Climate Foundation, the answer is right beneath our fins.

On land, deserts occur where there is a lack of water, which is quite emphatically not a limiting factor for life at sea. Instead, offshore subtropical oceanic surface is mostly empty because it is depleted of key nutrients, which are constantly sinking down from the surface in the form of dead plankton, animals, or even fecal matter. But not far below the surface, just past the reach of sunlight, you can find almost unlimited cool, nutrient-rich water. It rarely mixes with the surface waters—and thereby replenishes its depleted nutrients—for the simple reason that cooler water is denser than warmer water. Most scuba divers are well aware of just how stark the layering of the ocean can be, having regularly descended through a shimmering transition point, or "thermocline," where the water abruptly transitions from a sun-heated, freely mixing layer at the surface to a much cooler layer below. The shimmer is a result of the different densities of the two layers of water, which refract light at slightly different angles. Marine stratification gets stronger with higher ocean surface temperatures. In fact, one pernicious consequence of global heating is an increase in marine stratification and an attendant loss in ocean productivity, as profoundly occurred during the Permian Mass Extinction.

Some seaforesters, even those who imagine drawing down many gigatons of carbon, see their main job as drawing up cold, nutrient-rich deep water to the surface using renewable energy. They aim to create oceanic oases in vast expanses of empty ocean by using renewable energy to restore natural upwelling. As it turns out, the key to using seaforestation to pump gigatons of carbon safely *down* into the deep is to restore natural upwelling to the surface. The Climate Foundation calls this process *irrigation*.

As absurd as watering the ocean may sound, they've taken the first steps to show that irrigation works in practice through proof-of-concept projects from the Philippines to Tasmania. Ultimately, they envision many one-kilometer-square platforms submerged just deep enough that their kelp and red algae forests barely touch the ocean's surface and ships can pass over them. They believe that with each platform, they can harvest kelp four times a year, each drawing down thousands of tons of carbon out of the atmosphere and storing it safely in the deep sea. The exact number remains to be proven, but the Climate Foundation anticipates as much as three thousand tons per platform per year.

As if regenerating a healthy climate in a generation, making livestock production more sustainable, and preventing toxic algal blooms were not enough on their own, seaforestation may actually be able to help us achieve even more. Seaforestation via irrigation has some pretty spectacular perks, even beyond those of regular seaweed farming.

For example, it could restore the Great Barrier Reef, off Australia. Networks of submerged seaforestation platforms could potentially pump up enough cold water to blunt marine heatwaves that have already negatively affected more than half of all corals globally and killed almost a third on the Great Barrier Reef.

Sea forests could boost declining fish stocks as well.

Some scientists have projected that the protein and energy needs of ten billion humans could be met by seaforesting anywhere between 4 and 9 percent of the ocean's surface. It turns out that in the open ocean, fish need shelter to hide from predators, a major reason why healthy kelp forests are associated with strong commercial fish harvests. A smaller distributed area would be enough to sink and store gigatons of atmospheric carbon dioxide, helping humanity reach drawdown.

A million square kilometers of submerged, storm-tolerant seaforests are easier to summon up in a spreadsheet than in reality, of course. It would take an enormous, civilization-defining effort to achieve such a large transformation, not to mention additional improvements to the many technological milestones already achieved in seaforestation development. But most of the technical hurdles are being addressed to sustainably grow and harvest seaforests at scale.

Much like the swaying kelp forests of Monterey Bay, it is an alluring vision—one worth trying our hardest to achieve. ●

Harbor seal (*Phoca vitulina*) in the kelp forest (*Macrocystis pyrifera*), Santa Barbara Island in the Channel Islands, Santa Barbara, California.

Mangroves

A rib of land separates ocean and barrier beach from the red-colored tidal estuary and wetlands area where the river runs toward larger waters. The river is full of earth that it is red and shallow. In its marshy places, plants grow from its clay. This red estuary is alive and breathing, moving with embryonic clay and silt. Mangroves, a network of tangled roots and twisted branches, are a part of creation and renewal in this land. Coastal plants, they live in the divide between land and water. Both marine and terrestrial, these plants are boundary-bridgers that have created islands and continents. Consuming their own fallen leaves, they are nurturers in the ongoing formation of the world, makers of earth, and the mangroves have a life force strong enough to alter the visible face of their world.

—From *Dwellings*, by Linda Hogan,
Chickasaw poet and novelist

Mangroves live at the nexus of sea, soil, and sky and are one of the most carbon-rich ecosystems on the planet. They are also one of the most endangered.

Mangrove forests can be found in 123 nations and territories, totaling thirty-four million acres of land and water, principally in tropical areas. They can live in fresh or salt water but usually prefer something in between. There are eighty different types of mangrove trees, ranging from dwarf mangroves in Florida to three-story forests in Colombia. The giant mangroves on the coast of Gabon are among the tallest trees in the world. Mangroves often live in extreme conditions. They endure constant wave action, tidal ebb-and-flow, frequent flooding, and occasional battering by hurricanes. The waterlogged soil beneath them contains low levels of oxygen. They tolerate levels of salt that would kill 99 percent of all other plant species on earth. And yet mangroves have thrived for 200 million years, originating on the edge of an ancient sea.

Mangrove forests provide a unique habitat for a rich diversity of plants and animals, including fish, birds, reptiles, key deer, sea turtles, crocodiles, manatees, and black clams. The Sundarbans, a huge mangrove forest in the

Ganges River delta of Bangladesh, is critical habitat for the endangered Bengal tiger. The Sundarbans are also used for agriculture. Worldwide, millions of people depend on mangrove forests. Their dense web of roots act as nurseries for a wide variety of sea life. The trees supply local communities with seafood, material resources, and income. Mangroves shield vulnerable populations from natural disasters, such as hurricanes and tsunamis, and protect land from erosion. They buffer the impact of sea-level rise. Mangroves trap sediment, nutrients, and pollutants, thus acting as a natural water purification system. This is particularly important in developing nations where coastal cities have limited sewage systems and extensive shipping operations that often pollute large areas. Mangrove forests also have important recreational value, earning income from nature-based tourism while also playing an important role in traditional customs and modern cultural practices.

Mangrove ecosystems play a vital role in addressing climate change. They are long-term carbon sinks, storing carbon in the trees themselves as well as soils underwater (where reduced oxygen levels slow the rate of carbon dioxide release). Mangrove forests remove up to four times more carbon from the atmosphere per acre than terrestrial forests and can store almost twice as much carbon. Mangrove ecosystems currently hold between 5.6 and 6.1 billion tons of carbon worldwide, mostly in the submerged soils. Although the total area occupied by marine vegetation, which includes mangroves, seagrasses, and salt marshes, is relatively small—less than 1 percent of the planet's terrestrial surface—they account for 50 percent of all carbon sequestered in marine sediments. This is one reason why researchers, conservationists, and policy-makers frequently refer to marine vegetation as *blue carbon*, a term coined in 2009 to distinguish it from green carbon, which is associated with plants and trees farther inland. The term has proven to be a handy concept for use in the protection, restoration, and management of these important marine habitats.

There is a sense of urgency for blue carbon ecosystems. The world has lost nearly 50 percent of mangrove forests since 1980, with most of that loss occurring in Southeast Asia due to landscape conversion as a result of aquaculture projects, illegal logging, and industrial and urban coastal development. These impacts are expected to continue and be exacerbated by climate change and population growth. Other major losses have also been reported in Central America and Africa, making mangroves one of the world's most threatened ecosystems. When mangroves are degraded or destroyed, carbon stores that took millennia to accumulate are released in a matter of years, flipping an important carbon sink into a significant carbon source. In Indonesia, 618,000 acres of abandoned shrimp aquaculture ponds that were once pristine mangroves now emit up to 7 million tons of carbon dioxide per year. Restoring these abandoned ponds back to mangrove habitat would not only halt these greenhouse gas emissions but also absorb as much as 32 million tons of carbon dioxide annually. Taken together, the restoration of mangroves could result in 3 billion tons of removed or avoided greenhouse gas emissions by 2030.

By halting ongoing mangrove loss and restoring mangrove forests globally, we have a significant opportunity to combat climate change *and* help millions of people adapt to its impacts. It is critical to mobilize governments, institutions, communities, and individuals to protect this "super" ecosystem. Under the historic 2015 Paris climate agreement, many countries included mangroves in their commitments to reducing greenhouse gas emissions. While this is an important start, we now must turn these commitments into action for one of the earth's most important ecosystems. ●

River and mangrove forest in the Sarawak Mangrove Reserve, Borneo, Malaysia.

Tidal Salt Marshes

The ebb and flow of tidal waters inside the salt marsh carry life in and sweep it back out again. Raccoons that take advantage of low tide to fill their bellies with crabs and mussels are forced to beat a hasty retreat across the mudflats as the tide comes in. Minute by minute, the waters creep up through the low marsh, where the *Spartina alterniflora*, or smooth cordgrass, grows tallest—up to seven feet high—and the ribbed mussel, with whom it has a symbiotic relationship, burrows below, enhancing nutrients in the nearby soil. These helpful bivalves also provide a habitat for fiddler crabs, which take advantage of mussel mounds for their burrows. As the water rises into the high marsh, the cordgrass gradually gives way to glasswort, sea lavender, spike grass, black grass, and marsh elder. Here the shy saltmarsh sparrow feasts on insects to feed her chicks, hidden from sight in a nest built of woven marsh grass, just barely above the high tide mark. As the tide climbs still higher, it laps against the upland border, where marsh elder, switchgrass, and reeds grow.

Below the surface, plants decompose slowly in the largely anaerobic, salty conditions, and the regular influx of saltwater prevents the formation of methane. Meanwhile, the nutrients washed up in the tidal waters leave deposits, forming a thick layer of marsh peat that continuously grows with each tidal cycle. For this reason, scientists believe saltwater marshes store more carbon per acre than tropical forests. Globally, that amounts to nearly a ton of carbon sequestered per acre annually.

rising by one-fourth of an inch annually. As a result, in 2019 global sea level was 3.4 inches above its average height in 1993. Higher sea levels cause more damage as stormwater advances farther inland, but they are also problematic for tidal salt marshes.

Over the course of a year, as sediment from the incoming tide gets deposited and as root materials pile up on the marsh floor, the marsh gains up to one-third of an inch of elevation, with surface sediments alone holding between 10 and 15 percent carbon. Until recently, most tidal marshes have been able to grow fast enough to keep pace with sea-level rise. For those that have not, restoration methods include manually increasing the marsh's elevation by spreading a fine layer of sediment over the entire area. Another problem caused by a combination of sea-level rise and seawalls or levees is the inability for tidal salt marshes to migrate inland as needed for survival. By identifying areas where barriers can be removed or altered to accommodate marshland, we create an opportunity to save these tidal wetlands and allow them to regenerate in order to continue to capture carbon and protect coastlines.

Regenerating salt marshland requires a focus on re-creating historical tidal patterns in a given area. If there is too much water from sea-level rise and blockage from coastal development, the marshland is reclaimed by the ocean and "drowns." When there is insufficient water, invasive species take root, forcing the native plants out and dramatically altering the ecosystem upon which fish and birds rely. Restoring a marshland's connection to life-giving tides requires a number of techniques, from dredging to removal of man-made drains to modifications of water-control structures like tide gates. Many tidal marshlands are able to recover once tidal flows are brought back.

One of the marvels of tidal salt marshes is how its inhabitants have adapted to thrive in difficult conditions. Inundated by salt water one moment and drying in the scorching sun the next, the species that call it home are hardy. A reflection of its inhabitants, the marsh is also hardy. Up until this point, tidal marshlands have survived pollution, development, and rising sea levels. They may be at the tipping point, however, and without conservation support, marshlands will be squeezed out by roads, houses, and tidal seawalls. Every year in the United States, 7.3 million hectares of coastal wetlands capture 6.7 million tons of greenhouse gases. By making room for marshland and supporting their ecosystems, the great blue heron will continue to fish, carbon will be stored in its soil, and the marsh will provide protection for the shoreline that no human-made structure can recreate. ●

Tidal marshes become carbon dioxide sources when they are converted into roads, houses, or agricultural land. Almost two million acres of coastal wetlands disappear annually around the world, contributing about 500 million tons of carbon dioxide to the atmosphere. In the United States, more than half of tidal wetlands have been drained and converted into farmland, dissected by highways, or cut off from tidal waters by seawalls and levees. However, rising sea levels and an increase in the frequency and severity of hurricanes have been forcing people, especially those in coastal communities, to recognize the critical role of tidal marshlands.

According to satellite data from the National Oceanic and Atmospheric Administration, sea level is currently

———

Reeds and mudflats on the North Sea in Holland.

Seagrasses

Seagrasses include many varieties of grasses, lilies, and palms, a group of plant species that can live entirely immersed in seawater. Like their terrestrial relatives, seagrasses have leaves and roots and produce blossoms and seeds—the flowering plants of the sea.

Seagrasses have been called the forgotten ecosystem, perhaps because they seem so tame when compared with richly endowed mangroves and spectacular coral reefs. They cover the shallow slopes of coastlines from the tropics to the Arctic. Like grasses found on land, seagrasses can form dense underwater meadows, some large enough to be seen from space. These immense areas provide habitat to millions of animals, from tiny shrimp and seahorses to large fish, crabs, clams, turtles, manta rays, and marine mammals such as dugongs and sea otters. Sea grass meadows are essential to fisheries, coastal protection, and maintaining water quality. However, the most unknown quality of seagrasses may be their most important. While sea grass meadows occupy less than 0.1 percent of the ocean area, they provide nurseries for 20 percent of the largest fisheries on the planet, and 10 percent of the carbon buried in ocean sediment each year.

There are seventy-two known species of seagrass with a great diversity of sizes, shapes, and habitats, ranging from eelgrass, with its long, ribbon-like leaves, to spoon grass, with paddle-shaped leaves that form lush, low meadows. The tallest seagrass species—*Zostera caulescens*—has been found to reach thirty-five feet tall in Japan. Like all plants, seagrasses depend on light for photosynthesis; hence, they are most common in shallow depths where sunlight is brightest, but deep-growing seagrass has been found at depths up to 190 feet.

Seagrasses have been used by humans for more than ten thousand years. People fertilize fields, insulate houses, weave furniture, and thatch roofs with seagrasses. By providing critical habitat, they support fisheries and biodiversity. They also trap and stabilize sediment, which not only maintains water quality but also reduces erosion and buffers coastlines against storms. These benefits make seagrasses one of the most valuable ecosystems in the world.

Despite their critical importance, seagrasses are highly endangered. While the current global area of seagrasses is between 45 and 150 million acres, more than 12 million acres of seagrass meadows have been lost in the past century. The rates of loss have increased over that time, from 1 percent per year prior to 1940 to 7 percent per year since 1980. Globally, 24 percent of seagrass species are now classified as threatened or near threatened on the International Union for Conservation of Nature Red List. While seagrasses are impacted by coastal development, poorly managed fisheries, and aquaculture, the greatest threats to seagrasses are coastal pollution and poor water quality. Seagrass loss is occurring at rates similar to tropical rainforests, coral reefs, and mangroves.

The accelerating loss of seagrasses globally is a significant contributor to climate change. Seagrass meadows are natural carbon sinks and can be more effective at sequestering carbon than forests. They absorb carbon from the water and then bury it in the sediment below for up to millennia. The world's oldest living organism is a patch of Mediterranean seagrass, *Posidonia oceanica*, which is estimated to be 200,000 years old and still absorbs and deposits carbon into the thirty-six feet of carbon-rich soil on which it grows. On average, every acre of seagrass buries 0.5 tons of carbon per year (which adds up to 80 million tons per year of carbon captured by seagrasses and kept out of the oceans and atmosphere). Ongoing loss of seagrasses is not only reducing their capacity to remove carbon from the atmosphere; the carbon stored in the soil below the seagrasses can be released once the plants that were holding it in place are degraded or removed. Current rates of seagrass loss are potentially releasing 300 million tons of carbon into the oceans and the atmosphere every year.

Seagrass meadows in Virginia were wiped out by an ocean-borne pandemic and hurricane in the 1930s and never recovered. Over the past two decades, in four coastal bays, 74 million eelgrass seeds have been broadcast into 536 restoration plots that had been barren for nearly a half a century. The grasses have spread to nine thousand acres and have become the largest eelgrass habitat between Long Island and North Carolina. Once established, the grasses clear the water and moderate the waves, providing seafloor stability and sufficient light for the plants to thrive and reseed naturally. The areas seeded are part of the forty-thousand-acre Volgenau Virginia Coast Reserve, protected from boat anchors, propellers, and pollution. What the reserve cannot do is protect seagrasses from rising ocean temperatures. Halting ongoing losses and conserving the world's seagrasses must be a priority for addressing climate change. If we can ensure their future, these remarkable and ancient plants will continue to absorb carbon while also protecting and nurturing the people and rich biodiversity of the world's coasts.

One seed of hope is the scallop, a saltwater clam once ubiquitous along the eastern shores of the United States. In the Volgenau Virginia Coast Reserve, scallops have been found twenty miles from the closest spawning cages, an indication that scallop larvae drifted down the coast into the reserve and began a process of regeneration that was neither predicted or expected. Marine ecologist Mark Luckenbach compares the advent of a harvestable bay scallop population to the introduction of the gray wolves into Yellowstone. It's the beginning of regeneration. ●

Green turtles can travel thousands of miles in their lifetime, traversing entire oceans. They read the earth's magnetic field perfectly to guide them in their migrations. They return unerringly to the beach where they hatched.

Azolla Fern

Nearly 50 million years ago, atmospheric carbon dioxide levels were at least double, if not triple, current levels. However, carbon dioxide levels quickly declined to what they are today. There may be many explanations for this transition, including changes in the position of the continents. One partial explanation is that rapid blooms of a diminutive fresh-water fern helped lower carbon dioxide levels. This small fern, *Azolla arctica*, is related to modern azolla species, *Azolla filiculoides*, which have a huge potential to sequester carbon dioxide while replacing fossil fuel–intensive fertilizers, providing animal feed, and/or creating feedstock for biofuel.

In 49-million-year-old sediment cores from the Arctic Ocean, scientists found layers rich in azolla spores and organic matter from the surrounding shores. At the time, the Arctic ocean was mostly land-locked, and many fresh-water rivers flowed into it. This allowed azolla to bloom, and this tiny fern contributed to substantial amounts of organic carbon burial over the span of about 800,000 years that its spores are still visible in the ancient Arctic sediments. This carbon burial likely contributed to at least part of the carbon dioxide drawdown and global cooling in that era.

Azolla, also called water velvet or fairy moss, is not your typical fern. Unlike its larger, frond-producing cousins, azolla grows as a coin-size rosette, lying almost flat on the surface of freshwater. Its delicate roots dangle, making no effort to find the soil, even when floating in an inch of water. It places its microscopic seed in an air bag that floats away. It has more protein than soybeans. Azolla contains a special kind of bacteria, held in tiny, oxygen-free pockets called heterocysts. Called *Anabaena azollae*, the species is unique to the azolla fern and has become completely dependent upon it for its survival, having transferred some of its genes to the fern. *Anabaena* is a type of blue-green cyanobacteria that sequesters inert nitrogen from the air, which allows azolla to self-fertilize. And because of that, it can grow at blazing speeds, doubling its coverage on the surface of a water body in as little as 1.9 days.

Azolla could once more play an important role in sequestering carbon dioxide from the atmosphere through the deliberate actions of humans. Recent research outlines multiple potential impacts for azolla, including regenerative agriculture, green fuel, clean water, and, most of all, a livable climate. More than a thousand years of agricultural practice have witnessed azolla's promising role in rice cultivation. The earliest written account of the use of azolla to increase rice production dates back to AD 540, when Chinese scholar Jia Si Xue described how rice farmers inoculated their paddies with azolla in the book *The Art of Feeding the People*. What Jia could not have known, though, was how azolla accomplished its job.

Azolla acts as a "biofertilizer," a living organism that provides critical nutrients to its surroundings. To some extent, azolla is able to transmit nitrogen directly into the water it grows in and makes bigger contributions when bits of it die and get incorporated into the soil where rice plants are rooted. By draining azolla-filled rice paddies at the right time, farmers create a significant pulse of nitrogen as well as a full complement of nutrients that maturing rice plants need in order to maximize their output. By growing azolla alongside rice, farmers in settings where fertilizers are scarce or expensive can increase yields of their paddies by 50 to 200 percent. In more affluent rice-growing regions, azolla can dramatically reduce or entirely eliminate the need for chemical fertilizers.

Boosting yields and displacing energy-intensive chemicals are not the only ways azolla can help with climate change. It draws down carbon dioxide directly from the atmosphere into its tissues. By one estimate, if azolla were added to all the flooded rice paddies in the island nation of Sri Lanka, it would draw down more than 500,000 tons of carbon dioxide each year.

Japanese farmer Takao Furuno has even taken the system further. Furuno's book *The Power of Duck* relates his decades of experience perfecting a farming system that incorporates fish and ducks alongside azolla and rice. Azolla feeds the ducks, providing a steady diet that enables them to pick off the vast majority of the eggs of invasive snails as well as insect pests that attack the rice plants. The duck feces then fertilize both the paddy's soil and phytoplankton. The latter then provide food to loaches, the eel-like fish that Furuno harvests. By turning his rice paddies into a polyculture that mimics a natural wetland ecosystem, Furuno is able to sustain high productivity in the absence of chemical fertilizers, herbicides, or pesticides. By occasionally drying the paddies to grow organic vegetables, he is able to sustain rich and healthy soil indefinitely without external inputs.

Growing azolla in dedicated ponds can create a renewable supply of cheap "green manure" for fertilizing other crops. To date, researchers have shown that azolla mulch can boost yields of wheat, taro, soybeans, and mung beans, and we have every reason to believe it would be a boon for most of the plants we grow.

Azolla can be used as a protein- and oil-rich superfood for domesticated animals. A bevy of studies have shown that replacing 5 to 40 percent of the feed of dairy cows, pigs, chickens, tilapia, and rabbits with azolla either

leads to a boost in the animals' growth rates or a decrease in total cost of feed per unit of meat produced. In the latter cases, azolla is replacing protein-rich feeds like soybean meal, the production of which is usually chemical-intensive and a major driver of land clearing in places like the Amazon rainforest. It may also also safe for human consumption and has even been suggested as an ideal food for astronauts. Not only that, but eggs made from hens eating azolla are omega-3 eggs, containing EPA and DHA, critical to our cognitive health span. Other types of food products made with azolla would be rich in the omega-3 fatty acids and that could transform the health spans of people around the world.

Beyond agriculture, azolla has shown promise as a feedstock for biofuel production. Azolla isn't grown on land, so it wouldn't compete for space with carbon-storing terrestrial ecosystems like rainforests or grasslands, as do corn, sugarcane, and oil palm. In large artificial ponds that replaced agricultural fields in the temperate climates that the fern favors, they would be as efficient as current biofuel crops grown at the same latitudes. Early trials suggest that one acre of azolla could concurrently produce nearly as much ethanol as an acre of corn and as much biodiesel as an acre of palm oil. Azolla wouldn't require energy-intensive nitrogen fertilizer as do those other crops, thanks to *Anabaena*. If azolla were grown for the use of its proteins and carbohydrates as animal feeds, its omega-3 fatty acids EPA and DHA would improve the nutrient content of the foods made from these feeds, while other oils could be fractioned off for use as a carbon-neutral fuel for the tractors, trucks, and machinery involved in animal husbandry.

Great care would need to be taken if azolla were introduced into settings where it is currently absent, whether in the name of growing biofuel feedstocks or otherwise. Some species of azolla have become invasive when introduced outside of their native ranges. But there is nothing preventing the sustainable harvest of the fast-growing fern from its existing habitats or from artificial ponds in controlled agricultural settings. When azolla freezes, it usually dies, so that presents one way to manage its proliferation.

Less controversial is azolla's potential for phytoremediation, a term that refers to the use of plants for environmental cleanup. Its ability to soak up phosphorus and even excess nitrogen from waterways is clear, reducing eutrophication of waterways. Not only that, but azolla has a remarkable affinity for all manner of pollutants, including heavy metals like lead, nickel, zinc, copper, cadmium, and chromium, as well as certain pharmaceuticals, and even for the salt ions that cause salinization of some agricultural lands. By concentrating these elements and compounds in its tissues, azolla can be used to clean up mine tailings and fly ash, or even as a way to clean wastewater so it can be used for irrigation. Depending on the type of cleanup operation, the azolla could potentially be harvested for use as green manure or used for biofuel production.

Azolla has already had a positive impact on the earth's geological history and Asian agriculture. With more research and financing, it may once again change the world. ●

A Morelet's crocodile (*Crocodylus moreletii*) amid profuse Azolla fern in Laguna Catemaco, in Los Tuxtlas Biosphere Reserve at the center of the Sierra de los Tuxtlas, Veracruz, Mexico.

Forests

Forests are crucial to our well-being. They are water-sheds, habitat, and refuge. They cleanse the air, cool the air, and create the air. Forests cover nearly 30 percent of the earth's terrestrial surface—some 10.8 million square miles (reduced from approximately 23 million square miles at the end of the last Ice Age). There are 60,065 known tree species. Brazil, Colombia, and Indonesia have the greatest number, with more than 5,000 species each. There are thousands of different forests, and all store significant amounts of carbon, amplified where there are carbon-rich soils such as peatlands and wetlands. Forestlands contain most of the planet's terrestrial plant and animal species, with tropical forests by themselves containing at least two-thirds of species diversity, possibly as much as 90 percent. Forests are critical to freshwater supplies, as they help regulate and maintain the aquatic environmental conditions and related habitat resources their ecological communities require.

In the past two decades, thanks to scientists such as Suzanne Simard of the University of British Columbia, our understanding of forests has transformed. Just as research on the human microbiome revolutionized our understanding of health and disease, the biological interactions among trees, fungal networks, microorganisms, and unrelated plant species paint a new picture of forests and what occurs within them. Our old image of trees competing for water, sun, and nutrients has been replaced by research showing how primary or intact native forests are social creatures, interacting, sharing knowledge, and taking care of their community. Trees learn—they visually sense animals near them (including us)—retain memory, and accurately anticipate future weather. Trees in a forest behave like a living organism rather than a collection of parts. The forest community includes bacteria, viruses, algae, archaea, protozoa, springtails, mites, earthworms, and nematodes, collectively numbering in the trillions in a single handful of soil.

However, trees are a commodity. Tree trunks and forest acreage are totaled up like a purchase order. This is misguided. While some forest products, such as timber, are useful, the wanton deforestation that usually comes with tree harvesting is both short-sighted and dangerous. It imperils the critical role forests play in regulating our climate. An estimated 2,200 billion tons of carbon are stored in forests, distributed among the three major forest biomes: approximately 54 percent occurs in tropical forests, 32 percent in boreal forests, and 14 percent in temperate forests. Boreal forest ecosystems have the highest carbon densities, and more recent estimates for boreal forests suggest that the total ecosystem stock of carbon, including biomass and soil carbon, is larger than the carbon stock in tropical and temperate forests combined. Any diminishment, including "sustainable" logging, reduces the amount of stored carbon in a forest while raising the risk of wildfire.

To solve the climate crisis, forests are crucial. Our top priority is protecting primary forests, sometimes called old-growth, from destruction. They are the largest, most resilient, most carbon-rich forests on earth. Letting them continue to grow, and to sequester additional carbon, is one of the most effective strategies we have to reverse climate change. Another priority is reforesting land that has been cleared or degraded by human use. But it must be done in the right way. Many climate schemes propose bioenergy projects that would plant and combust more than a hundred million acres of fast-growing trees in the global South, capture the carbon emissions from the incinerators, and bury it. These tree "crops" would be machine-harvested and burned to provide so-called clean energy.

Scientists calculate that planting a trillion trees will help attain timely carbon goals. It sounds logical. Tree-planting targets are a mainstay of nearly every climate pledge and proposal. Yet the adage "not seeing the forest for the trees" has never been more apt. Tree planting has a role to play, but we must be careful. Trees are being planted around the world in order to offset corporate emissions and achieve so-called net-zero goals. But emissions goals need to be achieved by reducing *actual* emissions. Forest plantings and expansion should be done to restore native species, water, and human rights in addition to drawing down carbon from the atmosphere. Too often, however, tree planting requires taking over land in the South to offset emissions created by a more affluent and industrialized North. In this sense, it is no different from prior colonization visited upon Africa, South America, and Asia over the centuries. The urgency for climate solutions can lead to seeing the world in ways that are less complex—less confusing, if you will, as if a tree could be viewed in isolation.

Because much of the world's deforestation has occurred on Indigenous lands, a global forest strategy makes no sense unless it includes protecting the people who have cared for the forests for thousands of years. Mainstream conservation is science-based and analytical, but it often doesn't acknowledge that the forestlands being studied are stolen lands. Restoring forests begins with restoring human rights. Racist expulsion of Indigenous people continues to this day. The more than five hundred Indigenous nations represent some of the most knowledgeable and motivated people in the world when it comes to repairing, restoring, and regenerating their land. A critical step will be to define what a forest is exactly. At the moment, destructive eucalyptus plantations are called forests, but these are just monoculture plantings that destroy native diversity and cause forced evictions of native populations.

There is great potential for restoring degraded forests. At the same time, because of global warming and logging we are seeing more megadroughts and megafires. The drying causes insect infestation and tree death. Tinder combined with dry conditions make reforestation a formidable challenge in some areas of the world. The Bonn Challenge, an initiative launched by Germany and the International Union for Conservation of Nature in 2011, calls for 350 million hectares (1.35 million square miles) of forest restoration by 2030 and looks to governments and the private sector to make pledges toward these objectives. As of February 2021, about 210 million hectares have been pledged, mostly by governments. The Convention on Biological Diversity Aichi Target 15 calls for restoring 15 percent of degraded ecosystems around the world—about 500 million hectares (1.15 million square miles), assuming about three billion hectares of degraded lands globally. Similarly, the United Nations Sustainable Development Goal 15 requires nations to achieve a land degradation–neutral world by 2030.

In this section, we explore the ways these goals can be accomplished. ●

Douglas fir in the Cascade Range, whose woodlands contain the greatest amount of biomass of any forest on the planet.

Proforestation

Environmental scientist William Moomaw coined the term *proforestation* when he realized that protecting intact forests as well as letting degraded forests recover and mature would have a greater impact on global emissions than any other land-based solution. A 2021 University of West Virginia research project that studied changes since 1901 showed that as carbon dioxide levels in the atmosphere rise, the capacity of trees to take up more carbon dioxide increases, though eventually this capacity plateaus. This reversed a long-standing belief that the stomata, the minute pores in the epidermis of the tree leaves that allow movement of gases in and out, became more constricted in the presence of additional carbon dioxide. It turns out to be the opposite: even more carbon is being taken up by the trees than was conventionally understood, which is confirmed by Moomaw's studies.

Forestation is rightly promoted as a prominent drawdown solution for climate change, but practices differ.

Afforestation is planting trees where none grew before; *reforestation* replaces trees where they were previously grown. Both are activities implemented by humans that result in the sequestration of carbon over the lifetime of the trees. However, the contribution of a newly planted tree to carbon removal is limited during the first decades of its life as it grows to maturity. To meet 1.5°C climate goals, afforestation and reforestation would require 3.7 million square miles of land, an area greater than the landmass of China.

Proforestation is different—the required land is already forested, whether as intact, old-growth forest or as degraded forest that needs simply to recover and grow. The world's total yearly emissions of carbon are about 11 billion tons. However, the net annual increase in elemental carbon in the atmosphere is about 5.4 billion tons, because 5.8 billion tons are sequestered by land, plants, and oceans. Of the three, forests are the greatest remover of carbon dioxide on the planet, and primary, mature forests

For example, boreal forests sequester nearly double the amount of carbon stored in tropical forests, yet they are threatened by logging, fires, pests, and mining. Trees from the boreal forests of Canada are harvested and pulped to make luxury two-ply toilet paper.

A practice that does damage to growing forests is the burning of wood pellets. Called bioenergy, it has been falsely championed as a carbon-neutral energy source. The belief is based on the proposition that the carbon from the incineration of one tree will be canceled out by the growing of a replacement tree. Employing that logic, coal could be considered renewable. Diverse, species-rich woodlands in North America, especially in the Deep South, are being cut down because the European Union considers wood pellets to be renewable energy. Burning today's trees is being justified by the planting of tomorrow's trees, even though it is not economical. Solar and wind energy are far less expensive forms of energy today. But subsidies from the European Union and the UK make wood pellets popular. To compound the loss, wood-pelleting factories in the U.S. are built in lower-income African American communities, creating high rates of asthma and lung disease. Wood-pellet plants produce unacceptable amounts of dust and particulates, and this is being done in the name of renewable energy and carbon neutrality.

Protecting the world's remaining intact forests is made more difficult by the fact that the United Nations defines forests as a catchall term. It considers tree plantations, timber forests, and mature, intact forests as equals. A twenty-year-old loblolly pine plantation and the 130-million-year-old Taman Negara forest of Malaysia are hardly equivalent. In defining forest conservation, the UN uses forest cover as a blanket metric and ignores characteristics like the forest's ability to sequester carbon, the age of the trees, species diversity, and ecosystem functionality. Every older forest offers complexity, connectedness, differentiated habitat, greater biodiversity of plants and animals, water retention, air purification, flood control, and—if that is not enough—beauty, wonder, and awe. Degeneration is the breakdown of connections within any living system. Regeneration honors and protects those connections, which are also the means to most effectively sequester carbon and protect us from climate disruption. When practicing proforestation of intact forests, all aspects of life are protected, enhanced, and supported. ●

are responsible for the majority of that sequestration. Until recently, the scientific community assumed that older trees sequestered carbon marginally, if at all. Now we know trees accumulate significant amounts of carbon to almost the end of their long lives. Proforestation would have a forty times greater impact between now and 2100 than newly planted forests.

Intact forest landscapes (over 50,000 hectares in size) where there is no habitat fragmentation and that contain diverse populations of native species are the top priority for protection. They exist throughout the world, from Russia to Gabon to Suriname to Canada. Intact forests comprise only 20 percent of tropical forest area, yet they account for 40 percent of the total aboveground tropical forest carbon. Despite this critical role they play in carbon sequestration, only 12 percent of intact forests are protected, leaving them vulnerable to exploitation, most commonly for timber harvesting and agricultural expansion.

Scientists measuring the health and caliper of the largest trees on earth, the giant Sequoia of Sequoia National Forest in Tulare County, California. The larger trees are over 250 feet high and up to 102 feet in circumference at the base, more than the distance between home plate and second base in a baseball field.

Boreal
Forests

The boreal is home to the largest intact forest systems in the world. They comprise diverse communities of pine, larch, spruce, fir, shrubs, bushes, mosses, lichen, and animals, a green swath that sweeps around the Northern Hemisphere from Alaska and Canada across to Scandinavia, Russia, and the northern tip of Japan. It is the largest terrestrial biome on the planet—known as the *taiga* in Russia—a complex and mysterious habitat replete with quagmires, marshes, and bogs, an endless swath of coniferous trees where hunters get lost and are never seen again. Underneath shaded canopies, black lichen hangs like matted hair. It is the lair of wolves, grizzlies, wood bison, caribou, and moose, along with a profusion of small carnivorous mammals—lynx, martens, minks, stoats, sables, wolverines, badgers, and weasels. It is home to more than six hundred Indigenous communities who know and understand the land, forests, and waters like no one else. And most important, there is more carbon under the lakes, in the forests, and throughout the peatlands of the boreal latitude than is in the atmosphere itself.

The North American boreal is the largest unbroken expanse of contiguous forest, peatlands, and wetlands on the planet, interlaced by streams, rivers, lakes, and ponds, forming an area of approximately 1.2 billion acres. One to three billion birds fly to the North American boreal for their summer refugia, migrating from as far away as Patagonia. In the fall, three to five billion birds fly back with their hatchlings to wintering sites. These include birds we see in our backyards, parks, fields, and forests—warblers, sparrows, ducks, waxwings, and ravens, but also the endangered whooping crane.

Summers are cool and short, winters long and cold. In many areas, the soils are thin, sandy, and toxically acidic, due to the constant deposition of needles, resins, oils, and chemicals from the trees. In areas where light can get through, there are fairy-tale patches of salmonberries, blueberries, and red and black currants. In bogs and fens, carnivorous sundews and pitcher plants trap and digest unsuspecting insects and spiders. The conifers that dominate the boreal are dark green to maximize light absorption, form perfect pyramidal cones to shed the heavy winter snow, and create antifreeze resins in the needles to keep from freezing solid.

The boreal has the highest carbon density of any region on earth, with more carbon below the ground than intact tropical forests have above the ground. The forests hold 1,140 billion tons of carbon in soil and biomass, 50 percent more than what is in the atmosphere. The damp, cold conditions in the boreal prolong decay and create carbon-rich bogs and peatlands. When boreal forests are harvested and clear-cut, the disturbance dries out the soil, which creates carbon emissions greater than the loss of the trees. If half of the boreal forest and its carbon stock is lost, carbon dioxide in the atmosphere would be over 500 parts per million.

There are urgent campaigns to stop the logging of the boreal forest, especially when one of the main products is virgin fiber for single-use paper products, including luxury toilet paper. Using Charmin toilet paper made by Procter & Gamble, or Kimberly-Clark's Quilted Northern Ultra Plush, flushes away, piece by piece, the greatest forest and carbon sink in the world, as well as its attendant biodiversity. Along with Georgia-Pacific, these companies are engaged in what is known as the "tree-to-toilet" pipeline. They have countered by saying they support sustainability because they "cut one and plant one"—reasoning that does not comport with forestry science. The old-growth trees being cut contain greater amounts of carbon than the newly planted trees will capture in forty years. Once cleared, boreal forests do not spring back to life. It takes trees decades to regrow in cold northern climes. Animals that once occupied the dense forests, such as the caribou, do not return to the single-species forests replanted by toilet paper companies. Converting boreal forests into tissue, paper towels, junk mail, and shopping bags defies common sense. There are many alternative fibers and recycling possibilities. A spokesman for a company employing boreal-sourced fiber stated that the company uses only "waste" wood created by other logging companies, logic that is equivalent to taking the feet of elephants after they have been killed by poachers.

The boreal forest is being scraped, polluted, and destroyed in ways that will require hundreds, if not thousands, of years to recover. On the Athabasca tar sands of Alberta, Canada, oil companies are undertaking the largest construction, land-clearing, and mining project on earth. Situated on both sides of the Athabasca River, the operation, if completed, will cover an area the size of Ireland. It is a vast area containing a thick, tar-like substance called crude bitumen. Extracting, heating, and refining the bitumen requires 2.8 gallons of fresh water for every gallon of gasoline produced. Afterward, the toxin-laden water, which is no longer of use, is placed into unlined storage ponds that so far cover eighty square miles. The ponds contain nickel, lead, vanadium, cobalt, mercury, chromium, cadmium, arsenic, selenium, copper, silver, and zinc. As yet there is no proposal to detoxify the tailings and water left behind. The hot water that is used in the deeper areas for steam extraction of bitumen remains underground and can contaminate groundwater for hundreds if not thousands of square miles.

European brown bear (*Ursus arctos*) reflected in a forest pond in evening mist, Finland.

Because it is warming faster than the rest of the world, the boreal is facing consequences ranging from insect infestations to wildfires. Temperatures in the Arctic are rising at twice the global rate, and as disproportionate warming moves climate zones northward, it outpaces the speed at which trees are able to migrate. The warming poses a significant risk to the substantial quantities of carbon currently locked into the soil. An earlier and more extensive melting of the snowpack causes the forest to dry out in advance of the fire season, leading to wildfires that will be increasingly hard to control. Using tree rings and lake deposits of soot, scientists have been attempting to trace the boreal's fire history. The recent fires in Alaska's boreal seem to be the most severe in eight thousand years.

Today, less than one-third of the boreal forests remain primary forests. The logging that continues to occur is not sustainable, but it could be mitigated. If governments would commit to setting those lands aside and allow them to fully regenerate back to primary forest, average per-acre carbon storage would be triple the storage in temperate or tropical forests. It demonstrates that the earth is inherently a regenerating system when we cease cutting, scraping, burning, or poisoning it.

Canada's First Nations people are taking the lead in creating protected and conserved areas all across that country, as monitors and stewards of their lands and the natural cycles of nature witnessed in the global flocks of birds that visit their homelands. A comprehensive plan to end the climate crisis depends upon the protection of the boreal forest and requires an end to the ecological degradation caused by resource extraction. A significant effort to turn back the exploitation of the boreal forest was undertaken by four Anishinaabe First Nations starting in 2002. The nations signed an accord that would protect their lands through a UNESCO World Heritage designation. The Anishinaabe have stewarded the land for at least seven thousand years. In 2018, Pimachiowin Aki was designated as Canada's nineteenth World Heritage site. In a local Ojibwe language, Pimachiowin Aki means "the land that gives life." Situated on both sides of the Manitoba-Ontario provincial boundary, the 11,212-square-mile site contains two provincial parks, a conservation reserve, 3,200 lakes, more than 5,000 freshwater marshes and wetland pools, and 285 archaeological sites. Biological surveys count more than 700 different vascular plants and some 400 mammal, bird, amphibian, reptile, and fish species. Included are loons, leopard frogs, lake sturgeon, black ash, jack pine, and wild rice. The archaeological sites reveal a longtime pattern of resource rotation that shifted the hunting, fishing, and harvesting areas to ensure full recovery of the land and waters.

One of the Indigenous leaders who is making a significant impact is Steven Nitah, of the Łutsel K'e Dene First Nation. He was the chief negotiator in creating Canada's newest national park, the 10,200-square-mile Thaidene Nëné National Park Reserve in the Northwest Territories, which is designed to integrate Indigenous people instead of displacing them, as has occurred in so many places in North America. The Łutsel K'e will continue to hunt, fish, and manage the forests, lakes, and rivers as they have always. In working with government, citizens, and local officials, Nitah has proposed a new way of understanding what are called nature-based solutions, land-use projects that sequester carbon. Nitah proposes a different approach: Land Relationship Planning. "Using" land is what people have been doing for centuries in the world. Reimagining our relationship to land is what Nitah proposes—the relationship humans have to the farmlands, grasslands, wetlands, and forestlands. It describes how people are connected to place, the interrelation and bonds that exist or are intended.

Recently, several proposals have been put forth to set aside 30 to 50 percent of nature for nature. Professor E. O. Wilson's Half-Earth Project works to conserve one-half of our planet's intact land and seas in order to preserve most of life on earth. Advocates call for 30 percent to be set aside by 2030. This is important for climate change. When we destroy nature, the intricate carbon fabric that connects all of life frays and shreds. The carbon in dying soils, trees, wetlands, and animals oxidizes and rises into the atmosphere as carbon dioxide. The boreal is not just a forest or habitat crisis. It is an extinction crisis, a civilizational crisis. As large, unbroken habitats shrink into fragmented islands, the number of species they support does not fall proportionately. Wilson believes we may still be able to save 85 percent of all species if we protect 50 percent of the earth. It is a fascinating proposal that has been a long time in coming. If ever there was an idea that deserved the title "Earthshot," this would be it. In the past few years, climate scientists and corporations speak increasingly about nature as an overlooked means to address climate change and "save the earth." The world is coming to the conclusion that in order to save nature, we need to, well, save nature. Exactly. Boreal and tropical forests deserve the highest priority with respect to the life (and carbon) in and on our terrestrial lands, but all forests are worthy of protection for now and always. ●

(*Left*) Snow-covered taiga forest in Finland.

(*Below*) Tailings pond at the Syncrude mine north of Fort McMurray, Alberta, Canada. Tailings ponds in the tar sands are unlined and will leach toxic chemicals into the surrounding environment for decades.

Tropical Forests

One of the most consequential actions we can take to end the climate crisis is the protection of tropical forests—and we must do it quickly. The world's forests are a critical sink for atmospheric carbon dioxide. Between 2001 and 2019, they sequestered twice as much carbon dioxide as degradation and deforestation caused them to emit—a net difference of nearly eight billion tons each year. That's more than the total amount of carbon dioxide emitted by the United States in 2019. Tropical forests are key. Until recently, they sequestered more carbon than any other forest type. But today they are being destroyed at a rate that may soon reverse their role as a carbon sink and cause the release of the large amounts of carbon that have been stored for decades in their branches, trunks, and roots. The reason for this loss is land degradation caused by logging, mining, urbanization, and conversion to agricultural use. Combined, they eliminate a football-field-size patch of tropical forest every 4.5 seconds. The rate is increasing. In 2020, despite a global pandemic and an economic downturn, 12 percent more tropical forests were destroyed than the year before, generating a fresh loss of 2.6 gigatons of carbon dioxide.

Of the three largest rainforests on the planet, only one, in Africa's Congo basin, remains a carbon sink and retains most of its accumulated carbon. A second, in Indonesia, has become a net source of carbon due to degradation and deforestation, and the third, in the Amazon basin, is teetering. The challenge is not just carbon. A recent study concluded that the amount of other greenhouse gases being emitted from the Amazon, including methane and nitrous oxide, likely exceed the climate benefit provided by the forest's uptake of carbon dioxide. Continued logging and deforestation will amplify this condition. Logging opens the door. Mature, intact tropical forests won't burn typically, but road construction and the elimination of trees dries the forest, creating conditions for fire. Tree combustion releases stored carbon, and clearing the land afterward causes carbon in the soil to decompose and return to the atmosphere.

One of the simplest solutions is to let tropical forests *grow*. Protect them from degradation and let them be. Older, larger tropical trees can asorb as much as three times more carbon than younger, fast-growing trees in similar environments. Specific types of trees matter. Large hardwood trees—the most attractive commercially—can take hundreds of years to mature because they grow only a few millimeters in diameter a year. As a result, they absorb a great deal of carbon, particularly in the last third of their lives. Large trees contain nearly half of the aboveground carbon. If an old-growth forest is converted to a plantation—palm oil, for example—the carbon loss can be significant. When logged and burned, tropical forests can lose

dropped to 6 percent. While deforestation activities by humans are the principal cause of the decline, physical changes to forests provoked by rising temperatures, extended droughts, and other effects of climate change are contributing as well. Rates of tree mortality have risen, primarily due to higher-than-normal temperatures and changing rainfall patterns. Scientists estimate there is a temperature threshold for tropical forests beyond which they cease to function properly. According to model projections, the Amazonian forest will reach this threshold by 2040. Reducing global greenhouse gas emissions is essential to the future of tropical forests.

The integrity of tropical forests cannot be separated from the Indigenous peoples who live within their borders, and any strategy to protect these critical natural resources must honor and support Indigenous rights. Research demonstrates that Indigenous reserves have low rates of deforestation and the highest levels of biological diversity. The lands that store nearly a quarter of all the stored carbon in tropical forests around the world are managed by Indigenous peoples, and one-third of the land has no formal protection or acknowledgment of Indigenous rights. This carbon is at risk, as are the people who live on the land. In Brazil, for example, forests within the Menkragnotí Indigenous reserve are a carbon sink, soaking up 11 million tons of carbon dioxide more than they produce, while the land outside the reserve has become a net source of carbon dioxide due to deforestation caused by mining, pasturing, and agriculture. During the devastating fires in the Amazon basin in 2019, fragmented and degraded forests facilitated the spread of the wildfires, while intact, protected forests on Indigenous reserves, such as the Kayapo in southeastern Brazil, were largely spared. Ensuring Indigenous rights to their land is a critical solution to the climate crisis.

Deforestation in the tropics is largely driven by the production of four commodities—cattle, soy, palm oil, and wood—much of which is exported. Reducing industrial demand for these products will lower the pressure on tropical forests and allow them to be protected. Working with Indigenous peoples and supporting them financially while implementing national economic policies that stop degradation and deforestation will help ensure that tropical forests and their significant reservoirs of organic carbon continue to support one of the most extraordinary and irreplaceable webs of life on earth. ●

between 35 percent and 90 percent of their carbon stocks. Recovery by natural regrowth can take forty years, but replenishment of carbon stocks can take one hundred years or more. For climate purposes, that is too long and too late.

Intact tropical forests are resilient. They resist fires unless fragmented or degraded, are less prone to drought, and are less susceptible to invasive species. They are cooler and wetter and recover more quickly from natural disturbances. Tropical forests shelter at least two-thirds of all species on earth, including a huge diversity of animals, plants, birds, insects, fungi, and countless soil microbes. Tropical forests are biological treasure chests. Many species are unique and have features adapted to microclimates within patches of rainforest. The health and productivity of tropical forests is inseparable from the web of life below their canopies. If wildlife populations are damaged, then the ecological integrity of the entire ecosystem begins to decline. Trees can regenerate in a few decades, but ecosystem recovery takes much longer and for certain species may not happen for hundreds of years. From a carbon and climate perspective, tropical forest ecosystems should be the focus of our efforts, not simply trees.

Time is short. The rate at which the world's tropical forests sequester carbon has dropped by roughly one-third since the 1990s, when they removed 17 percent of global carbon dioxide emissions. By the 2010s, this total had

Sunrise over lowland rainforest in the Danum Valley in Borneo. The Danum Valley Conservation Area encompasses 171 square miles. It is one of the most diverse forests in the world, with over 200 species of plants per hectare, 270 bird species, and 124 species of mammals, including the Bornean rhinoceros and pygmy elephant. The rainforest is said to be 130 million years old. In 2019, the tallest tropical tree in the world was discovered there, a 331-foot Yellow Meranti Tree.

Afforestation

Afforestation is the deliberate introduction of trees to land not previously forested. It can happen on a scale from small to large. The practice can also be applied to areas that were forested historically but are no longer, such as highly degraded land. It is different from *reforestation*, which is the regeneration of forests that have been damaged or cut down. Planting trees is an important natural solution to climate change because they sequester atmospheric carbon in their leaves, branches, trunk, bark, and roots for the duration of their lives. As trees grow and mature, the quantity of carbon stored can be significant. Additional reasons to plant trees include improving degraded ecosystems by stabilizing watercourses and hillslopes; improving soil health; integrating into agroforestry; creating living barriers that can slow the spread of deserts; creating habitat for wildlife; and reducing the damage caused by floods.

In 2015, scientists determined that there were approximately three trillion trees on earth, which sounds like a lot except it represents a *decrease* of nearly 50 percent since the early days of human civilization. More alarming, researchers calculated a net loss of ten billion trees each year due to deforestation, pest outbreaks, and wildfire. The land surface of the planet has absorbed about 30 percent of the carbon dioxide emitted by the combustion of fossil fuels over the past three decades, with the

majority of the sequestration occurring in forests. What if we afforested more land? A separate study estimated that more than two billion acres worldwide are available to grow as many as a trillion additional trees. The researchers excluded agricultural fields in their analysis but included grazing land, noting that a few trees could benefit livestock. "The potential is literally everywhere—the entire globe," said Thomas Crowther, who was an author on the study. "In terms of carbon capture, you get by far your biggest bang for your buck in the tropics but every one of us can get involved."

And we are getting involved. Large-scale tree planting campaigns have been announced recently, including the World Economic Forum's plan to have *one trillion* new trees in the ground by 2030. Planting projects have become popular with national governments, wealthy individuals, and major corporations,.

These efforts raise crucial questions. What sorts of trees should we plant? And where? On whose land? To what end? Have we consulted with the traditional Indigenous custodians and caretakers of the land? And what unintended consequences might afforestation have? While it is clear that afforestation can have a big role to play in the climate crisis, careful site and species selection must be considered. Some scientists, for example, object to the possible inclusion of lands that are neither deforested nor degraded and would be harmed ecologically by tree planting, such as Africa's vast grasslands and savannas. Many wildlife species are evolutionarily adapted to these landscapes and could suffer if their habitats are altered. New trees might adversely affect groundwater supplies. Nonnative trees could spread, displacing native ones. There are ethical questions too, such as whether low-carbon-emitting nations in the developing world should bear the ecological brunt of tree-planting projects on behalf of high-emitting countries with industrial economies. Social considerations are critical as well. Traditional livelihoods can be disrupted by the introduction of trees. Or the wrong type of tree might be selected. An afforestation project in Ireland that employed a nonnative spruce generated indignation from nearby residents. For climate purposes, tree planting projects need to endure for many years in order to maintain their carbon storage. This is a challenge for one of the most popular afforestation methods: tree plantations. Often composed of fast-growing nonnative species with high commercial value, plantations lose their sequestration benefit when the trees are harvested.

There are good examples we can follow. Botanist Akira Miyawaki developed a nature-based process for growing small, dense forests on degraded land using a diverse mix of native species adapted to local conditions. Tree species successfully planted include pines, oaks, Japanese chestnut, neem, mango, teak, guava, and mulberry. Miniforests have become popular in urban settings. In Kenya, Wangari Maathai founded the Green Belt Movement in 1977 in response to women who told her that streams were drying up and firewood was harder to find. Her organization worked with women to plant trees in local watersheds to store rainwater in the soil, boost crop production, and provide fuel, while receiving a small stipend for their effort. Fifty million trees have been planted and tens of thousands of women have been trained in forestry and food processing. Jadev Payeng, the "Forest Man of India," single-handedly planted a rich forest spread over 550 hectares to prevent erosion on Majuli, a river island in the river Brahmaputra.

China is home to more afforestation than the rest of the world combined. After calamitous flooding along the Yangtze River in 1998 exposed rampant deforestation and land degradation in the region, the government doubled down on a massive tree-planting and forest conservation program across the nation. The Three-North Shelterbelt Program includes approximately 50 billion trees and aims to create a three-thousand-mile barrier—dubbed the Green Great Wall of China—along the Gobi Desert to halt its encroachment and reduce the occurrence of sandstorms. However, there have been setbacks. The use of tree species not adapted to local conditions, including poplar and willow, and grown in vast monocultures has had limited success. China is exploring new approaches. The 160,000-acre Shandong Ecological Afforestation Project, south of Beijing, engaged local residents to plant a multistory forest using a mix of native and commercial species. Over a five-year period, hillside vegetation increased from 16 percent to nearly 90 percent, soil erosion was reduced by two-thirds, and water infiltration into the soil increased by about a third.

In California, arborist David Muffly has tackled a difficult question: What trees do we plant in shifting and unpredictable conditions caused by climate change? Native oak trees in the Bay Area, including the blue oak, are struggling to survive. A very small percentage of oak seedlings grow to maturity. Muffly's answer is to seek out similar tree species that have proven to be resilient and adaptable to hotter and drier conditions, such as the island oaks found on the Channel Islands off Los Angeles. These native oaks were once much more widespread, when California was drier thousands of years ago. Another candidate is the Engelmann oak, a native of northern Baja and the hills near San Diego.

Planting trees strikes a chord that resonates deeply in humans. It is a reason why more than thirty nations formally recognize an Arbor Day, including the United States. Done correctly, afforestation is a satisfying and cost-effective strategy for maximizing carbon sequestration. ●

(*Left*) "Johnny Pineseed" near Williams Lake, British Columbia, Canada.

(*Below*) Prickly tea tree seedling planted as one of fifty-seven thousand native seedlings placed in Mitchell River National Park in Gippsland, Victoria, Australia, by Greenfleet Australia.

Peatlands

Peatlands are wetland ecosystems where organic matter decomposes slowly and builds up large stores of thick, black carbon in the form of acidic peat. They are found on 3 percent of the earth's land surface yet comprise up to one-tenth of the earth's carbon stock. They account for 50 to 70 percent of the world's freshwater wetlands and are known by many names—swamp forests, fens, heaths, bogs, or mires. Except for Antarctica, peatlands are found everywhere in the world, from Arctic permafrost regions to the humid tropics, coastal regions, and boreal forests. Sogginess is a defining characteristic of waterlogged peatlands. Lack of oxygen creates an acidic, low-nutrient environment in which plants and wood decompose, forming cumulative layers of acidic peat. Fungi, sphagnum moss, reeds, and heather can all thrive in bogs, eventually forming a thick mat across an entire area.

More than 50 percent of the world's peatlands are located in Indonesia, including Sumatra, Kalimantan (on the island of Borneo), and the Malaysian state of Sarawak. They are unusual forests in every way, vastly different from the peatlands in boreal forests or Scotland, which are composed primarily of sphagnum moss, sedges, heather, and scrub. In the tropics, trees in lowland swamp forests can grow to the height of a twenty-story building. The forests are flooded in the rainy season and later dominated by dark, tannin-colored pools in the dry season.

The swamp and adjoining lowland forests of Indonesian Borneo contain numerous cultures whose livelihoods and food sources have been eliminated. Dayak communities used a shifting cultivation practice for millennia. There are more than 70 million Indigenous people belonging to one thousand ethnic communities in Indonesia. Deforestation has created poverty, human rights abuses, and sudden and enduring loss of traditional lands. It has likewise impacted plant and animal species. There are more than fifteen thousand plant species in the lowland forests. Bird species include hornbills and the operatic white-rumped shama, one of the most melodious birds in the world. Mammals include the clouded leopard, the Sumatran rhinoceros, the iconic orangutan, bearded pigs, yellow muntjac deer, gibbons, otters, macaques, langurs, and the proboscis monkey. Virtually all the island's species of

WEST
KALIMANTAN

EAST
KALIMANTAN

BORNEO

CENTRAL
KALIMANTAN

SOUTH
KALIMANTAN

mammals, reptiles, and birds are in decline, with some on the verge of extinction.

Indonesia aggressively opened up its rainforests starting in 1996. The government built 2,500 miles of irrigation canals and channels in order to create massive agricultural fields for rice cultivation. The result was the opposite of what was intended. The canals drained and dried out the peat forests. Logging moved in, followed by fire. Major paper and palm oil companies employed illegal large-scale slash-and-burn methods to clear the land. Peatland forests burn differently from conventional forests, creating what are known as zombie fires that are nearly impossible to extinguish, as ignition can burrow deep underground and tunnel sideways, erupting in unexpected and unintended locations. When the deeper layers of peat ignite, they consume a belowground fuel supply built up over five to ten thousand years. In the droughts of 1997 and 2002, forest fires spread throughout the peatlands, causing the emission of an estimated three to four billion tons of carbon. The 1997 fires caused the greatest single-year increase in atmospheric carbon dioxide since direct

measurements began in 1957. The Indonesian fires of 2015 consumed 6.4 million acres and covered most of Southeast Asia in a yellow haze. In addition to the environmental loss, the smoke created a healthcare and economic nightmare. An estimated one million people were stricken with respiratory diseases. The cost to quell the fires and mitigate impacts from the air pollution totaled more than $35 billion. Only four years later, in 2019, Indonesia was struck by another round of fires, with emissions nearly equal to those in 2015. Total carbon emissions were double those of the Amazon fires widely reported that year.

The conversion of a single acre of peatland emits twenty-eight to forty tons of carbon dioxide per year from

(*Left*) Early morning mist rising from the canopy of the lowland forest in the Danum Valley, Sabah, Borneo.

(*Above*) A NASA satellite image taken on September 24, 2019. By the end of 2019, 2.23 million acres had been burned in Kalimantan. Half of everything you buy in a supermarket has palm oil in it.

the drying peat. Former swamp forests are now used for palm oil plantations and have to be further drained as land subsides, creating a continuous flow of carbon emissions. Because the peat can run more than fifty feet deep, emissions will continue for decades. If we do the math, a ton of palm oil produced on former peatlands emits twenty times as much carbon dioxide than a ton of gasoline. Not surprisingly, Indonesia is the fourth-largest emitter of carbon dioxide on the planet. Ironically, palm oil is shipped in oil tankers to Europe, where it is classified as a renewable, sustainable biofuel under EU standards. Palm oil is also used throughout the food industry, for cookies, crackers, chips, candy, cereals, and chocolate, and in the cosmetic world, for shampoos, conditioners, body oil, lotions, lipstick, shaving gel, and other skin-care products. Pulp and paper made from single-species tree plantations on former peatlands carry a similar carbon impact. Their output creates newsprint, packaging, shipping boxes, and glossy catalogs.

The 2015 conflagration drew international condemnation and embarrassed Indonesian president Joko Widodo, who responded by establishing the Peatland and Mangrove Restoration Agency, committed to recovering more than 6.4 million acres of forested peatland by the end of 2020. Additionally, the government ordered plantation and forestry companies to rehabilitate another four million acres of peatland, a project that the agency is responsible for managing. It was charged with damming and blocking one hundred thousand miles of canals to rewet the land. Thus far it has restored five hundred thousand acres of former peatland, 8 percent of its 6.4-million-acre goal. The main obstacle faced by the Peatland Restoration Agency is the question of what to do with the restored lands. Very few crops grow in swampy or waterlogged lands. It is a branch of agriculture called paludiculture, unknown to most institutions, including agricultural colleges. Because of growing consumption, the efforts may flounder unless cash crops can be introduced for smallholders on the restored lands.

Out of the need to reverse course in Indonesia, the Katingan Mentaya Project was created, which preserves and rehabilitates 370,000 acres of original peatland forest, saving invaluable tracts of peatland from conversion into a plantation. It sells carbon credits, and with this funding the group has been able to install automated fire-monitoring towers, dam old drainage canals, perform carbon emissions research, and employ more than 1,500 local Indonesians to help protect the forest from seasonal fires. In order to cultivate local support and engagement in the project, families living near the forest are given seedlings to plant in some of the more degraded areas of the forest, and are compensated yearly for their participation. In Denmark, peatland restoration is supported as a way to reverse global warming. Not so long ago, the Danes were doing what the Indonesians were doing. They dug ditches, put in pipelines, and drained bogs to create new farmland. The Danish government, as part of its commitment to reduce greenhouse gas emissions 70 percent by 2030, is encouraging farmers to regenerate peat production by flooding their land, also known as "rewetting the swamp." Farmer Henrik Bertelsen is integral to Denmark's goal of becoming a carbon-neutral country by 2050. He floods 220 acres, nearly three-quarters of his land, to do it. Bertelsen's fields will avoid an estimated 2,700 tons of carbon dioxide equivalent emissions annually. His plan to flood his farm is part of a larger movement in Denmark to restore boglands that once made up 25 percent of the country's landmass. By the late 2000s, that number had dropped to 4.7 percent. Wetlands need to be wet, yet it can be difficult to achieve wetting in peatland restoration projects. Former bogs are often located on sites that have been drained, with irrigation canals and ditches put in place to lower the water table so that commercial crops can be grown.

Scientists recently found what is likely the "largest tropical peatland complex in the world." Greta Dargie and her team spent several years in punishing and difficult field incursions into exceptionally remote territory of the Democratic Republic of the Congo, known as the Cuvette Centrale, a large, level region in the heart of the biggest rainforest in Africa. Dargie's PhD thesis for Leeds University, in the UK, published in 2017, outlined how she, senior colleagues, and her graduate adviser, Simon Lewis, had discovered undocumented peatlands greater than the size of England. Her paper "rattled" the assumptions of many scientists who had known about the flooded conditions in the Cuvette Centrale (cuvette means "bowl" in French) but did not know that the soils were peat and thus could contain 33 billion tons of carbon.

The future of peatlands' spongy carbon sinks depends on a combination of factors: the technology to locate and monitor peatlands, efforts to wean the world off palm oil, and the willingness to preserve and protect remaining intact peatlands. Focusing global efforts on 3 percent of the land on earth means keeping 650 billion tons of carbon out of the atmosphere. For example, if the Cuvette Centrale peatlands in the Congo caught fire, it could triple annual global emissions of greenhouse gases. Indonesia has allowed 250,000 square miles of peatland to be converted to plantations, an area the size of France, for palm oil. The Cuvette Centrale is, unfortunately, the perfect habitat for *Elaeis guineensis*, which is the source of the most prevalent variety of palm oil cultivated today, because it originated in the Congo. Investors are already planting there. ●

(*Left*) Palm fruit being moved into-high pressure steam chambers to remove impurities. The Sapi palm oil plantation is said to be the largest palm oil trader in the world. One hectare produces six tons of palm oil (soy oil produces one ton per hectare).

(*Below*) A proboscis monkey in the Kinabatangan Wildlife Sanctuary, Sabah, Borneo, one of the many endangered species due to forest destruction by palm oil companies. They are arboreal monkeys that rarely venture onto land, but with habitat fragmentation they need to travel by land to find food and thus become prey to jaguars and native people as bushmeat.

Agroforestry

Chickens originated in the forests of Southeast Asia, evolving from ancestors called jungle fowl. This bit of knowledge is key to the vision of an innovative, poultry-centered agriculture that involves trees. Fowl graze freely in the shade of hazelnut trees, feasting on insects in the cover crops planted between the rows of trees. The chickens fertilize the soil while receiving leafy protection from overflying predators. The trees generate a commercially useful product. This is the basic practice of agroforestry, an agricultural method that combines varied and dynamic combinations of trees, woody perennials, annuals, and livestock.

Agroforestry can scale from less than an acre to thousands of acres supporting entire communities of plants and people. The diverse combination of plantings honors the ecological needs of the land as well as the economic needs of the farmer, creating a resource management system that has been widely touted by scientists and practitioners as nearly ideal, especially in areas of the world vulnerable to climate fluctuations and soil erosion. Trees provide shelter from wind, slow down water erosion, and reduce the evaporation of moisture from soil and crops. Tree shade reduces soil temperature and creates microclimates favorable to crops that might not have grown in open sunlight. Tree roots anchor the soil; falling leaves, pruned branches, and decomposing bark mulch the ground. Water infiltration is improved, and the soil is continuously enriched with organic matter. Shrubs and blossoming trees provide pollen for beneficial insects, including pollinators. The shade, moisture, and organic matter provided by trees support a wide variety of soil microbes that enhance nutrient uptake and build soil structure.

Many of the trees used in agroforestry fix nitrogen, an essential nutrient for plant health and growth. In general, commercial crops cannot access atmospheric nitrogen directly. Instead, they form symbiotic relationships with a type of fungi in the soil that makes nitrogen and phosphorus available to plant roots in exchange for carbon. These fungi acquire nitrogen from bacteria who get it from the roots of certain plants, shrubs, and trees that can absorb the gas from the atmosphere through their leaves. In conventional, monocropped agriculture, the absence of nitrogen-fixing plants on the farm means the nutrient must be applied as an artificial fertilizer, with well-documented negative consequences. In agroforestry systems, the fertilizing nitrogen is provided naturally by plants and trees. As a bonus, much of the carbon exchanged between fungi and roots, which originated as carbon dioxide in the air, will remain sequestered in the soil.

In America, crop agriculture and forestry were historically viewed as separate disciplines, particularly in the West, where forests were for saw logs, and the Midwest, where the Great Plains were for annuals—soy, corn, and wheat, for pigs, cows, and people, in that order. However, if seen from a drone, most watersheds in the nation are a mosaic of crops, trees, livestock, and natural areas, each vitally important to the culture and economy of local communities. In the mid-1990s, the U.S. Department of Agriculture began to create standards that would encourage landowners to integrate trees with crop and livestock practices, dividing agroforestry into three farming systems.

Alley cropping: food crops grown in passages created by rows of trees or bushes. Alleys can be narrow or wide, and the rows of trees can be straight or curve to follow contours on the land. The trees can be fruit, nut, softwood for timber, or hardwood for specialty use. The annual or perennial crops grown in alleys can be any type, including cover crops that can be grazed by livestock. Different types of trees and crops grown together can diversify farm income sources while creating ecological benefits. Shade protects livestock, which fertilizes the soil; bushes shelter pollinators; leaf mulches provide food for earthworms and microbes. Alley cropping systems are dynamic. As trees and shrubs grow, their canopies and roots expand and their nutrient and water needs change, affecting crop production. Farmers use this dynamic to their advantage, growing more crops as soil fertility rises, for example, or planting new trees. The system's crop versatility buffers a farm operation against market and climate fluctuations, making it more resilient.

Silvopasture: the intentional integration of trees and livestock on the same parcel of land. The Latin word *silva* means "woods, forest, or jungle." Its combination with *pasture* signifies agroforestry managed for trees and animals. Year to year, livestock are carefully managed to ensure the sustainability of the pastures, often employing short-duration grazing. Long-term, cultivated trees supply a steady stream of valuable products, including edible leaves for livestock. Some silvopasture systems emphasize animals, which can include cattle, sheep, goats, llamas, horses, pigs, chickens, ostriches, yak, caribou, and deer. Other silvopasture systems focus on fruit, nut, and wood products, with grazing as a complementary activity. Sheep grazing in vineyards, for instance, can help keep weeds and other vegetation under control. Pigs and boars clear forest understory, and their rooting, carefully managed, can stimulate grass growth. In turn, the shade provided by trees provides relief to grazers and wildlife on hot days.

Forest farming: growing crops under the protection of a managed tree canopy. Forest farms are usually small-scale operations and are most frequently found in tropical environments, where they are also called home gardens. While some food is wild-harvested, most forest farms are planned and cultivated. Particular attention is paid to vertical space

and the arrangement of tree species of varying heights. The interplay between crops and trees as they grow is the key to successful forest farming. Some forest farms focus on high-value woodland products, such as mushrooms, berries, nuts, and herbs used for medicinal, spiritual, and cultural purposes. Seeds from wild plants are collected and sold, providing another source of income. Livestock other than fowl are generally excluded from forest farm systems. By intermixing crop production with the ecology of a natural forest, farmers reap the best of both worlds.

Research confirms what practitioners have long experienced: integrating trees on farms creates important benefits, including carbon sequestration in soils and plants, greater crop yields, less wind and water erosion, and diverse enterprises for rural economies. On average, a shift from row crops to agroforestry increased soil organic carbon by 34 percent.

In India, the integration of trees and farming is a centuries-old tradition and is currently deployed on 33 million acres, providing 65 percent of the country's timber. India was the first nation in the world to formally adopt a national policy for agroforestry. In Africa, examples of agroforestry include shade-grown coffee plantations in Ethiopia, multistory home gardens in Tanzania, woodlots in Kenya, and savanna-like systems in the Sahel region, south of the Sahara Desert. In Southeast Asia, fruit and nut trees are planted within and along the edges of rice fields to create shelter for seedlings as well as provide an extra source of income for farmers. In Indonesia, agroforestry practices are successfully being implemented to restore land severely degraded by decades of rainforest clear-cutting.

In Niger, West Africa, a farmer-led movement is restoring highly degraded land using an effective, low-cost woodland management technique that grows indigenous trees and shrubs from stumps, roots, and seeds. Called farmer-managed natural regeneration (FMNR), this agroforestry system had its origins in the severe famines that struck the Sahel region in the 1970s and 1980s. Faced with chronic poverty and a steady loss of vegetation and other natural resources due to desertification, farmers and agricultural specialists discovered an "underground forest" of living tree stumps, the remnants of native woodlands that had been cut down years earlier to clear the land for

A bird's-eye view of organic agroforestry at Serendipalm, an organic, fair-trade palm oil project funded and created by Dr. Bronner's in Asuom, Ghana. Seen here are the intercropped oil palm, cocoa, bananas, cassava, citrus, and various timber trees. A fair-trade premium is paid to farmers and is held in a fund managed by the farmers, employees, and Serendipalm's management. Fair-trade income has funded drinking-water wells, materials and uniforms for schools in Asuom, residential homes for the nearby hospital, public toilets, five thousand mosquito nets, a computer lab, and scholarships, among other things.

crops. They coaxed the stumps to resprout, starting the process of regeneration, and employed traditional pruning methods to encourage vertical growth. Today, the numerous trees and shrubs are integrated into existing agricultural activities, including livestock grazing, creating a dynamic set of relationships that have resulted in a rise in soil fertility, soil moisture, and crop yields. They also provide fruit and firewood. A recent survey revealed that more than ten million acres of Niger, roughly 50 percent of its farmland, have been regreened by FMNR, increasing food security and helping to build resilience to weather extremes. FMNR has caught the attention of farmers, development agencies, and NGOs in other parts of Africa and around the world.

In Ghana, an innovative agroforestry enterprise began in 2007, when the Dr. Bronner's corporation partnered with local farmers and agricultural workers to produce organic, fair-trade palm oil to be used in the company's bar soap. Palm fruit oil is a popular substitute for partially hydrogenated oils, which contain unhealthy trans fats, and is used in a wide variety of foods, cosmetics, cleaning agents, and biofuels. However, palm plantations are highly destructive to the environment, and industrialized oil factories can have detrimental effects on local communities. To provide a regenerative alternative, Dr. Bronner's launched the Serendipalm project, which buys palm fruits from nearly eight hundred small organic farms (five to seven acres) and processes the oil in the town of Asuom, using a mill that employs more than two hundred people at well-above-average living wages. The leftover palm kernels are sent to a fair-trade palm oil project in neighboring Togo, and the remainder of the fruit is mulched into fertilizer to be spread on the farms, returning nutrients to the soil.

In 2016, Serendipalm began helping its organic palm farmers, who often grow cacao, fruit, and other tree crops on their land as well, to adopt additional agroforestry practices, including mixing different types of trees in stratified layers to mimic natural forests. The combination of different tree heights, leaf sizes, and canopy densities optimizes the collection of sunlight. This photosynthetic boost improves plant growth and vigor. Among other advantages, healthy trees are less susceptible to pressure from pests. Leaf litter, branch clippings, and the different types of plant roots combine to stimulate diverse populations of soil microbes. The result is higher crops yields for the farmer, higher-quality products for Serendipalm's markets, improved food security for local communities, greater ecological benefits for plants and animals, and higher levels of carbon sequestration. Demand for palm oil and organic cocoa from the project continues to rise and Serendipalm is planning an expansion of its work with farmers to include fair-trade cassava flour, turmeric, and fruit purees.

In Polynesia, agroforestry was widespread throughout history and provided most of the raw materials that residents needed for subsistence, craft, construction, and rituals. The particular type of agroforestry practiced creates vegetation patterns called novel forests. They include mixes of introduced and native species that were initially perceived by non-Polynesians as natural landscapes but were, in fact, intentional systems based on Indigenous knowledge. Novel forests are low-labor, multigenerational

resources that comprise a particular form of biotic landscape, often without permanent infrastructure such as terraces, canals, or field systems. They can replace some of the ecological functions of natural forests and may exceed previous levels in certain areas. Trees are grown in permanent multistory home gardens with understory crops of banana, taro, and yam. Expansive patches of novel forests contained nineteen species of tree crops, including Tahitian chestnut, mountain apple, and candlenut, though the dominant crops are breadfruit and coconut.

In Europe, there is a long tradition of integrating pastoral and forest enterprises with wildlife conservation and local food production. One example is called a *dehesa*, a diverse mosaic of trees, shrubs, perennial grasslands, annual crops, and grazing animals, all part of a highly managed silvopasture system tailored to land and water limits. Occupying nearly eight million acres across southwestern Spain and eastern Portugal (where they are called *montados*) the savanna-like *dehesas* are the result of careful cultivation. Over the years, land was cleared of evergreen trees, shrubs, and other woody species in favor of native grasses and oak trees, including helm and cork. Helm oaks grow an acorn prized by pigs. The thin soils and dry climate of the region meant the land was used primarily for livestock grazing at low densities. Over time, agricultural uses expanded to include browsing goats, the cropping of certain plants, mushroom-gathering, beekeeping, and the production of natural cork (to serve the burgeoning wine industry). Recently, there has been a focus on olive and grape groves and raising heritage livestock breeds such as the Spanish fighting bull and the Iberian pig. *Dehesas* are also home to a variety of endangered species, including the Spanish imperial eagle, the Spanish lynx, and the black vulture.

A dehesa, like all agroforestry systems, is an intentional landscape requiring human management. Without intervention, brush and other woody vegetation grow back, reducing the benefits of the savanna effect. Oak regeneration requires pruning, nurturing, and protection from livestock when the trees are young. As a traditional landscape with deep roots in the region, *dehesas* have enduring relationships with local farmers and communities. Generational knowledge based on long experience has been critical to their success. Modernization efforts, including the adoption of industrial food-production practices, have largely been resisted. Researchers have identified the carbon sequestration benefits of *dehesas*. The carbon and nutrient cycles remain largely self-contained within their boundaries. The low-intensity production of *dehesas* combined with the diverse sources of carbon, including leaves and manure, means they are effective at capturing and storing organic matter. According to one study, these agroforestry practices improve soil carbon storage while reducing the carbon dioxide flow from the soil to the atmosphere. In another study, higher amounts of soil carbon were measured below the tree canopies than in the pastureland, suggesting that maintaining or increasing tree cover in *dehesas* may increase long-term storage of carbon in their soils.

There is potential for agroforestry systems in arid environments as well. Agaves, for instance, have been cultivated as sources of food and fiber by Indigenous societies for thousands of years. Best known for their role as the source of tequila, two hundred varieties of agaves grow around the world, including on degraded lands that are no longer suitable for annual crop production. They thrive in hot climates, require very little water, are drought-tolerant, grow all year, and live as long as a century, making them ideal for a world of rising temperatures under climate change. Certain varieties of agave average forty tons per acre of dry weight each year, which is very high and could have a big impact on climate change while providing regenerative incomes for people in regions beset by a legacy of degraded land. Agaves are often grown in association with nitrogen-fixing trees, including mesquite, ironwood, huisache, and acacia.

Agroforestry includes cacti. The prickly pear cactus is being actively cultivated as a source of food and fuel. The edible parts of the iconic desert plant, called nopal in Mexico and featured on the country's national flag, are used in salads, salsas, soups, tamales, and casseroles, or ground into flour for tortillas. The inedible parts, usually thrown away as waste, are being distilled into a source of biofuel for vehicles and biomethane gas for electricity generation. This double use of a plant that grows bountifully in areas inhospitable to most crops has the potential to support human populations in lands vulnerable to climate extremes.

Over the past century, trees were removed from agricultural fields around the world as mechanized fence-to-fence production took hold in industrialized nations. Deforestation is prevalent in nearly all developing nations, often at the expense of traditional food systems. Today, agroforestry systems are making a comeback, spurred by their ability to meet a variety of pressing environmental challenges, including carbon sequestration, while sustainably producing food, fiber, and forage. ●

The Serendipalm agroforestry project supports many adjoining smallholder partners and trains them to create the most effective maintenance programs for best yield and impact. All the cuttings from pruning help increase soil fertility. Here, a group of farmers are opening cocoa pods to extract the beans. The harvesting farmer invites neighboring farmers and friends for the breaking of the pods. Food is prepared for all as an act of community reciprocity. The wet beans in the pans are then fermented.

Fire Ecology

Fire can either destroy or regenerate a landscape. In Western industrialized nations, for more than a century fire has been viewed as destructive. However, Indigenous people have actively used fire for thousands of years to cultivate abundant, productive forests and grasslands around the world that were immune to destructive conflagrations. When colonists arrived in North America in the 1500s, the forests were lush with plant life and healthy, magnificent trees. Many forests were so spacious that colonists could drive their wagons between the trees. European settlers didn't realize that they were experiencing tended forests, not an ecosystem in its wild state. For millennia, Native tribes along the coast of western North America have used ancestral burning practices to lower the risk of extreme fire events while promoting desired landscapes and abundant game. Rotating sections of forest and grasslands were burned over the years to reduce fuel loads of brush and second growth. Regenerated, the forests grow food, including huckleberries, acorns, lily bulbs, and mushrooms, and useful products such as hazel shoots, used to weave baskets.

The word *indigenous* is the adjective for the noun *indigene*, which refers to people inhabiting or existing on lands from the earliest times, or at least before the arrival of colonists. The majority of Indigenous cultures have occupied their homelands for thousands of years. The cumulative knowledge derived amounts to what could be called local or observational science—centuries of insight about how to interact with their homeland for the greatest benefit to the living systems they depended upon. What emerged and developed was the understanding of biophysical cycles and how they impacted the landscape. Indigenous people increased the abundance of the land—and fire ecology played a crucial role. That was not the case with the colonists. The European hunger for timber saw fire as a threat to assets, holdings, and property. In the United States, lighting fires on public lands was criminalized in 1911. Fire suppression became Forest Service policy—all fires were ordered to be put out by 10:00 a.m. the day after their discovery. National forests grew unchecked as a result, acquiring densities of small trees and understory vegetation that have fueled intense, destructive fires.

Global warming and overgrown forests have created ideal conditions for uncontrollable fire. Fire seasons are longer now in many parts of the world, the number of fires have increased, the amount of land burned is more extensive, and the cost of firefighting has risen dramatically. In 2019, more than eight billion tons of carbon dioxide were released globally by forest fires, nearly one-fourth of total carbon dioxide emissions. In the 2019–20 fire season, Australian fires released 491 million tons of carbon dioxide, effectively doubling the nation's total greenhouse gas emissions for the year and vaulting it from the fourteenth highest emitter to sixth.

The basic principle of Indigenous fire management employs carefully timed, low-intensity fires to remove underbrush and regenerate important grasses and perennials, especially in fire-prone environments. The key is timing. Fires are lit in certain areas depending on the season, air temperature, weather, and wind. Some criteria are more subtle. They come from continuous observation and a close connection to the land, such as whether there is heavy dew in the morning and the afternoon. At the beginning of autumn in western North America, Native tribes burned after the first rain because they could maximize impact on pest populations while minimizing risk to the mature trees. Native tribes in California understood that some species took longer to recover after a fire, so they burned segments of the forest in a specific rotation. This allowed time for certain plants to mature. In the first year after what is known as a cultural burn, grasses and shoots of hazel could be gathered for basketry. In the second and third years, berries are abundant on the shrubs.

A similar technique is employed by Aboriginal people in Australia, but the timing of their burns is mandated by the monsoon. As sections of land dry out after the rainy season, Indigenous Rangers light and monitor hundreds of small fires. Today, the majority of defensive burning by Aboriginal people takes place in Australia's Northern Territory. Aboriginal fire prevention methods have reduced dangerous wildfires by half in northern Australia. Wildfires are burning 57 percent fewer acres, and greenhouse gas emissions have gone down by 40 percent. Because Australia employs a cap-and-trade system whereby emitters compensate those who sequester or avoid emissions, the organizations using ancient burning techniques have earned $80 million, which they have reinvested into their communities for better education and hundreds of jobs.

Native fire management techniques are gaining momentum in the United States. The Klamath River basin has been home to Indigenous people for thousands of years. Until the 1900s, salmon were plentiful, as were their predators, including black bears and bald eagles. Today, the Klamath is no longer teeming with salmon, sturgeon,

or steelhead. The Megram Fire consumed 125,040 acres inside Six Rivers National Forest in 1999. Karuk and Yurok tribe member Frank Lake realized that in order to restore salmon and the watershed, he needed to bring Native fire management practices back to his ancestors' land. First he had to convince the U.S. Forest Service to switch from fire suppression to Indigenous methods employing fire, which could reduce the number and potency of the wildfires in the region.

After a decade of collaboration and advocacy, the U.S. Forest Service is now partnering with tribal groups and nonprofits to use Karuk fire practices to burn several thousand acres within the Klamath and Six Rivers National Forest. If all goes well, Frank Lake hopes the partnership will one day maintain the more than one million acres of Karuk tribal lands. For now, he is part of a growing community of fire ecologists, researchers, environmentalists, and government officials who understand the value of Indigenous fire management. Increasingly, management tactics are shifting from waiting for lightning to strike to proactive, regular burns. It's a change of policy welcomed by Indigenous peoples.

Members of the North Fork Rancheria Mono tribe in Northern California call targeted burns "good fire." Historically, they used fire to maintain the health of the woodland ecosystem that provided their food, water, wood, and fiber. Good fire practices created significant cultural bonds with the land. Today, as Indigenous wisdom is being recognized by state and federal authorities, tribes are bringing good fire back. Small areas are being burned with low-intensity fires managed by tribal members. But it's not simply a matter of employing proper management practices. Good fire is an attitude. Instead of "fighting" fire as if it were an enemy, the Mono and other tribes view fire as a partner in the critical work of regenerating and stewarding our shared planet. This partnership is woven into the lives of Indigenous peoples. Land and fire can't be separated from culture and history. They can't be managed abstractly; they are part of a spiritual, social, and ecological whole. Although Indigenous fire knowledge is being respected once again, there is a risk that it will be oversimplified and measured by a narrow set of outcomes. It is the whole that matters. Effective, long-lasting solutions must build on Indigenous knowledge, support Native communities, and protect their rights. ●

A Yurok firefighter manages the boundary of an Indigenous-prescribed burn near Weitchpec, California, during a fire training exchange, or TREX. In recent years, Indigenous fire ecology has received attention as more wildfires have raged destructively across the West Coast of the United States.

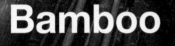

Bamboo

Bamboo is delicious. It's food for people and for pandas, and medicine for both. Some species can grow thirty inches a day. It can reach maturity in five years. When harvested, it regenerates without reseeding. It doesn't require irrigation, pesticides, or fertilizers. It's a grass but looks like a forest. It may provide a third more oxygen to the atmosphere than a similar-sized stand of trees. It sequesters carbon, grows in degraded land, loves sunlight and shade, and can be a crop in agroforestry projects. Native to five continents, bamboo is used by two and a half billion people worldwide. It is attractive and has deep cultural value. It is lightweight and yet incredibly strong. Anything made from wood can be made with bamboo. Its fibers are soft, durable, flexible, and absorbent. It can be made into charcoal for cooking fuel. Thomas Edison used a bamboo filament in his lightbulbs. Its shoots are a popular staple in meals. It can be turned into rugs, drapes, beds, chairs, tables, cups, utensils, plates, toys, jewelry, art objects, motorcycle helmets, and musical instruments.

Bamboo makes great toilet paper. The process is similar to the one that converts virgin wood into conventional toilet paper, breaking down fibers into pulp with heat, water, and pressure. The difference is the source. Virgin wood can originate in the boreal forests of Canada. Slow-growing, carbon-dense trees are clear-cut, damaging a critical carbon sink and destroying wildlife habitats for decades. By contrast, bamboo requires less land to grow, regenerates quickly, and is adaptive to multiple climatic conditions.

Bamboo can sequester significant amounts of atmospheric carbon. The speed at which bamboo grows, the height it can reach (over ninety feet), its extensive root system, and its ability to resprout after harvest add up to a large amount of carbon storage above and below ground. During the early years of growth, a bamboo stand can sequester more carbon than a comparably sized group of fast-growing trees. A forest of giant bamboo has been shown to store roughly 134 tons of carbon per acre over a sixty-year period. Part of bamboo's advantage can be found in its roots. Some of the 1,662 species of bamboo grow in clumps, but many grow from rhizomes—horizontal roots, also called runners, that spread quickly and can reach ten feet in length in a single season. While the average life span of an individual bamboo culm is less than ten years, the roots can live for decades, storing carbon. The key is cultivation. Unmanaged bamboo doesn't grow as robustly as managed stands, storing less carbon as a result. Trimming and selective harvesting give bamboo room to grow, enhances the carbon dioxide–capturing capacity of a given stand, and doesn't harm the root system.

Silica structures called phytoliths are abundant in the cells of bamboo plants. Phytoliths seal carbon inside themselves. The silica is highly resistant to degradation, so when bamboo trunks or leaves fall to the ground, the stored carbon continues to be sequestered, often for thousands of years, long after the bamboo itself has decomposed. The climate impact could be significant. The sequestration potential from phytoliths is estimated to be 750 million tons of carbon dioxide per year if half of the 10 billion acres of potentially arable land were used to grow bamboo. A recent study indicates that this total may be higher when phytoliths found in belowground bamboo trunks and root systems are included.

Bamboo's carbon continues to be sequestered when the harvested material is turned into a long-lasting product, including timber for the construction of a home, office building, or bridge. Remarkably, bamboo has greater tensile strength than steel, and greater compressive strength than concrete. Bamboo can replace hardwoods, such as mahogany, to create flooring, furniture, and other household items, thus reducing pressure on endangered forest systems and critical wildlife habitat. If treated to protect against rot and insect infestation, bamboo can last fifty to one hundred years. Advances in manufacturing enable bamboo to be cut with straight edges, cross-laminated, and arranged into shapes that serve a range of structural needs, increasing its utility and further enhancing its role as a global carbon sink. New technology focused on digital design and the fabrication of superstrong composites is allowing bamboo to become a substitute for steel, cement, and other widely used building materials, avoiding the greenhouse gas emissions normally generated by their manufacture, and instead locking carbon into the built environment for decades.

Bamboo is an important tool for restoring degraded ecosystems. Its root systems, particularly runners, can stabilize damaged land, protecting it from wind and water erosion. Bamboo can be grown on steep slopes and in thin soils. China has been planting bamboo for years to slow desertification, restore degraded farmland, and create windbreaks. In Malawi, rows of giant bamboo are being planted to offset the destruction caused by decades of deforestation while providing a replenishable, quick-growing source of timber and fuel for local communities. In Nicaragua, a group called EcoPlanet Bamboo has restored thousands of acres of degraded forest by planting clumps of Guadua bamboo, a native species, interspersed with standing trees. The resulting bamboo fiber, certified by the Forest Stewardship Council for sustainability, is exported for manufacture. In India, an area of former farmland devastated by kiln owners who mined the topsoil for industrial brickmaking has been planted with groves of moringa, bamboo, guava, banana, mango trees, staple crops, and vegetables. Leaf litter has accumulated on the ground over the ensuing years, increasing rainwater infiltration and boosting microbial populations in the soil as it decomposed. The depleted water table has risen by fifty feet in twenty years.

Bamboo has important economic and social benefits. As a crop, its rapid growth and regeneration provides steady income for farmers. It is a source of charcoal for cooking stoves and home heating. It burns more cleanly than traditional wood-sourced charcoal. Trees in many parts of the world require decades to mature, so the use of bamboo as a fuel reduces a significant source of deforestation. Bamboo can be converted into pellets or biogas to be used in the production of electricity. It can be a source of biochar, which sequesters its embedded carbon for long periods. Bamboo's hardiness and versatility mean it is well suited to meet changing ecological and financial conditions brought on by global warming.

Bamboo isn't perfect. Like many grass plants, it can spread invasively if not managed properly. Care must be taken to ensure that it is integrated with other types of vegetation so it doesn't outcompete its neighbors. If grown as a plantation monoculture, it can have a detrimental effect on native species, including wildlife. Bamboo has become a popular source of fiber for clothing, but its use requires scrutiny. Rayon, for example, is a fabric made from tree cellulose that has undergone a production process involving toxic chemicals that have been linked to human illnesses. Its by-products create air and water pollution. Some manufacturers use (and promote) bamboo as a feedstock for rayon, often without explaining that all the original fiber will be chemically dissolved. If the process of creating bamboo rayon can employ green chemistry, it can become a sustainable clothing fiber.

Bamboo is a multipurpose natural climate solution, valuable for agroforestry, food production, building construction, land restoration, rural economic development, wildlife habitat protection, and atmospheric carbon sequestration—all rolled into one. ●

The sinuous trunk of a Korean pine (*Pinus koraiensis*) being overtaken by Moso bamboo (*Phyllostachys heterocycla f. pubescens*) in the Gyeongsang-do walled town in Korea.

Patricia Westerford in *The Overstory*
Richard Powers

When Suzanne Simard moved from Canada to study forest ecology at Oregon State, she found a different world than the one she left. Having grown up in the temperate rainforests of British Columbia, where she was raised by a family of horse loggers, she was accustomed to seeing trees that were selectively cut and skid along the ground one by one. Cedar logs would be deposited in roadside collection areas, with little damage to the forest floor. The disturbances that did occur could help nuts and seeds germinate, sometimes with fertilization courtesy of the horse. In Oregon, logging was going full-bore. Checkerboard sections of forestlands were clear-cut, scraped clean with bulldozers, and sprayed with herbicides. The "cleared" land was then planted with rows of seedlings. Gone were the cedars, maples, bitter cherry, hazel, alder, spruce, and ash. Tree plantations were free of competition, properly spaced, well watered, and bathed in sunlight. However, it was obvious that the replanted Douglas fir seedlings were not as robust or healthy as the native fir. Failure seemed to coincide with the uniform plantings and the barren industrial landscape, devoid of brush and native trees.

For her doctoral thesis in forestry, Simard went to the woods. That may sound obvious, but at that time, forestry students largely stayed on campus, performing genetics research in labs or with cuttings in greenhouses. She wanted to work out why diverse, dense, old-growth forests were more productive and healthier than Weyerhaeuser and Plum Creek second-growth forests, plantings that had been optimized using current forest science. She was pretty sure the cause was underneath the trees, in the soil, among the roots—changes in forest ecology caused by wiping out the diverse plant life and replacing it with one species.

She returned to the forests of British Columbia for her research. Starting with injections of two types of carbon into Douglas fir and paper birch, she discovered that the Douglas firs were receiving carbon (sugars) from the paper birches when the fir trees were more shaded in the summer. When the seasons reversed and the birches were leafless, the Douglas firs were sending carbon to the paper birches. This unexpected, paradigm-busting phenomenon didn't have a scientific name, at least not in the world of forestry; interdependence, symbiosis, reciprocity, altruism, and generative are some descriptors. Through the filamentous, whitish, fungal underground network known as mycorrhizae, the trees were exchanging food, water, and chemical signals to support multiple species of trees and plants. The plantation trees were essentially orphans. No network, no support, no "family." When her research was published in Nature in 1997, the editors referred to her paper as the "The wood wide web." It caused a sensation in the press and pushback from male colleagues. Since then, Simard's work has been replicated by scientists many times over. Her fundamental discovery is that an intact forest is a community, not a grouping of competitive species. Scientists do not question her methods, techniques, or results. It is her thesis that is criticized, by scientists who cannot forgo the Darwinian legacy of unbending competition as the main determinant of species evolution. Curiously, Darwin's writings were informed by economist Adam Smith's theories of competitive advantage, and that interpretation has informed capitalist ideology ever since. Simard sees it differently. She uses the words mother trees in her work, to describe old-growth trees that support hundreds of other plants and trees. Simard was looking for another quality in her trees and plants, their ability to interrelate, not impede.

In Richard Powers's Pulitzer Prize–winning 2018 novel The Overstory, one of the main characters, Patricia Westerford, is fashioned after Simard. In the novel, Dr. Westerford publishes a paper saying, "The biochemical behavior of individual trees may make sense only when we see them as members of a community." In the following months, she is sharply, if not cruelly, criticized by colleagues and university faculty. The combination of shunning and ridicule drums her out of the academy. She travels west and disappears into the wilderness for many years, becoming a guide. More than two decades later, she resumes a career at a forest research station in the Cascades. Her original findings are eventually confirmed, and she is greatly sought out after her book The Secret Forest is published. She is invited to give a keynote at a prestigious climate conference in Silicon Valley, where executives and corporate leaders gather to discuss the plight and future of the world. Her talk is entitled "The Single Best Thing a Person Can Do for Tomorrow's World." She is nervous and flustered, and struggles to get all the words out. At one point in her talk, people walk out. —P.H.

The Sir Isaac Newton coast redwood is located in Prairie Creek Redwoods State Park, on the unceded lands of the Yurok Nation. It is the third largest single-stem coast redwood, just shy of 300 feet in height and nearly 70 feet in diameter. The burl seen to the left weighs 40,000 pounds. It stores the genetic code of the parent tree. In the forest are Douglas fir, Sitka spruce, and western hemlock.

The auditorium is dark and lined with redwood questionably obtained. Patricia looks out from the podium on hundreds of experts. She keeps her eyes high above the expectant faces and clicks. Behind her appears a painting of a naïve wooden ark with a parade of animals winding up into it.

"When the world was ending the first time, Noah took all the animals, two by two, and loaded them aboard his escape craft for evacuation. But it's a funny thing: He left the plants to die. He failed to take the one thing he needed to rebuild life on land, and concentrated on saving the freeloaders! The problem was, Noah and his kind didn't believe that plants were really alive. No intentions, no vital spark. Just like rocks that happened to get bigger.

"Now we know that plants communicate and remember. They taste, smell, touch, and even hear and see. We, the species that figured this out, have learned so much about who we share the world with. We've begun to understand the profound ties between trees and people. But our separation has grown faster than our connection.

"In this state alone, a third of the forested acres have died in the last six years. Forests are falling to many things—drought, fire, sudden oak death, gypsy moths, pine and engraver beetles, rust, and plain old felling for farms and subdivisions. But there's always the same distal cause, and you know it and I know it and everyone alive who's paying attention knows it. The year's clocks are off by a month or two. Whole ecosystems are unraveling. Biologists are scared senseless. Life is so generous, and we are so . . . inconsolable.

"You see, a lot of folks think trees are simple things, incapable of doing anything interesting. But there's a tree for every purpose under heaven. Their chemistry is astonishing. Waxes, fats, sugars. Tannins, sterols, gums, and carotenoids. Resin acids, flavonoids, terpenes. Alkaloids, phenols, corky suberins. They're learning to make whatever can be made. And most of what they make we haven't even identified.

"Dragon trees that bleed as red as blood. Jabuticaba, whose billiard-ball fruits grow right out of the trunk.

Thousand-year-old baobabs, like tethered weather balloons loaded with thirty thousand gallons of water. Eucalypts the color of rainbows. Bizarre quiver trees with weapons for branch tips. *Hura crepitans*, the sandbox tree, launching seeds from its exploding fruit at 160 miles per hour.

"At some time over the last four hundred million years, some plant has tried every strategy with a remote chance of working. We're just beginning to realize how varied a thing working might be. Life has a way of talking to the future. It's called memory. It's called genes. To solve the future, we must save the past. My simple rule of thumb, then, is this: when you cut down a tree, what you make from it should be at least as miraculous as what you cut down.

"My whole life, I've been an outsider. But many others have been out there with me. We found that trees could communicate, over the air and through their roots. Common sense hooted us down. We found that trees take care of each other. Collective science dismissed the idea. Outsiders discovered how seeds remember the seasons of their childhood and set buds accordingly. Outsiders discovered that trees sense the presence of other nearby life. That a tree learns to save water. That trees feed their young and synchronize their masts and bank resources and warn kin and send out signals to wasps to come and save them from attacks.

"Here's a little outsider information, and you can wait for it to be confirmed. A forest knows things. They wire themselves up underground. There are brains down there, ones our own brains aren't shaped to see. Root plasticity, solving problems and making decisions. Fungal synapses. What else do you want to call it? Link enough trees together, and a forest grows aware.

"We scientists are taught never to look for ourselves in other species. So we make sure nothing looks like us! Until a short while ago, we didn't even let chimpanzees have consciousness, let alone dogs or dolphins. Only man, you see: only man could know enough to want things. But believe me: trees want something from us, just as we've always wanted things from them. This isn't mystical. The 'environment' is alive—a fluid, changing web of purposeful lives dependent on each other. Love and war can't be teased apart. Flowers shape bees as much as bees shape flowers. Berries may compete to be eaten more than animals compete for the berries. A thorn acacia makes sugary protein treats to feed and enslave the ants who guard it. Fruit-bearing plants trick us into distributing their seeds, and ripening fruit led to color vision. In teaching us how to find their bait, trees taught us to see that the sky is blue. Our brains evolved to solve the forest. We've shaped and been shaped by forests for longer than we've been *Homo sapiens*.

"Trees are doing science. Running a billion field tests. They make their conjectures, and the living world tells them what works. Life is speculation, and speculation is life. What a marvelous word! It means to guess. It also means to mirror. Trees stand at the heart of ecology, and they must come to stand at the heart of human politics. Tagore said, *Trees are the Earth's endless effort to speak to the listening heaven.* But people—oh, my word—people! People could be the heaven that the Earth is trying to speak to.

"If we could see green, we'd see a thing that keeps getting more interesting the closer we get. If we could see what green was doing, we'd never be lonely or bored. If we could understand green, we'd learn how to grow all the food we need in layers three deep, on a third of the ground we need right now, with plants that protected one another from pests and stress. If we knew what green wanted, we wouldn't have to choose between the Earth's interests and ours. They'd be the same!

"To see green is to grasp the Earth's intentions. So consider this one. This tree grows from Colombia to Costa Rica. As a sapling, it looks like a piece of braided hemp. But if it finds a hole in the canopy, the sapling shoots up into a giant stem with flaring buttresses. Did you know that every broadleaf tree on Earth has flowers? Many mature species flower at least once a year. But this tree, *Tachigali versicolor*, this one flowers only once. Now, suppose you could have sex only once in your entire life. . . .

"How can a creature survive, by putting everything into a one-night stand? *Tachigali versicolor*'s act is so quick and decisive that it boggles me. You see, within a year of its only flowering, it dies.

"It turns out that a tree can give away more than its food and medicines. The rain forest canopy is thick, and wind-borne seeds never land very far from their parent. *Tachigali*'s once-in-a-lifetime offspring germinate right away, in the shadow of giants who have the sun locked up. They're doomed, unless an old tree falls. The dying mother opens a hole in the canopy, and its rotting trunk enriches the soil for new seedlings. Call it the ultimate parental sacrifice. The common name for *Tachigali versicolor* is the suicide tree.

"I've asked myself the question you brought me here to answer. I've thought about it based on all the evidence available. I've tried not to let my feelings protect me from the facts. I've tried not to let hope and vanity blind me. I've tried to see this matter from the standpoint of trees. What is the single best thing a person can do for tomorrow's world?" ●

Twisted limbs of a bristlecone pine tree (*Pinus longaeva*) at dawn in the Ancient Bristlecone Pine Forest, Inyo National Forest, California. *Pinus longaeva* are more than five thousand years old, the oldest living organism on earth.

Wilding

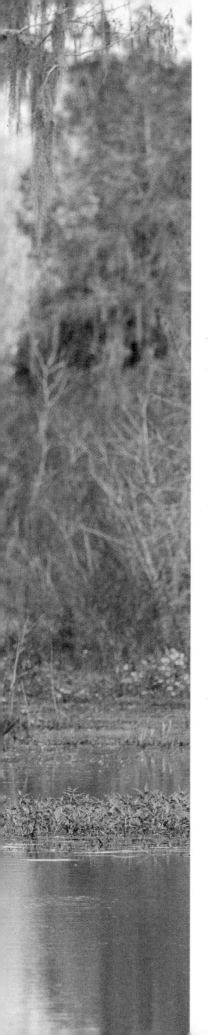

One can spend a lifetime in a city and never encounter wilderness. But is that really true? The adjective *wild* distinguishes a plant or animal growing in nature from one cultivated by human beings. From an individual's viewpoint, the distinction is valid—breakfast bacon is not wild boar. However, from the perspective of the living world, it makes no sense. If you live in Paris, the river Seine is wild, even if polluted. Wildness grows in the cracks of sidewalks. The human body is suffused with a vast system of microbiota, known and unknown organisms that outnumber our human cells. You could say that we are mostly bacteria learning to be human. Each of us is a culture. Our organs, gut, skin, and follicles contain unique sets of bacteria, different from those in every other person on the planet. And we liberally share them with others: the daily activities of touching, shaking hands, pecks on the cheek, and eating food together exchange microbes and create a web of interconnectedness said to harmonize interactions with our family and the environment.

Although our innate human biodiversity has not been fully identified or quantified, research suggests that there are more genes in the human microbiome than stars in the universe, and approximately half are what are called singletons, genes that are unique to each individual. The science-based climate movement proposes that there are nature-based solutions, which this book also puts forward; however, the proposition reveals a fundamental misunderstanding. Nature is not an "it" out there. It is us. The belief that we are individuals with distinct parts and functions acting as a unit gives way to bodies as ecosystems inextricably connected to expansive sweeps of spores, alga, bacteria, pollen, and viruses. It turns out that we truly are earthlings. There may not be any "individuals" in nature. Trees are connecting points within vast networks, interacting with mycelia, microbes, fungi, bacteria, nematodes, phages, and viruses. That pretty much describes a human being too.

When we think about how to end the climate crisis, we rarely consider wildness as being integral. The swamps, beetles, elephants, penny bun fungi termite mounds, sandhill cranes, and coral reefs fall under the rubric of biodiversity and threatened habitat. In this way, we separate our personal well-being from the well-being of the living world in all its mystery, majesty, and vastness. But we are not separate. Protection of the wild is crucial. Take away the bacteria inside your body and you will perish. Remove the micro- and macrobiota on earth and life as we know it comes to a halt. We damage our internal wildness when we take antibiotics, eat processed foods, or oversterilize our living environment. We destroy our external wildness when we plow wetlands, set bait traps for wildlife, glyphosate our soil, overfish, acidify oceans, set fire to forests. When we restore wildlife habitat, we restore resiliency, reproduction, viability, and evolution.

In this section, we detail several areas where people can make a difference with respect to pollinators, wildlife migration, wolves, salmon, beavers, and wildlife corridors. Also explored is the idea of bioregions, geographical areas organized and governed according to their unique biological attributes, as contrasted with political force and financial greed. An inspiring piece is an essay about the 3,500-acre Knepp Castle Estate, in Sussex, England, by Isabella Tree. She writes eloquently about rewilding, how she and her husband, Charlie Burrell, created the conditions for spontaneous regeneration of their degraded land, a newly born wilderness that astonished conservationists in the UK with its beauty and diversity. Nature repairs quickly, elegantly, and abundantly when harm ceases. In the case of the Knepp Castle Estate, it involved introducing a heritage species of pig, cattle, and horse. A deadened, money-losing farm gave birth to a living ark. The Knepp Castle Estate has become so wild that people pay to witness the mammalian, avian, sylvan, and insect life that has returned to Sussex. When life regenerates, complexity proliferates. Diversity burgeons. Productivity soars. Species reappear. And the climate responds. ●

Trophic Cascades

In 1926, with a final gunshot, the sole remaining member of the last wolfpack in Yellowstone National Park slumped to the ground. Then, as even today, wolves were viewed as dangerous predators that would consume you or your sheep if given half a chance. Famed writer and naturalist George Monbiot once posed a question. "Arrange these threats in ascending order of deadliness: wolves, vending machines, cows, domestic dogs and toothpicks. I will save you the trouble: they have been ordered already." Nearly 170 Americans are killed each year by swallowing toothpicks, and only one human death has been caused by a wolf in this century. The vending machines? People frustrated and upset by not getting their Cheetos shake the machine and it falls down and crushes them.

It was assumed that the heroic hunters who eliminated the northern Rocky Mountain gray wolf had removed a menace from Yellowstone. But within a few years, the park ecosystem began to unravel. The types and numbers of plants changed. Without their main predator, the elk and deer populations surged. The herbivorous appetite of those wild ungulates mowed down aspens, cottonwoods, willows, and maples. The trees were joined by the loss of huckleberries, currants, dogwoods, and wild rose, and then came the grasses and forbs: sweet clover, dandelion, salsify, cow parsnip. That was just the beginning. The bird population dropped for lack of seeds, nuts, bark-dwelling insects, and places to nest. The beavers lost their winter food source, willow. Without their dams, streams eroded. Fish were impacted by silted rivers. Without trees, bank erosion widened the rivers, creating shallows of warmer water. However, other species flourished. With less competition, coyotes multiplied. Running rampant in meadows and the diminished forests, they increased their intake of field mice and small mammals, which dramatically reduced the numbers of foxes, badgers, golden eagles, red-tailed hawks, peregrine falcons, and ospreys. Yellowstone National Park, presumptively a wild refuge for untamed nature, became the poster child for ecosystem collapse.

An ecosystem is an area and all the living organisms within it, as well as the physical components supporting it—streams and rivers, rainfall, rocks, minerals, and local climate. The living and nonliving interact to create a complex, tangled system that is created and energized by photosynthesis and plant foods. Plants are eaten by a variety of creatures and insects, from berry-eating brown bears to nectar-sipping butterflies. Plants are fed by microorganisms, minerals, fungi, water, and sunshine. Every ecosystem forms a cascade, upward or downward depending on your perspective, of what is eating what, from larger mammals down to the wisps of fungus dotting the forest floor and feeding off rotting leaves and pine needles.

Trophic describes the act of feeding, of obtaining nutrition. The parts, pieces, remains, or death of any plant or organism is food for another species in the food chain, which is roughly divided into four trophic levels according to the food consumed. It starts with the decomposers: fungi, worms, nematodes, and bacteria that break down organic debris. Above that are plants, from mosses to shrubs to trees, energized by sunshine, rain, and soil nutrients provided by the decomposers. Then come herbivores that eat the plants, from field mice, bluebirds, and squirrels to bison, deer, and elk. On top of the fourth and last level are predators: carnivores that eat herbivores—wolves, owls, and mountain lions.

How an ecosystem is assembled differs with place, flora, and fauna. However, there is one element all have in common: apex predators, animals that have no natural predator and thus sit on top of the food chain. The key word is *natural* predator. Surrounding Yellowstone is the human predator. For decades, ranchers told horrific stories of death and loss by wolf predation to justify why they wiped out the wolf population around and in the park. By 1945, the gray wolf had been extirpated throughout the northwestern United States. All but one species of *Canis lupus* was listed as an endangered species in 1975 by the U.S. Fish and Wildlife Service. The hoary tales of wolves savaging herds and flocks are not based on science or observation. Wolves are not fond of sheep or cattle. They favor elk and follow their migration, which oftentimes overlaps with ranches and farms. The predator myth of widespread livestock loss was propagated by fear, not data. Before their extirpation in the northern Rockies, wolves accounted for less than 1 percent of all livestock loss.

Between 1995 and 1996, thirty-one Canadian wolves were released back into the park. Seventy years of ecosystem deterioration began to reverse, astonishing biologists and the opponents of reintroduction. Rangers reported hearing a sound similar to the report of a fired rifle, but in this case it was the sound of beavers returning to the park, the slap of their paddle tail on the surface of a pond. Stands of willows and cottonwoods returned to riverbanks. Before the wolves were reintroduced, elk would winter over in the park and eat the willows down to the root stubs. The elk were now on the move in and out of the park. A characteristic engendered by apex predators is something called the ecology of fear. As the elk and deer became aware of the existence of wolves in the park, they reverted back to their instinct of constantly moving and migrating, which reduced the consumption of trees and shrubs in any one area. More willows meant more beavers. Although the wolf has been romanticized and given total credit for these changes, it wasn't the only influence. Three other apex predators grew in number at the same time: grizzly bears, cougars, and hunters who were killing elk outside the park.

Wolves, grizzlies, and cougars are more than apex predators. They are what is known as a keystone species, a type of organism that glues an entire ecosystem together. Other examples of keystone species are bees, hummingbirds, sea otters, and trees such as aspen groves in boreal forests. Even starfish. They are key because their lives are crucial to the lives of other species. When they are removed, ecosystems can deteriorate or be lost altogether. They lose resilience and can become prey to invasive species that were once held in check. It was long believed that an ecosystem was a pyramid limited by the availability of food on the bottom level. No soil meant no plants; no plants meant no herbivores, which meant no predators. The theory was logical: plants determined the number of herbivores—deer, elk, rabbits—which in turn determined the number of predators. That belief was turned upside down by legendary biologist Roger Paine, who coined the term *keystone species*. As an assistant professor of zoology at the University of Washington in Seattle, he would take his students to Makah Bay, near the tip of the Olympic Peninsula. In June 1963, he performed an experiment there to test his theory that a single species could be critical to the function of an entire ecosystem.

Paine identified a six-foot-high slab in a tidal pool that extended for twenty-five feet. The rocky, sea-washed slab contained goose and acorn barnacles, limpets, chitons, sponges, seaweed, sea urchins, anemones, black mussels, and a species he wanted to "toss out"—a purple-and-orange starfish known as *Pisaster ochraceus*. Twice a year, using a crowbar, he pried off the starfish and frisbeed them out to sea. He wanted to find out what happened when a particular animal was taken away from the tidal pool ecosystem. The tidal pool was his lab. It was observational science, the kind that informed Charles Darwin, Alfred Wallace, and Copernicus, and is at the heart of virtually every Indigenous culture. Paine was more modest in his

---

A coyote and a gray wolf member of the original Yellowstone Druid Pack pause while feeding on an elk carcass in the Lamar Valley, Yellowstone National Park, Wyoming.

description; he called it "kick it and see" ecology. By turning over the starfish before they were flung, he could see what they were eating, which was just about everything from limpets to chitons and mussels to barnacles. Within a year of the starfish removal, the tidal pool community had completely changed. Acorn barnacles spread over 60 to 80 percent of the wall at first, but they were soon displaced by the smaller, fast-growing goose barnacle. The four types of algae that had been there previously were nowhere to be seen, and the chiton and limpet had moved away. The anemone and sponge populations declined, while a tiny predatory snail increased more than tenfold. From fifteen species, the community had shrunk to eight. A few years later, the mussels dominated the entire wall and had eliminated virtually all the other species.

In architecture, an arch is held in place by a wedge-shaped stone called a keystone. It is not necessarily bigger than the other arch stones, just different in shape and function. Likewise, certain species determine the stability and diversity of ecological communities. When sea otters were hunted to near extinction by Russian, British, and American trappers in the nineteenth century, sea urchins, their principal food source, exploded in number and consumed the magnificent kelp forests described by Sir Francis Drake as a wonder of the world. Paine hypothesized that the disappearance of sea otters since the late 1800s could be responsible. When two students approached Paine wanting to study sea otters, he suggested they travel to two islands in the Aleutians, one with a robust sea otter population, the other without. Where there were otters, they found rockfish, kelp beds, eagles, and harbor seals. None were seen at the otter-free islet. Thanks to Paine's work, we know how one species can help preserve the populations of many other species. Simply by living, keystone species create a trophic cascade of effects on multiple levels, another term coined by Roger Paine.

The reintroduction of the wolf reversed the trophic cascade that had started after they were killed off in many parts of Yellowstone Park. The reduced elk population triggered widespread benefits for other species that had been affected by the disruption of the food chain. The seventy-year absence of the gray wolf caused damage and reshaped the Yellowstone ecosystem. Restoration has occurred, but not completely and not everywhere. In some places, willows have not returned, because the eroded streams destroyed willow habitat. Beavers cannot return to those areas to restore stream health without winter food. The undoing of our natural landscapes and ecosystems took centuries, and may take centuries to fully recover.

What do wolves have to do with the climate emergency? Their absence and subsequent presence remind us that we do not fully understand how ecosystems work. Each ecosystem is a repository of carbon stored above and below ground, systems of life that are unimagined in their full complexity. How we treat them determines whether they emit, retain, or sequester carbon in their soil and biomass. The respect and protection of biodiversity is not a sideshow to the climate dilemma. It lies at the heart of the solution. We continue to eliminate predators, species, plants, and wetlands and degrade the function of ecosystems worldwide, even though the health of global ecosystems will determine our future. This understanding of climate impact can be lost in the rousing headlines about wind turbines taller than the Eiffel Tower and electric trucks autonomously moving food across the country. Hydrogen-fueled cars and triple-glazed windows are ingenious technologies, but technology alone will not allow a return to a stable climatic environment.

There are three types of keystone species. There are predators that control the population and behavior of prey, ranging from sperm whales to eagles. The presence of such a predator changes prey behavior. Australian scientists observed that when tiger sharks are distant from seagrass beds, the grasses can be mowed down by concentrations of grazing sea turtles. When tiger sharks are present, sea turtles fan out over wide areas and do not cause damage.

Then there are ecosystem engineers, species that physically transform their environment, the beaver being the quintessential example. Another example is prairie dogs. Their earthen colonies have been called "coral reefs" in seas of grass. Approximately 150 native birds and animals rely on ecosystems created by prairie dogs for their sustenance and habitat. Their feeding, tunneling, and clipping of vegetation create habitat for burrowing owls and mountain plovers, engender better feed for grazing buffalo, prevent the onset of invasive brush, and increase microbial complexity and nutrients for grass and forb cover, and the rodents themselves provide food as prey to owls, hawks, and ferrets. Regrettably, prairie dogs are used to this day as live targets in organized shootings in Nebraska and other states.

The third are mutualists. Mutualism is about reciprocity, the recognition by living organisms that their survival depends on the well-being of other forms of life. For centuries if not millennia, the Western view of life was some variation of dog-eat-dog. "Survival of the fittest" was a misinterpretation of Darwin's assertion that survival was about being the most fit for a given environment. The African red-billed oxpecker is the classic example. They sit atop buffalo, hippos, and zebras eating insects, ticks, and parasites, as if the heads and backs of animals were lunch counters. When predators approach, the oxpecker hisses a warning cry to its host. Bees looking for nectar see ultraviolet reflectance in flowers that directs them to the nectar- and pollen-laden stamen at the center. When they leave, their footprint can be seen from a distance by other bees, which then avoid that flower until the nectar is replaced in the carpals. The flower benefits from efficient pollination, and the bee benefits from productive access to nectar.

If mutualism is when species act reciprocally, to the benefit of both, we may want to ask which one of these three types of keystone species humanity should choose to be, given the disorder and suffering in the world that we cause to the land, creatures, and one another. ●

———————

(*Left*) Three black-tailed prairie dog pups gobbling roots and shoots.

(*Below*) Hippopotamus with two red-billed oxpeckers in the South Luangwa National Park, Zambia.

Grazing Ecology

Grazers are among the great unsung carbon cyclers on earth.

Grazing is a natural process with an ancient history. We know from the fossil record that grass eating animals first appeared 55 million years ago. Extensive grasslands came into existence 30 million years later and supported growing populations of grazers over the ensuing millennia, coevolving together. Their diets originally included a mixture of grasses, forbs, and the leaves of woody species, but over time grazing became increasingly specialized. The first true grazers—animals that drew their nutrition primarily from grass all year round—emerged ten million years ago and included the ancestors of many herbivores familiar to us today: bison, wildebeests, zebras, water buffalo, sheep, elk, camels, llamas, horses, cattle, moose, and yaks, as well as rabbits, grasshoppers, and geese. Today, 27 percent of the planet's terrestrial surface is grassland, which makes it one of the largest reservoirs of carbon. That means grazers are a critical part of ecosystem health at big scales. Yet, while the role that trees, plants, and soil microbes play in the cycling of carbon on earth has been

well documented, the role of grazers, particularly wild ones, has been consistently underestimated.

Over eons, grazing animals developed the physiological ability to gain nutritional energy from grass by defoliating it with their mouths and fermenting the fibrous material in their stomachs. The digestive systems of grazers can be either of two types—foregut or hindgut fermenters. The former has a four-chambered stomach, including a rumen, where bacteria and other microbes break down the grass cellulose into fatty acids and proteins that are absorbed into the bloodstream. Called ruminants, these grazers include cattle, sheep, goats, elk, bison, yaks, antelopes, gazelles, and many others. Hindgut fermenters have a single, large stomach located near the end of their digestive tract. They include elephants, zebras, horses, rhinos, rabbits, sloths, and many rodents. Both types are adapted to grassland in different ways. Ruminants consume high-quality forage, which they efficiently process—including the chewing of cud—while hindgut fermenters eat low-quality forage, which they must consume in large quantities to get the nutrition they need. The difference in

their diets is one reason why so many grazing animals can coexist in a large landscape.

In the wild, some grazers congregate in vast herds and migrate long distances to meet their nutritional needs, while protecting themselves from predators. This behavior has created a long coevolutionary history between grazers and grasslands that is both sustainable and resilient. Historical examples include the vast bison herds of North America and the saiga antelope migrations across the Eurasian steppe. Today, the few remaining great migrations include the wildebeest and zebra herds in Africa and the caribou in the Arctic. Migrating grazers travel long distances because the quality and quantity of forage can vary greatly across a grassland ecosystem, and animals evolved to make instinctive decisions about how to maximize their nutrition, including which plant to bite and where to move next. The search for essential minerals available in forage also influences their behavior. In Africa, zebras and wildebeest, often numbering in the millions, migrate across the Serengeti grassland following the annual "green-up" as the rains arrive and the seasons shift from dry to wet. In their pursuit of young plants, zebras and wildebeests get along with each other partly because they eat different types of grass (zebras prefer taller and lower-quality varieties).

Grazers are not passive players in grassland ecosystems. One study in the Serengeti has shown that grasses of all types are more concentrated in grazed areas than ungrazed (fenced) ones, by an average of 43 percent. Grazing provokes grass regrowth by removing old, decayed, or dead plant tissue, allowing more sunlight to reach the base of the plant, where most growth takes place. Animal dung and urine supply natural fertilizers for the plants, including nitrogen. All of this increases solar capture and photosynthesis aboveground and water and nutrient uptake below the soil surface, improving plant vigor and causing

Blue wildebeest in the savanna of Etosha National Park in Namibia. 1.5 million wildebeest, along with gazelles and zebra, undertake the largest migration on earth through the Serengeti and the Masai Mara ecosystem.

roots to expand as new leaves reach for the sky. As a result, the grass becomes more nutritious, especially in the early stages of the growing season.

From the perspective of an individual grass plant, grazing is a form of disturbance that can have an intense effect. If the plant is dormant because of cold temperatures or a lack of water, its shearing by the mouth of a grazing animal probably won't cause much damage. When it is growing, however, the leaves are photosynthesizing, sending energy to all parts of the plant. A big bite by a hungry herbivore can substantially set back a plant's growth—but only for a while. The loss of greenery stimulates root growth and penetration, leading to more robust and mineral-rich grasses. And some green tissue almost always remains on the plant, especially near the base, and it begins growing again. It's ecological symbiosis—plant and animal both profit. Grasses have co-evolved with grazers. For example, saliva stimulates plant growth. The benefits grasses derive from the disturbance depends on many factors including the time of year the grazing occurs, the severity of the grazing event (how much green tissue was removed), and whether the plant has sufficient time to recover and grow before being defoliated again. This is why migratory grazers fit the bill—they graze and go, staying only briefly to eat the most nutritious grass and not returning again for a year or more.

Over millennia, grazers have become an intimate part of grassland ecosystems, whether as part of large migrating herds or smaller clusters of animals, becoming an essential component of life on earth. They play a direct role in the exchanges of energy, water, carbon, and greenhouse gases between land and the atmosphere. Changes in herbivore behavior or abundance can lead to dramatic consequences for plant composition, productivity, nutrient cycling, and other ecosystem processes. When grazers are displaced or removed from grasslands, the ecology of the system is altered significantly and degrades. In North America, for instance, the reintroduction of a herd of bison to a conservation reserve in a tallgrass prairie in Kansas during the 1980s allowed researchers to quantify the impact of their grazing on the land. They were particularly interested in comparing it with prescribed fire as a way of maintaining ecosystem health. The impact of grazing was highly positive, they determined, leading the scientists to call bison a keystone species and supporting their reintroduction to other parts of the Great Plains.

In Europe, grazing animals likely played an important role in the continent's ecology. Prehistorically, the forests in the region may not have been as dense as they are today. Instead, they were likely mosaics of large and small grasslands, scrub, solitary trees, and trees growing in groves. The indigenous wild grazers, such as deer, boars, and aurochs (ancestors to modern cattle), along with

seed-dispersing birds, were essential to creating this mosaic. Thorny shrubs unpalatable to grazing animals sheltered trees, which eventually formed a grove. When the trees grew high enough, the shade created by their canopy caused the shrubs to die back. Grazing pressure returned, preventing the growth of new trees outside of the grove. Eventually, the trees in the center of the grove, which are oldest, decayed and died. Increased sunlight now reached the ground, promoting grass growth, which in turn attracted more grazers, maintaining the grasslands. Recently, humans have used livestock to keep the landscape open, such as the savannah-like *dehesas* in southern Spain and *montados* in Portugal.

This dynamic process of natural disturbance contributes to diverse webs of life and enhanced carbon cycles. Advocates for the beneficial effects of grazer-grassland interactions argue that ecological goals can be accomplished by carefully managing herbivores in a way that mimics the behavior of their ancestors, including large landscapes such as the Tibetan Plateau.

Humans have a deep history of working with domesticated grazers, such as cattle and sheep, in sustainable and regenerative ways. One traditional relationship practiced by Indigenous peoples around the world is pastoral-

ism, in which herds of animals, including llamas, camels, goats, and yaks, are guided by people across a landscape to fresh forage and water. Pastoralists are attuned to dynamic interplay between weather, animals, and the environment. They know from long experience what their animals need to be healthy, when to move the herds in order to avoid overgrazing the land, where to go, and how to avoid conflicts with predators and other wildlife. Over the centuries, the land shaped the herds, and the herds shaped the land. Both shaped human cultures. As many as five hundred million people still practice some type of pastoralism and more than three-fourths of all nations have pastoral communities.

A well-known community are the Maasai, who live in Kenya and northern Tanzania. With over one million people, the Maasai have developed a sophisticated agro-pastoral way of life that has enabled them to survive and thrive in rough and arid environments. Their cattle are a source of milk, meat, and wealth. Some Maasai grow crops, including corn and beans. The Maasai have a strong conservation ethic and go to great lengths to integrate their activities with the needs of a diverse array of wild herbivores in the region, such as elephants, zebras, and buffalo. They know from experience that pastoralism is ecologically beneficial to the environment and consistent with wildlife conservation. There are conflicts, especially with lions and other predators, but the Maasai have developed ways to minimize trouble. The result are strong, time-tested relationships between people, grazing animals, and the land, supported by a deep well of Indigenous knowledge.

These bonds, however, are threatened by climate change and human encroachment. Rising temperatures and longer periods of drought imperil the health of the cattle and stress grasslands. The spread of urban areas and the privatization of formerly communal lands means there is less room for the Maasai to roam. Fences are going up, interrupting the ebb-and-flow of grazing animals across the land. Wildlife tourism, while profitable economically, has impacted the traditional Maasai way of life. Still, the Maasai are resilient, as is the land that sustains them. ●

Bison on the Blackfeet Indian Reservation in northern Montana moving to their fall pasture. The Blackfoot Confederacy tribes are working to restore bison on the Northern Great Plains, which could regenerate the tall grass prairies.

Wildlife Corridors

Let's start indoors. Let's start by imagining a fine Persian carpet and a hunting knife. The carpet is twelve feet by eighteen, say. That gives us 216 square feet of continuous woven material. Is the knife razor-sharp? If not, we hone it. We set about cutting the carpet into thirty-six equal pieces, each one a rectangle, two feet by three. Never mind the hardwood floor. The severing fibers release small tweaky noises, like the muted yelps of outraged Persian weavers. Never mind the weavers. When we're finished cutting, we measure the individual pieces, total them up—and find that, lo, there's still nearly 216 square feet of recognizably carpetlike stuff. But what does it amount to? Have we got thirty-six nice Persian throw rugs? No. All we're left with is three dozen ragged fragments, each one worthless and commencing to come apart.

—David Quammen, *Song of the Dodo*

David Quammen's oft-told analogy describes what happens when ecosystems are fragmented by roads, fencing, overgrazing, farming, suburbs, and development. Between 2001 and 2017, approximately 24 million acres was lost to development in the U.S. Highways and freeways that connect suburban tracts pose existential challenges to animals that need to cross. Anyone who has tried to walk or run across a busy thoroughfare at night can imagine the risk for a bear or deer. The result of habitat fragmentation is slow-motion devolution and species loss, a pathway to eventual extinction. What might this have to do with climate and global warming? Everything. Let's start with the true carpet.

Whether it is flocks of sandhill cranes fleeing Wisconsin's White River Marsh in mid-October or the great ungulate migration of 1.5 million wildebeest, as well as hundreds of thousands of zebras and gazelles, moving through the Serengeti, the movement of animals across plains, over mountains, and through the skies is crucial to the preservation of native habitat worldwide. Ecosystems are classified by communities of plants and animals that

on whether it is a hummingbird or a tortoise. They connect patches of isolated habitat and allow species to flow and migrate. Without them, species can't follow food and water sources and are at risk of genetic isolation and local extinction. Up to half of the planet's plant and animal species in the world's most biodiverse regions could be extinct by the end of the century due to global warming. Corridors that lead to a cooler climate provide species a lifeline to outrun rising temperatures. Reconnecting habitat corridors in the western United States on 75 percent of land area provides routes for animals to move and adapt to climate disruption. As birds, insects, reptiles, and mammals travel through corridors, plants and trees get pollinated, and their seeds get dispersed. Habitat connectivity increases genetic diversity and resilience within a population, which increases the ability of a species to adapt to changes caused by global warming.

We do not need a reason to save the elephant, wolf, tiger, whooping crane, and hundreds more species. That they exist at all is reason enough: these creatures embody extraordinary evolutionary qualities, intelligence, and beauty. What we might miss is that when we "save" the elephant, or virtually any species that depends on an intact ecosystem, whether it be a beetle or a bear, we benefit. It saves us—literally. Protecting the viability and resilience of planetary ecosystems is crucial to human existence and the effort to stabilize greenhouse gas emissions and reverse global warming. Wildlife corridors, habitats, and biodiversity are not a separate problem with respect to global warming. The biosphere creates the atmosphere and the atmosphere creates the biosphere. They are inseparable.

Globally, land conversion, human population growth, physical barriers, and agriculture have strained habitats from the wetlands of the Midwest to the rainforests of Asia and the boreal forests of Canada and Russia. Indonesia's rainforests, home to rhinos, tigers, elephants, and orangutans, is fragmented because of palm oil production for potato chips and Halloween candy. The border wall between Mexico and the United States blocks bighorn sheep, wolves, and ocelots from their ranges, and cut through Native American gravesites, six national parks, five conservation hotspots, six wildlife refuges, and numerous other wildlife and conservation areas, including the National Butterfly Center, which is home to two hundred species of butterflies.

When species disappear, ecosystems deteriorate. As unraveling progresses, more species are lost, until you have scattered biological fragments that fail to support the inherent complexity of a living community. In a given

dwell within. Every community is different, all are complex, and all are threatened or at risk.

Terrestrial and coastal ecosystems contain more than three billion tons of carbon, nearly four times the amount of carbon contained in the atmosphere. Three decisive actions are required to stop global warming. First, reduce and eliminate emissions from fossil fuel combustion. Second, sequester carbon into the soil by photosynthesis in grasslands, forests, farmlands, mangroves, and wetlands. Third, protect the carbon here on earth. If we forgo habitat protection and focus solely on fossil fuel replacement, it will be for naught. A 15 percent loss of terrestrial carbon could result in a one-hundred-parts-per-million increase in atmospheric carbon dioxide. Establishing wildlife corridors is crucial to preventing that loss.

Wildlife corridors are pathways in water, land, and air in which birds, mammals, invertebrates, reptiles, and insects migrate, feed, drink, and move to and from connected habits. Corridors preserve sufficient habitat so that a given species can complete their cycle of life, a lifespan that can range from two to one hundred years, depending

Family of African bush elephants, the largest living terrestrial animal, weaving through the reed banks of the Linyanti Swamp in Chobe National Park, Botswana.

ecosystem, birds, mammals, frogs, bats, reptiles, trees, plants, fungi, fish, and marshes form active relationships, networks that are dynamic, symbiotic, and intricate. When creatures large and small go missing, plants falter. The loss of diversity in plant and animal communities accelerates losses in the other. When ecosystems diminish, declining biomass releases its carbon stores. As plant communities wither, wetlands disappear, pollinators recede, soil becomes parched, and more carbon is emitted. The carbon oxidizes—it becomes carbon dioxide.

As of the mid-2000s, 58 percent of elk migratory routes, 78 percent of pronghorn migratory routes, and 100 percent of bison migratory routes were blocked by human development in the Greater Yellowstone Ecosystem. Stuart Schmidt of South Dakota is a fourth-generation farmer and rancher. By managing his cattle to mimic the natural herd movement of the Great Plains, he encourages native grasses and flowers to cover the topsoil across his pastures. The increase in diversity of plants and grasses has resulted in more diverse insects, leading to more diverse birds and mammals. The nature-mimicking grazing leaves residual grass for deer and pronghorn to sustain themselves through the winter migration. Managing the grass and restoring it to its natural state allows the Great Plains' species to share the grassland and migrate seasonally.

Migratory songbird populations traversing back and forth from the United States to Mexico have plummeted by more than 50 percent since the 1960s. There are many reasons for the decline, but habitat loss is foremost, in their nesting grounds, wintering grounds, and everywhere in between. There are three billion fewer songbirds in North America than in 1970. The greatest losses have befallen the birds that breed in at-risk grasslands, where more than seven hundred million individuals constituting thirty-one species have been lost.

Aquatic corridors are crucial for the health of marine ecosystems. The California Current, flowing along the west coast of North America from southern British Columbia to the southern tip of Baja, Mexico, is one of the most important marine corridors on the planet. About thirty species of fish, marine mammals, seabirds, reptiles, and invertebrates that depend on the current are threatened, including the southern resident population of orcas, Pacific leatherback turtles, southern sea otters, short-tailed albatross, and chinook, chum, and coho salmon. Their migratory pathways are crucial not only to their own survival but to the entire ecosystem of the Western Pacific.

The math is clear. We have lost five billion acres of forestland. Tropical forests, once 12 percent of landmass, are now only 7 percent. Marine corridors are threatened by warming, plastic, shipping lanes, pollution, and overfishing. America's grasslands are disappearing faster than the Amazon rainforest. The Great Plains lost 2.1 million acres in a single year in 2018. Almost half of the world's temperate grasslands have been converted for agricultural or industrial uses. The double-glazing of the atmosphere with carbon dioxide deservedly is receiving a great deal of attention. The loss of intact ecosystems, their release of greenhouse gases, and the impact upon the atmosphere deserves equal publicity. This is not a problem looking for solutions; they already exist. It is a problem looking for awareness. ●

A FEW MIGRATORY SPECIES

Whooping crane	Cheetah	Leatherback turtle	Kouprey
Black skimmer	Lion	Sperm whale	Gazelle
Wood thrush	Tiger	Manatee	Elephant
Redhead	Margay	Free-tailed bat	Vicuña
Wolverine	Vaquita	Wildebeest	Oryx
Shearwater	Antelope	Monarch butterfly	Caribou
Osprey	Sturgeon	Lemon pansy butterfly	Black stork
Cerulean warbler	Tarpon	Emigrant butterfly	Flamingo
Black rail	Red-flanked duiker	Blue tiger butterfly	Red-breasted goose
Striped bass	Black-backed jackal	Dragonfly	Reed warbler
Bluefin tuna	Bull shark	Ladybird beetle	Addax
Steller's sea lion	Whale shark	Yak	Goral
White-beaked dolphin	Green turtle	Mountain gorilla	Elk
Menhaden	Kemp's ridley turtle	Jaguar	Banteng
Leopard	Loggerhead turtle	Zebra	Black rhinoceros

Cape buffalo at a water hole with its companion, the red-billed oxpecker, in South Luangwa National Park, Zambia.

Wilding
Isabella Tree

When Charlie Burrell inherited the 3,500-acre Knepp Castle
Estate from his grandparents at age twenty-three, he also inherited a money-losing dairy and arable farm. There are thirty different Sussex words for mud—slub, gawm, gubber, sleech,
and pug are some—all of which make farming miserable when
the notorious Sussex Weald clay ends up like concrete in the
summer and porridge in the winter. In order to try and turn a
profit on their poor soils, Burrell, with his wife, Isabella Tree,
intensified agricultural methods on the farm with bigger, new
machinery and expensive artificial fertilizer and pesticides, and
invested in infrastructure and state-of-the-art dairy parlors.
They improved their farm yields, and the dairy was rated one of
the top ten in the country, but the Sussex clay prevented them
from getting onto the land all winter, sometimes for as long as
six months. The profits never materialized. In 2000, they made
a decision that has reverberated throughout the agriculture and
conservation worlds ever since: they let the farm go wild. They
let nature decide what the land, once a farm, wanted to become.
Following the advice of Dutch ecologist Frans Vera, they introduced free-roaming herbivores to act as proxies for the mega-
fauna that once roamed the UK and Europe, such as the auroch,
and the tarpan, the Eurasian wild horse. They ring-fenced the
property and removed miles of interior fencing. The results have
been spectacular and surprisingly profitable. Native plants,
wildflowers, and wetlands abound with thorny scrub, butterflies, Exmoor ponies, Tamworth pigs (the closest relative to a
wild boar the government would allow), red and fallow deer,
and old English longhorn cattle. The farm has one employee for
the animals. Because they cannot bring back wolves or other
apex predators, the animals need to be culled, producing what
Isabella calls the "most ethical meat" in the UK. The Knepp
Castle Estate has become one of the most biodiverse holdings in
all of the UK, rich with nightingales, the rare purple emperor
butterfly, the nearly lost turtledove, and five species of indigenous owls. They now run a profitable ecotourism business, with
glamping, camping, and wildlife safaris. The wildlife can be seen
or heard on eighteen miles of public trails that crisscross the
land. This is the story of regeneration, the primordial basis of
nature, and the latent abundance that arises when humans align
with life. —P.H.

It's a still June day on Knepp Castle Estate in West Sussex.
We can call it summer now. This is a moment we've been
waiting for, not sure if we dare expect it. But there it is—
from the thicket that was once a hedgerow, that unmistak-
able purring: soothing, inviting, softly melancholic. We
tread quietly past an eruption of saplings of oak and alder,
billowing with skirts of blackthorn, hawthorn, dog rose,
and bramble. The thrill of recognition is tinged with relief
and, though neither of us tempts fate by expressing it, a
hint of triumph. Our turtle doves have returned.

For my husband, Charlie, their gentle burbling takes
him back to the African bush, to his infancy running
around on his parents' farm. This is where the doves have
come from—their tiny flight muscles pumping 3,000 miles
from deep in West Africa, from Mali, Niger, and Senegal,
across the epic landscapes of the Sahara Desert, the Atlas
Mountains, and the Gulf of Cadiz; over the Mediterranean,
up the Iberian Peninsula, through France, and across the
English Channel. They mostly fly under the cover of darkness, covering between 300 and 450 miles every night at a
maximum speed of 40 miles an hour, usually making landfall in England around May or early June. Like their fellow
African migrant, the nightingale, they are famously timid. It
is their call that tells us they are here. Like the cuckoo and
the nightingale—who generally arrive here first—they have
come to breed, to raise their young far from the predators
and competitors of Africa, and to take advantage of the

long daylight feeding hours of the European summer. For most people our age, born in the 1960s, who have grown up in the English countryside, turtle doves are the sound of summer. Their companionable crooning is lodged forever, somewhere deep in my subconscious. But this nostalgia, I realize, is lost to generations younger than ours. In the 1960s there were an estimated 250,000 turtle doves in Britain. Today, there are fewer than 5,000. At the present rate of decline, by 2050 there could be fewer than 50 pairs, and from there it would be a hair's breadth to extinction as a breeding species in Britain. Now, at Christmas, when we sing of the gifts my true love gave to me, few carolers have ever heard a turtle dove, let alone seen one. The significance of its name, derived from the lovely Latin *turtur* (nothing to do with the reptile; all to do with its seductive purring), is lost to us. The symbolism of "turtles," their pair-bonding an allegory of marital tenderness and devotion, their mournful *turr-turr*-ing the song of love lost, the stuff of Chaucer, Shakespeare, and Spenser, is vanishing into the kingdom of phoenixes and unicorns.

As its territory shrinks to the southeast corner of England, Sussex is one of the turtle's final redoubts. Even so, numbers for our county are reckoned to be at best 200 pairs. Trouble on the migration route is undoubtedly partly responsible: periodic droughts, changes in land use, the loss of roosting sites, increasing desertification, and hunting in Africa—and the stupendous challenge of crossing the firing squads of the hunters of the Mediterranean. In Malta alone the slaughter claims 100,000 turtle doves every season. Around 800,000 a year are killed in Spain. Yet these impacts, considerable though they are, are not enough to explain the almost complete collapse of the population in Britain. In France, where hunters still shoot the birds on their return passage to Africa after the breeding season, numbers have decreased 40 percent since 1989—a significant loss, but nothing compared with ours, where, in recent times at least, we have opted not to shoot

Three fallow deer browsing at the Knepp Castle Estate wildlands. Native to most of Europe during the last interglacial, they survived the Ice Age in the Middle Eastern, Sicilian, and Anatolian refugia. In the Levant, they provided a source of meat going back to 420,000 BCE. They were introduced to southern England in the first century AD.

them. Across Europe, turtle-dove numbers have declined by a third over the past sixteen years to fewer than 6 million pairs—leading, in 2015, to a change in the bird's status on the International Union for Conservation of Nature Red List of Endangered Species from "Least Concern" to "Vulnerable," the start of a worrying downward slide.

But compared to the angle of European decline, the trajectory of the UK numbers is an almost vertical dive. The turtle dove's predicament in Britain is rooted in the almost complete transformation of our countryside—something that has come about in just fifty years. Changes in land use and, in particular, intensive farming have altered the landscape beyond anything our great-grandparents would recognize. These changes have taken place at all scales in the landscape, from the size of fields that now cover entire valleys and hills to the almost total disappearance of native flowers and grasses from farmland. Chemical fertilizers and weed killers have eradicated common plants like fumitory and scarlet pimpernel, on whose tiny, energy-rich seeds the turtle doves feed; while the wholesale clearance of wasteland and scrub, the ploughing of wildflower meadows, and the draining and pollution of natural water courses and standing ponds has wiped out their habitat.

In lowland England what tiny fragments of nature remain, whether left by accident or by design, are like oases in a desert, disconnected from natural processes—the interactions and dynamism that drive the natural world. We lost more ancient woods—tens of thousands of them—in the forty years after the Second World War than in the previous four hundred. Between the beginning of the war and the 1990s we lost 75,000 miles of hedgerows. Up to 90 percent of wetland has disappeared in England alone since the Industrial Revolution. Eighty percent of Britain's lowland heathland has been lost since 1800; a quarter of the acreage in the last fifty years. Ninety-seven percent of our wildflower meadows have been lost since the war. This is a story of unremitting unification and simplification, reducing the landscape to a large-scale patchwork of ryegrass, oilseed rape, and cereals, with scattered, undermanaged woods and remnant hedgerows the only remaining refuge for many species of wildflowers, insects, and songbirds.

The transformation of our countryside has impacted not just on turtle doves but on birds in general. In 1966, according to the RSPB, there were 40 million more birds in the UK than there are today. Our skies have emptied. In 1970 we had 20 million pairs of what are known as "farmland birds," such as quails, lapwings, grey partridges, corn buntings, linnets, yellowhammers, skylarks, tree sparrows, and turtle doves—most of them songbirds that depend on insects for their chicks and copses or hedgerows for their nests. By 1990 we had lost half of them. By 2010 that number had halved again.

Familiar, conspicuous in our skies and in our landscape, birds are, in a very real sense, our canaries in the mine—casualties connected to far greater and less visible losses. Preceding them, and following in their wake, are all the other species—including the less glamorous forms of life like insects, plants, fungi, lichens, bacteria—that share their fate. As the American biologist E. O. Wilson explained just thirty years ago, life's diversity is dependent on a complex web of natural resources and inter-species relationships. In general, the more species living in an ecosystem, the higher its productivity and resilience. Such is the wonder of life. The greater the biodiversity, the greater the mass of living things an ecosystem can sustain. Reduce biodiversity, and biomass may decline exponentially; and the more vulnerable individual species collapse. In *The Song of the Dodo*, David Quammen describes an ecosystem as being like a Persian carpet. Cut it into tiny squares, and you get not tiny carpets, but a lot of useless scraps of material fraying at the edges. Population crashes and extinctions are the signs of an ecosystem unravelling.

Against this background of almost inconceivable loss the turtle doves' appearance at Knepp seems little short of a miracle. Our patch—3,500 acres of former intensive arable and dairy farmland, just forty-four miles from central London—is bucking the trend. The turtle doves are here now because we have turned our land over to a pioneering rewilding experiment, the first of its kind in Britain. Their arrival has taken us and all those involved in the project completely by surprise. We began to hear turtle doves, only ever recorded here in ones and twos, just a year or two after the project began—three in 2005, four in 2008, seven in 2013, and by 2014 we reckoned we had eleven singing males. In the summer of 2018, we counted twenty. Occasionally, over the last couple of years, we've chanced upon a pair out in the open, sitting on telephone wires or on a dusty track, their pink breasts touched by the glow of evening, the tiny patch of zebra stripes on their necks a hint of Africa—a reminder that, just a few weeks earlier, these birds would have been flying over elephants. Their colonization of Knepp is one of the few reversals in the otherwise inexorable trend to national extinction; possibly the only optimistic sign for turtle doves on British soil.

The key to Knepp's success, conservationists are beginning to realize, is its focus on "self-willed ecological processes." Rewilding is restoration by letting go, allowing nature to take the driving seat. In contrast, conventional conservation in Britain tends to be about targets and control, doing everything humanly possible to preserve the status quo, sometimes to maintain the overall look of a landscape or, more often, to micromanage a particular habitat for the perceived benefit of several chosen species, or just a single, favored one. In our nature-depleted world this strategy has played a crucial role. Without it, rare species and habitats would have simply disappeared off the face of the earth. Such nature reserves are our Noah's Arks—our natural seed banks and repositories of species. But they are also increasingly vulnerable. Biodiversity continues to decline in these costly and micromanaged oases, sometimes even threatening the very species these areas are designed to protect. Something drastic needs to happen, and happen soon, if we are to halt this decline, and perhaps even reverse it. Knepp presents an alternative approach—a dynamic system that is self-sustaining and productive, as well as far cheaper to run. Such an approach can work in conjunction with conventional measures. It can be rolled out on land that on paper, at least, is of no conservation importance. It can add buffers to existing protected areas, as well as bridges and stepping stones between them, increasing the opportunities for species to migrate, adapt, and survive in the face of climate change, habitat degradation, and pollution.

When we began rewilding the estate nineteen years ago, we had no idea about the science or the controversies surrounding conservation. Charlie and I embarked on the project out of an amateurish love for wildlife and because we would have lost an impossible amount of money if we had continued to farm. We had no idea how influential and multifaceted the project would become, attracting policy makers, farmers, landowners, conservation bodies, and other land-management NGOs, both British and foreign. We had no idea Knepp would end up a focal point for today's most pressing problems: climate change, soil restoration, food quality and security, crop pollination, carbon sequestration, water resources and purification, flood mitigation, animal welfare, and human health.

Knepp is but a small step on that road to a wilder, richer country. But it shows that rewilding can work, that it has multiple benefits for the land; that it can generate economic activity and employment; and that it can benefit both nature and us—and that all of this can happen astonishingly quickly. Perhaps most exciting of all, if it can happen here, on our depleted patch of land in the over-developed, densely populated southeast of England, it can happen anywhere—if only we have the will to give it a try. ●

An entirely different ecosystem has emerged from ending the intensive industrial farming and dairy practices at the Knepp Estate. Without any farming at all, it now produces 75 live-weight tons of organic, wild-pasture fed, free-roaming meat every year. Rewilding has resulted in huge gains in terms of soil restoration, carbon sequestration, water storage and water purification, flood mitigation, air purification, and habitat for rare species and other wildlife, including pollinating insects—a space for nature, contributing to human health and enjoyment.

Grasslands

Grasslands account for about 15 percent of global terrestrial carbon storage, likely exceeding the carbon stored in temperate forests. Compared to the world's forests, grasslands store more of their carbon—up to 91 percent below ground, where it is secure from fire and other processes that release carbon. In drought- and fire-prone regions, grasslands are more secure carbon sinks than forests.

To understand why grassland carbon is below ground, it helps to think of these ecosystems as lands of frequent fires and large herbivores, including bison, elephants, wildebeest, and many extinct beasts such as mammoths, glyptodonts, and giant ground sloths. Over millions of years of fire and plant consumption above ground, grassland plants evolved to protect their biomass below ground. Although ecologists traditionally describe grasslands as occurring in regions intermediate in precipitation between forests and deserts, most grasslands receive more than enough rain to support forests or dense shrublands. For

these disturbance-dependent grasslands, fire and large herbivores prevent tree invasion and maintain herbaceous plant diversity. While it is true that fire releases carbon dioxide and herbivores eat plant biomass, any carbon loss is temporary, since fire and herbivores stimulate abundant plant regrowth that absorbs carbon dioxide from the atmosphere. Further, dung produced by herbivores and charcoal produced by fires contribute to soil carbon storage.

The ways that plants in grasslands store carbon is fascinating. Many prairie grasses have exceedingly long roots, containing much more carbon in below ground biomass and surrounding soil organic carbon than their above ground parts. Especially in tropical and subtropical grasslands, most grasses have the vast majority of their biomass under ground. If you were to dig up these ancient, seemingly tiny herbs, you would find that many are effectively underground trees with just the tips of their branches above ground. Many grassland herbs and shrubs rely on underground storage organs, which include multipurpose

structures, such as rhizomes, and highly specialized organs, such as lignotubers. These underground structures store high concentrations of carbohydrates that permit plants to re-sprout after fire, grazing, drought, or other disturbances remove their above ground stems and leaves.

Tropical grasslands store approximately 30 tons per hectare as biomass and 77 tons per hectare as soil organic carbon. Temperate grasslands store around 9 metric tons per hectare as biomass and 156 metric tons per hectare as soil organic carbon. Assuming 20 million square kilometers of tropical grasslands and 10 million square kilometers of temperate grasslands worldwide, plus flooded and mountain grasslands, the total global grassland carbon stock is 470 billion metric tons of carbon, out of a total terrestrial carbon pool of 3,300 billion metric tons.

Grassland conservation is a global priority. In many regions worldwide, grasslands have declined more than forests because they are easier to clear for agriculture and because natural fires have been excluded or suppressed.

Many grasslands are biodiversity hotspots with large concentrations of endemic species. Unfortunately, efforts to increase carbon storage in grasslands by planting trees put the carbon storage capacity and biodiversity of ancient grasslands at risk. An additional risk of planting trees is that dark forest canopies absorb more heat than highly reflective grasslands, meaning that tree planting for carbon storage in grasslands can be both bad for biodiversity and counterproductive for mitigating climate change. It is critical that grasslands be cherished for their biodiversity, as well as for their ability to support the livelihoods of pastoralists and hunters, at least as much as for their carbon storage capacity. Maintaining and restoring natural grasslands—and avoiding inappropriate tree planting—is critical if we are to promote both carbon storage and biodiversity conservation. ●

Sunlight on grasslands in the Tibetan Plateau at an elevation of over 13,000 feet.

Rewilding Pollinators

In 2008, designer Sarah Bergmann transformed a mile-long stretch of parking medians along Columbia Street, between Seattle University and a patch of urban forest called Nora's Woods, into a pollinator-friendly corridor of flowering plants. Concerned about the decline of populations of bees, butterflies, moths, and other native pollinators due to urbanization and pesticides, Bergmann decided to create a continuous pathway that allowed the insects to travel between the two isolated green spaces. After consulting with entomologists, gardening experts, and city planners, Bergmann engaged homeowners along the corridor and raised a small army of students and volunteers to help establish beds of low-maintenance plants in the parking medians as a regenerating source of food for the pollinators. The project quickly became popular, attracting the attention of conservationists, ecologists, planners, and landowners throughout Seattle and across the nation.

As Bergmann understood, pollinators are under siege around the world. While colony collapse disorder among honeybees and the precipitous decline of Monarch butterfly populations have grabbed headlines, pollinators everywhere are struggling under the rapidly expanding threats of habitat destruction, agricultural pesticides, invasive species, and climate change. According to a worldwide assessment, 16 percent of vertebrate pollinator species, such as bats and hummingbirds, are threatened with extinction, while more than 40 percent of invertebrate species face a similar fate. Not only do nearly 90 percent of all wild flowering plants depend at least to some extent on pollination, 75 percent of total global food production does as well. It is estimated that one in every three bites of food consumed in the United States depends directly on pollination, including apples, carrots, avocados, asparagus, cherries, blueberries, pumpkins, onions, sunflowers, citrus, and nuts. In the past fifty years, the volume of agricultural production dependent on pollinators has increased by 300 percent, including crops that are used for biofuels, medicine, and fiber for our clothes.

Migratory pollinators that travel long distances are especially vulnerable to threats, including drought, pesticides, and habitat degradation. Restoring and protecting migratory corridors—sometimes called nectar trails—is crucial to the survival of these species. Research can reveal landscape-level connectivity between habitat preserves, which often protect pollinator breeding grounds, and identify weak links in the corridors, while conservation and educational activities can do the necessary grassroots work to protect them. One important source of untapped connectivity is the thousands of miles of roads, electricity transmission corridors, and rights-of-way that crisscross

the nation, estimated to total around seventy million acres. Currently, many of these corridors are mowed and sprayed with herbicides for safety and weed suppression, but with a change in management practices they could become pollinator-friendly habitat, full of wildflowers, native grasses, and forbs. For example, the nonprofit Save Our Monarchs Foundation is partnering with the Omaha and Nebraska Public Power Districts to restore native milkweed habitat on hundreds of acres of transmission rights-of-way in that state.

Along the borderlands of the American Southwest, a binational effort is underway to restore native flowering plants along nectar corridors and heal damaged riparian areas, including creeks and wetlands, so they can once again provide reliable sources of food and water for pollinators. Ethnobotanist and author Gary Nabhan, who says regenerating nectar trails is critical to ecosystem health, calls this "bottom-up food chain restoration." Wild pollinators depend on diverse, vibrant plant communities, which, in turn, play a part in productive water and mineral cycles and the creation of soil carbon. One key is the reestablishment of native perennial shrubs, often grown in hedgerows along the edges of cropland and pastures, which serve as ideal habitat for a wide variety of pollinators and other beneficial insects. The ultimate goal is to create a corridor that shelters a sequence of flowering plants necessary for north-to-south (and reverse) migrations, including the sugars and other nutrients needed by pollinators to complete their long journeys.

Many pollinators, such as the lesser long-nosed bat and the monarch butterfly, are keystone species in larger ecosystems, linking plants and animals together in mutualistic relationships. If populations fall too low to remain viable, then a series of cascading negative ecological effects could happen, making them the proverbial canary in the coal mine for planetary well-being. Fortunately, we can take action. Whether it's a mini-garden of flowering plants in a street median, a ten-acre prairie restoration project under a power line, or the protection of a critical wetland along a migratory corridor, many options exist for restoring and protecting pollinator pathways. ●

_____

(*Left*) Julia heleconia (*Dryas iulia*) butterfly above a Yacare caiman (*Caiman yacare*) in Pantanal, Brazil. Butterflies often land on caiman to drink the salt from their eyes.

(*Above*) Clouded apollo butterfly (*Parnassius mnemosyne*) in the French Pyrenees.

Wetlands

For centuries, wetlands were drained and dried for land and fuel. Midwest farmers have been eliminating the prairie potholes and mires to create more arable land since the eighteenth century. Peat derived from wetlands has been harvested in Northern Europe for more than a thousand years and fuels power plants in Ireland and Russia to this day. Until 1953, wetlands were less studied because the term did not exist. They fell in the cracks between terrestrial and aquatic ecosystems, because they were neither and both. Today we know that wetlands are extraordinarily valuable, the most diverse ecosystems on the planet, one of the two greatest vaults of terrestrial carbon. They cover 4 percent of the land and contain six times more carbon per acre than grasslands. Peatlands alone protect a massive amount of carbon—around 650 billion tons. The atmosphere currently holds 885 billion tons of carbon.

More than 300 million years ago, plants swathed a hot and steamy earth. Growth was lush and profuse; sea-sons were barely noticeable, and the humid air was fecund with giant flying insects, including *Meganeura*, a dragonfly cousin with a nearly thirty-inch wingspan, while on the ground slithered five-foot-long millipedes. Swamp forests dominated entire regions of North America, China, and Europe. It was a photosynthetic bonanza known as the Carboniferous Period, which produced much of our fossil fuels. Gardens and woodlands retain some surviving plants from that era—horsetail, mosses, and ferns, the shrunken remnants of their giant predecessors.

Wetlands remain the most diverse and productive habitat on the planet, repositories of carbon, diversity, and life. Wetland variations are endless, depending on soil, clime, depth, and ecosystem. They can be seasonal or permanent, freshwater or saline, and come in a myriad of shapes, forms, and places. There are fens, bogs, peatlands, seeps, moors, mires, muskegs, mudflats, mangroves, quag-mires, marshes, bayous, tidal marshes, swamps, billa-bongs, floodplains, oases, and oxbows. Wetlands range from the countless bogs strewn across the summer Arctic to the 54,000-square-mile Pantanal, in South America. They are feeding grounds for migrating birds; sanctuaries for beavers, sloths, otters, and capybaras; nurseries for spawning fish, mollusks, and crustaceans; ponds for egrets, herons, and cranes; and refuge for alligators, frogs, snakes, and turtles.

In some of the largest wetlands, stagnant water resting upon saturated soil created low-oxygen environments where sphagnum moss and other plants slowly decomposed for centuries, forming peat. Some of the most extensive peatlands in the world are found in Canada, with more than 430,000 square miles of them, an area twice as big as France. Northern peatlands were formed by successive ages of intense glacial scouring that created vast regions of flooded depressions. There are different types of peatlands, from subalpine peat bogs in the Patagonian Andes to the peat forests of Indonesia. The Indonesian peat forests are more than fifty feet deep. Replacing peatland forests with palm oil plantations requires canals to channel the water away and dry the remaining soil, which causes extraordinary carbon emissions.

Where local rains are prolific and exceed outflows, wetlands perform a function rivers and lakes cannot. The waters move slowly, seep into the soil, and/or remain standing for days, months, or years. They perform unseen services of filtration and cleansing. Pollutants from upstream farms, including nitrates and phosphates, are taken up by plants, buffering downstream impacts to help prevent dead zones like those found in the Gulf of Mexico, where nitrogen-fed algae blooms kill marine life. In times of intense rainfall, wetlands help prevent extreme flooding and soil erosion. In droughts, their capacity to store water can maintain water flow in streams and rivers. In both cases, wetland restoration is less expensive and more effective than constructing steel and concrete infrastructure, including channels, dams, barriers, levees, and floodwalls. Water scarcity arises from lack of water capture as well as overuse.

Wetlands continue to be under threat and exploitation the world over. More than 65 percent were lost in the past century. However, there are international conservation institutions and organizations working to protect the world's remaining wetlands. The Convention on Wetlands was established in 1971 in Ramsar, Iran, where the treaty for the conservation of wetlands was signed. There is now a list of more than 2,400 wetland sites, known as the Ramsar List, that are considered to be of international importance. These are reserves, sanctuaries, barrages, lagoons, and lakes, among them the Okavango Delta, in Botswana, the Camargue, in France, the Wadden Sea, in the Netherlands, and the Everglades, in Florida. However, the math is still on the side of loss. When we abstract wetlands into numbers, types, biodiversity, and function, the human communities that rely on them mustn't be overlooked. Without the Pantanal, for example, tropical rains would quickly sheet off the land, causing downstream flooding. During the rainy season, about 80 percent of the Pantanal floodplains are submerged, nurturing a profusion of plant, animal, and bird life. At least 270 communities depend on the wetland, from Indigenous to local to touristic, but not all of those communities sustain the Pantanal. Many newer communities are degrading the wetland and surrounding lands through untethered cattle ranching, glyphosate-enabled soybean agriculture, and overfishing. In 2020, nearly one-third of the Pantanal—savanna, forest, and shrubland—went up in flames. Conservation areas such as the Encontro das Águas State Park, one of the world's greatest refuges for jaguars, was almost completely destroyed. The Brazilian government did almost nothing.

At the same time, wetland restoration is occurring worldwide, and it is one of the most fulfilling acts of conservation. In Illinois, the Wetlands Initiative is removing drainage tiles (underground pipes) from lands that were formerly lakes. In 2001, at the Dixon Waterfowl Refuge, which had been farmland for most of a century, volunteers and staff began removing a forty-mile network of clay pipe that drew off nine million gallons of water per day into the Illinois River. Once the drainage tile was removed, Hennepin and Hopper lakes, which had been drained for farmland, refilled in three months. Today, the Dixon Waterfowl Refuge serves as a migration site for thousands of birds. In 2018, citizen scientists and experts spent a weekend combing through the refuge and identified 915 species of plants, birds, invertebrates, insects, reptiles, amphibians, fungi, mammals, spiders, mites, and fish, a menagerie of diversity that was once a cornfield. This underlines a basic principle of regeneration: it is the default mode of nature. Once we stop burning, scraping, poisoning, cutting, and draining our environment, and create the conditions for life instead, life in its spectacular diversity regenerates without exception.

Among the inhabitants counted at the Dixon Waterfowl Refuge are white pelicans, with their nine-foot wingspans, the brilliant blue indigo bunting, and the almost prehistoric pileated woodpecker. White pelicans have been observed by airplane pilots and climbers gliding in spirals far above the 13,775-foot Grand Teton, in Wyoming. The birds rely on thermals to lift them to those heights, and after a while they glide down before starting over again. Why? No one knows. They seem to enjoy it. ●

Wetland area of Scots pine in the Abernethy Forest, Cairngorms National Park, Scotland.

Beavers

Nature's best water restoration technology comes with a sleek, oily fur coat and chisel-like incisors. Beavers are a keystone species that create watery habitats and wetlands for salmon fry, turtles, frogs, birds, and ducks. To be fair, they are also big rodents, and the clichés about being busy or eager as a beaver truly do describe this mammal. When it comes to modifying the earth for their own needs, they are second only to humans.

Beavers are refined and practiced engineers, tirelessly creating dams, canals, and family lodges. Before a beaver can build a lodge, it needs to build a dam. The dams create a deepwater refuge that provides protection from predators such as coyotes and bears. Their self-sharpening teeth and muscular jaws gnaw through trees, saplings, and felled branches, which they bury like skinny fenceposts into riparian outlets using their prehensile front feet. Once sticks and poles are placed, flexible branches and brush are intertwined in crisscross patterns. Then come the grasses, weeds, and mud caulking used to fill the openings and crevices and make the dams watertight. Their lodges are domelike, up to ten feet tall, with hidden underwater entrances, often situated in the middle of the pond. Inside is an extended family of young kits and yearlings, able to overwinter in iced-over ponds with nourishing stores of slender branches. The family feeds on the green cambium layer of bark, munching it like corn on the cob, with side dishes of leaves, twigs, lily tubers, and roots.

Beavers are notoriously nearsighted and, when seen up close, look as if they need glasses. They have transparent eyelids that fold over their retina like swim goggles, allowing them to see better underwater, but on land they rely instead on a keen sense of smell. Getting close to beavers is a rare occurrence, as they almost always avoid human contact and are gone before you see them. They dive quietly underwater, where they can hold their breath for up to fifteen minutes, or swim to their lodges. If startled or frightened, they will make a ruckus by smacking the surface of the pond several times with their scaly, paddle-like tales, an earsplitting sound that can be heard by beaver colonies far away. Hearing the sound, distant beavers will also submerge and disappear into their ponds and lodges.

But one predator was undeterred by ponds, lodges, and tail slaps: fur trappers. North American populations of beaver (*Castor canadensis*) are estimated to have numbered between 60 and 400 million individuals before European colonization. They were relentlessly trapped and traded from the seventeenth to the nineteenth centuries, when hats and raiment made of beaver pelt were fashionable in Europe. By 1900, no beavers could be found in the northeastern United States. The total population was estimated to be a mere 100,000 countrywide when the U.S. Fish and Wildlife Service interceded and prohibited further capture and killing. Two species of beaver were saved from extinction at that time: *Castor canadensis*, in North America, and *Castor fiber*, the Eurasian beaver, in Northern Europe, which was recently reintroduced to Scotland and the UK. Today, the beaver population in North America is estimated to be between ten and fifteen million.

For decades, beavers had been considered a destructive, tree-killing nuisance that required eradication. Even though extermination efforts still persist, the past two decades have seen a notable transformation in understand-

ing. Ecologists and scientists have come to the opposite conclusion: that beavers are restorers of habitat, floodplains, fish, aquifers, wildlife, and streams—whole ecosystems, to be concise. Benjamin Dittbrenner, executive director of Beavers Northwest, found that slowing down water flows from streams and rivers behind beaver dams increases groundwater recharging significantly. This process keeps streams running fuller and longer into the season. Further, as water moves through the sediment and the porous spaces beneath a streambed, it cools. The cold water, which eventually reemerges downstream as surface flow, is crucial for salmon fry and other aquatic invertebrates that rely on oxygen-rich water for their survival. As the snowpack in the Cascades decreases with climate change, scientists like Dittbrenner hope that the ecosystem services provided by beavers and their dam building might compensate for some of the water lost.

Researchers in the state of Washington are applying beaver conservation specifically to the cause of restoring salmon habitat. They are trapping so-called nuisance beavers from urban environments and repopulating them in upland river systems, then measuring the impacts they have on the ecosystem. Michael Pollock, an ecosystems analyst at the National Oceanic and Atmospheric Administration, estimated that at one time beaver ponds in the Stillaguamish River basin provided crucial winter habitat to some 7.1 million juvenile coho salmon, but with the loss of the beaver population, untended dams crumbled, ponds drained and disappeared, and the number of juvenile coho collapsed to 1 million. The coho lacked deep pools and rich riparian vegetation, which together provide juveniles with predatory protection and an abundant food supply. The beavers have been called earth's kidneys, owing to an unforeseen benefit of the dams. Silt builds up on upstream sidewalls and collects toxins, such as pesticides and fertilizers, which are then broken down and detoxified by the microbial populations. Thus, the benefits of beaver activity include increasing groundwater levels, decreasing and retaining stormwater runoff, creating habitat for species besides salmon, decreasing erosion and incision of streambeds, and increasing riparian vegetation. Ben Goldfarb, an environmental journalist who authored a definitive book on the secret life of beavers, writes, "Beavers, by capturing surface water and elevating groundwater tables, keep our waterways hydrated in the face of climate change–fueled drought. Their wetlands dissipate floods and slow the onslaught of wildfires. They filter pollution. They store carbon. They reverse erosion. And, whereas our infrastructure is generally inimical to life, they terraform watery cradles for creatures from salmon to sawflies to salamanders. They heal the wounds we inflict." ●

Bioregions

The concept of a bioregion was popularized by environmental activists and scientists in the 1970s to describe geographical areas that are defined by their unique landscapes and ecosystems, as opposed to the counties, states, and countries that constitute our political and financial systems. Bioregionalism took hold as a way to identify spatial and biological dynamics between ecosystems and physical systems. Bioregional maps are not overlaid with suburban names and abstract pastel colors identifying political borders. These features cannot be seen by birds or satellites. One of the first named bioregions was Cascadia, extending from northern Oregon to the Continental Divide and all the way up to the Copper River Delta, in Alaska, one of the most biodiverse and ecologically rich territories in North America. There are 185 distinct named bioregions according to researchers at One Earth. Bioregions were not intended to supplant politics. Rather, the understanding of bioregions creates political and cultural systems that are in accord with ecological stewardship. David McCloskey, then

a professor at Seattle University, was the first to describe Cascadia as a bioregion—the land of falling waters. Although the geography of a bioregion is delineated by both physical and biological factors, the preservation of culture, customs, and traditions is as crucial as saving native trees, plants, and animals.

The institution of bioregions has three goals. Foremost, it is intended to identify and inform inhabitants about where they live. Beyond street numbers and malls is a world that supports the life of the region that is largely unknown to cities and communities. Where does fresh water come from? What is a watershed? Where does fresh local food come from? What impacts do human communities have on the region, and are those impacts harmful or sustainable? Once these are identified, it is the goal of a bioregional community to maintain and regenerate degraded natural systems, whether forests, fisheries, or fauna.

Next is to teach and learn. How could human needs be met in ways that are in sync with the carrying capacity

about how you can't manage what isn't measured, bioregions set about establishing new sets of metrics. To this day, we do not have proper measures of what is happening on earth with respect to the biosphere. The useful metrics about climate, water, and biodiversity are meaningful when they are connected to where we live—that is where we can most readily act. Most of the impacts we have upon our local and greater world are hidden, occluded, or forgotten at the curbside, where our trash and recycling are picked up.

In 2019, Spencer Beebe, founder of the nonprofit Ecotrust, created Salmon Nation, which he calls a *nature state*, as contrasted with a *nation state*. By his own admission, it is both an idea and a place. A nature state is not meant to dissolve historical political boundaries, dysfunctional though they may be, but to create a state of mind and action that transcends political borders. Human life requires clean air and water, healthy soil, and a relatively stable climate if it is to continue. Learning from and acting on behalf of nature is not about ideologies, and neither are bioregions. Although Salmon Nation was delineated by the migratory pattern of one uniting keystone species: wild salmon. It extends from the coast and inland valleys of Northern California through western British Columbia and all the way to Attu Island, in the Aleutians, the westernmost point in the state of Alaska. And why salmon? They are the canaries of sea and stream, indicators of the health of the environment.

Salmon Nation is creating a network of "Ravens," self-organizing groups comprising thousands of community leaders bringing about positive change. They are both nonprofit and for-profit social entrepreneurs, including "artists, musicians, regenerative ranchers and farmers, restoration foresters, community fisherpeople, physicians, renewable energy and green building advocates, tiny-home builders, Indigenous leaders, philanthropists, investors, teachers, scientists and leading blockchain engineers." The purpose of Salmon Nation is to connect Ravens more cohesively, for the same reason most readers of this book seek to be part of something larger than their individual life with respect to the climate crisis. In times of chaos and disruption, we seek one another. Bill Bishop once wrote that "it used to be that people were born as part of a community, and had to find their place as individuals. Now people are born as individuals, and have to find their community." Spencer Beebe has consistently forged and recreated community in his forty-year dedication to the environment, promoting narratives that transform culture and unleash "the human imagination towards a new myth of people and planet." ●

of the region? Are deeper wells being drilled into depleting aquifers? Are dams impeding fish migration and habitat? Is industrial pollution destroying downstream marine environments? Is the coal-fired generating plant depositing mercury downwind on crops and people? By emphasizing community in the largest sense, bioregions can identify and understand how human needs can be satisfied.

The third goal is to explore the idea of reinhabitation. Rather than focus on saving what was left in the case of plants, animals, forests, wilderness, and the biosphere, the question was how to regenerate the environment with place-based ideas and initiatives so that the needs of all living systems are satisfied, honored, and increased. Essentially it is a question of how to create more—more life, more pure water, more fish, more trees, more grasslands, more wetlands, more wildflower meadows, more resilience.

Creating a bioregion requires understanding of the region as a whole, which means working as a community and making the commons home. With a nod to the adage

Russian brown bear on the Reka Zhupanova River, Kamchatka Peninsula, Russia.

Wild Things
Carl Safina

*Carl Safina's writings on ecology are considered literature—
eloquent, informed, accessible, poetic. And surprising. Two of
his most treasured works are* Beyond Words: What Animals
Think and Feel *and* Song for the Blue Ocean. *The test of a
great ecology book is that you see the earth and its creatures
with new eyes and that you feel empathy, a sense of connectedness that is crucial to the survival of the living world. As a
world-class ecologist who writes about animal cognition and
even precognition, Safina goes beyond the idea that animals
have emotions and thought—he sees them as conscious, feeling
beings, no less than we are, with levels of intelligence unknown
to us. Can the elusive veery, a small woodland thrush, predict
the intensity of coming hurricane seasons better than computer
models? After two decades of study, it seems the answer is yes.
The intensity of hurricane seasons is foretold by the timing of
veery migration from the eastern forests of North America to
their wintering grounds in South America. Safina proposes a
figure-ground reversal between our omniscience and the supposed lack of mental capabilities of the animal world. "There is
in nature an overriding sanity and often, in humankind, an
undermining insanity. We, among all animals, are most frequently irrational, distortional, delusional, worried." His gift of
revealing the humanlike qualities of nonhuman consciousness
transforms the dimension of what occurs when we shoot a wolf,
harpoon a whale (Japan, Norway, and Iceland still do), remove
primates from forests and consign them to a life in a concrete
zoo enclosure, or poison coyotes. His efforts have earned him
numerous awards and fellowships, from the MacArthur
"Genius" award to grants from Pew, Guggenheim, and the
National Science Foundation. His spirited descriptions of the
living world, bolstered by rigorous science, have awakened the
world to the importance of biodiversity because it provokes
wonder, awe, and fascination. —P.H.*

In June of 1976, as an undergrad, I drove all night to New Jersey's Island Beach State Park, arriving shortly before first light. Whip-poor-wills filled the pre-dawn with their name while I awaited two people and a cardboard box. We boated to a marsh island where my companions finally opened the box and I locked eyes with three slightly bewildered, downy young birds. They were peregrine falcons, part of the first captive-bred peregrine cohort scheduled for release in a grand attempt to reverse their species' DDT-induced disappearance across the United States. DDT and related pesticides had been banned four years earlier, making the environment less fatal for these and many birds. We placed the chicks in a specially erected tower. My job: tend them during their weeks till fledging. None of us knew whether re-wilding would work out. Or whether I would.

Things have gotten better, and things have gotten worse. A United Nations panel last year released a summary of an upcoming report, roughly extrapolating—based on the proportion of species that the International Union for Conservation of Nature has assessed as "threatened" or "endangered"—that a million species face extinction in this century. A million deaths, Stalin reputedly said, is just statistics. Even Mother Teresa said, "If I look at the mass I will never act." This emotional overwhelm, this paralyzing tsunami to the soul, has been termed "psychic numbing." Mother Teresa had added, though, "If I look at the one, I will."

If conservation and the environmental movement are remiss in anything, it is the inability to remember that mass statistics obscure real tragedies, and numbers numb us. Each species, individually, has scant voice to vocalize its tragic opera. But as troubles rise in chorus, they sing the woes of living things large and humble, no matter whether they darken skies or rustle grass or keep their peace among underwater boulders. Everywhere, trouble rumbles. Monkeys, elephants, tigers, lions, giraffes—the ones in every painting of Noah's Ark deemed worthy of salvation two by two—we send them to perdition one by one. Their flood is us as, in our billions, we rise to engulf the world.

Conservation's most troubling paradox is that the "most popular" species are heading toward extinction. All ten of the "most charismatic" animals—pandas, elephants, lions, tigers; the ones we paint on the walls of our babies' nursery rooms—are at risk of extinction in the wild. . . . Because species don't get onto endangered lists until they are rare, it is imperative that we wake up to the broad across-the-board declines that are happening.

Statistics are inescapable, but we can at least begin breaking down the numbing round statistic of a million species now imperiled, to more angular chunks of numbers. One-fifth of mammals are threatened with extinction.

Of bird species, more than 1,450—one in eight—are threatened. The highest number of threatened birds in any group is awarded to parrots; about half of roughly 400 parrot species are declining due to agriculture, logging, or capture for the cage trade, killing for food, and killing as crop "pests." Free-living African grey parrots face extinction in the wild and are down to 1 percent of former numbers in parts of their range. North America has lost nearly 30 percent of its birds since 1970, a trend likely to be roughly mirrored in Europe. Five birds extinct in the wild persist in captivity; but what can be their fate?

About that ark: It doesn't have enough beds. The formerly common are becoming rare. In North America alone, 20 common bird species—more than half a million individuals each—declined more than 50 percent in the last 40 years. Bobwhite quail (common in every woodlot in my youth) have declined more than 80 percent, even in good habitat. Nineteen American shorebirds have halved since the 1970s. Puffins and other seabirds worldwide have declined 70 percent since 1950. Whip-poor-wills like I heard in 1976 have dropped 70 percent.

A quarter of the world's thousand-or-so sharks and rays are considered "vulnerable" to "critically endangered," with just 23 percent considered safe—the lowest proportion in any vertebrate class. Hammerheads, makos, blue sharks—so abundant when I first began going to sea—I've watched them fade, for soup-thickener. Increasingly, "mass mortality events" kill thousands. In 2015, in Kazakhstan, 200,000 saiga antelope—60 percent of the world's population, died in a week when abnormal heat and humidity turned a harmless bacteria lethal. In Australia, the overall wildlife toll from the recent fires will likely prove entirely awful as iconic species like the koala and platypus decline sharply. In the last several years diminished food tied to warming water has mass-starved puffins, shearwaters, fulmars, kittiwakes, auklets, gulls, and hundreds of thousands of murres from Alaska to the U.S. West Coast. Scientists have documented well over 700 mass mortalities since the late 1800s, affecting more than 2,400 animal

Joseph Wachira saying goodbye to Sudan, the last male northern white rhino on the planet, as he died. The photo was taken by Ami Vitale at the Ol Pejeta Conservancy located in Laikipia County in central Kenya. "I saw Sudan for the first time in 2009 at the Dvůr Králové Zoo in Czechia (the Czech Republic). I can recall the exact moment. Surrounded by snow in his brick and iron enclosure, Sudan was being crate trained—learning to walk into the giant box that would carry him almost four thousand miles south to Kenya. He moved slowly, cautiously. He took time to sniff the snow. He was gentle, hulking, otherworldly. I knew I was in the presence of an ancient being, millions of years in the making (fossil records suggest that the lineage is over 50 million years old), whose kind had roamed around much of our world. On that winter's day, Sudan was one of only eight northern white rhinos left alive on the planet. A century ago there were hundreds of thousands of rhinos in Africa."

populations including mammals, birds, fishes, amphibians, reptiles, and ocean invertebrates. Many likely go undocumented.

Many of us easily remember summer streetlights holding halos of moths, often with bats dipping through the illuminated buffet. Last summer on Long Island a friend said, "Look at the street light." There was not a single insect visible. In Germany, scientists have documented flying-insect declines of roughly 80 percent, and in Puerto Rico's Luquillo rainforest researchers noted an astonishing 98 percent decline of ground insects and an 80 percent loss of tree canopy insects, with accompanying crashes of insect-eating birds, frogs, and lizards. Scientists are just now compiling documentation of steep declines in butter-

flies, bees, and insects worldwide due to farming and warming, with an extinction rate eight times faster than mammals and birds. With unusual urgency, scientists have called the implications "Catastrophic to say the least."

At this point in the history of the world, humankind has made itself incompatible with the rest of life on Earth. We're too much of a good thing. I don't think that's how we'd want to be remembered. Unless we see the big picture and care about our role in maintaining or destroying the miracle of living existence, we will continue to do the latter. But the big picture is exactly what can be numbing.

Fortunately, none of us has to tackle the big picture. For the last forty years I've had a quote from Gandhi pinned somewhere in my office through several moves. He said that "what we each do seems insignificant, but it is most important that we do it." It can be something little and local, or maybe something big. You can maybe help do something like putting falcons back into the skies or be head of the U.S. Fish and Wildlife Service; one woman—Jamie Rappaport Clark—eventually did both. In the humblest beginnings of individual effort, big things can incubate.

When we, collectively, do decide *not* to let animals be driven into that eternal void, it works. More than three-quarters of marine mammals and sea turtles increased significantly after gaining protection under the U.S. Endangered Species Act. Ospreys—nearly erased from the map in my youth because of DDT—now soar abundantly over our bays and rivers on their six-foot wings. Moths and butterflies increase when farms maintain bordering habitats. Conservation efforts have reversed near-certain oblivion for more than two dozen birds, various mammals—from rodents to whales—and dozens of others.

Around the world conservationists are endeavoring to stabilize ostriches, rhinos, big cats, bears, apes, cranes, deer, antelope, otters, musk ox, parrots, butterflies—and many more. Intensive protection of short-tailed albatrosses—once exterminated on their North Pacific nesting island for their feathers—has brought them from six rediscovered individuals to more than 4,000 birds. Rarest of all the world's cranes, North America's whooping crane hit a low of 15 adults in 1938. Today, following exhaustive conservation efforts, including captive breeding and restoration of several free-living populations, there are up to 250 mature individuals and a total population around twice that.

In 1985 I traveled from New York to California to see a California condor before they vanished; six remained in the wild. But condors breed well in captivity. Today, there are more than 300 free-flying condors in California, the Grand Canyon region, and Mexico's Baja peninsula. No condor would yet corkscrew into the sky had the U.S. Congress not passed the Endangered Species Act of 1973.

America's bald eagles recovered from roughly 400 breeding pairs south of Canada around 1960 to about 14,000 pairs today; they've been off the endangered list since 2007. Brown pelicans have increased more than 700 percent in 40 years, and the American alligator, listed as endangered in 1967, is abundant today. American bison, subjected to perhaps the most profligate slaughter of any creature in history, collapsed from 60 million to just 23 wild bison by 1900 in Yellowstone. More than 30,000 exist today.

Other recoveries, more painstaking, are less celebrated. The large North African antelope called scimitar-horned oryx fell from a population of tens of thousands in the 1930s to extinct in the wild by the early 1990s; they've recently been reintroduced in Chad. The black rhino had declined 98 percent from its 1960 numbers because of poaching; despite the loss of one subspecies, aggressive conservation has allowed their numbers to double to about 5,000. Gray whales were hunted to extinction in the Atlantic and hang on with perhaps just 150 in the Asian North Pacific, but they've recovered spectacularly along the West Coast of North America and are often seen from shore from Baja California to Alaska. Atlantic humpback whales have recovered so well that I often see them when I take our dogs to run on the beach on Long Island, New York.

No one worked on all of those successes. But someone worked on each of them, and that's what made the difference. It would help all of us, and the cause of the world's species, if we think more granularly; speak more specifically; focus on what can be meaningful; and stay observant of the many beauties remaining. Beauty is the single criterion that best captures all our deepest concerns and highest hopes. Beauty encompasses the continued existence of free-living things, adaptation, and human dignity. Really, beauty is simple litmus for the presence of things that matter.

Endangered species and wild things in the remaining wild places need us to care for them not selfishly but selflessly, for their sake, the sake of everything and everyone who is not us, for the sake of beauty and all it implies. As we make our habitual appeals to practicality, the argument we cannot afford to ignore, the one that must frequently be on our lips, is this: We live in a sacred miracle. We should act accordingly. ●

The whooping crane is the rarest and tallest bird in North America with a 7.5-foot wingspan, a red cap, and yellow eyes. Because of diminished wetlands due to agriculture and indiscriminate hunting for their meat and plumage, there were fifteen individual birds alive in 1938. The population now exceeds five hundred birds, both migratory and in captivity, one of the most successful restoration projects ever undertaken.

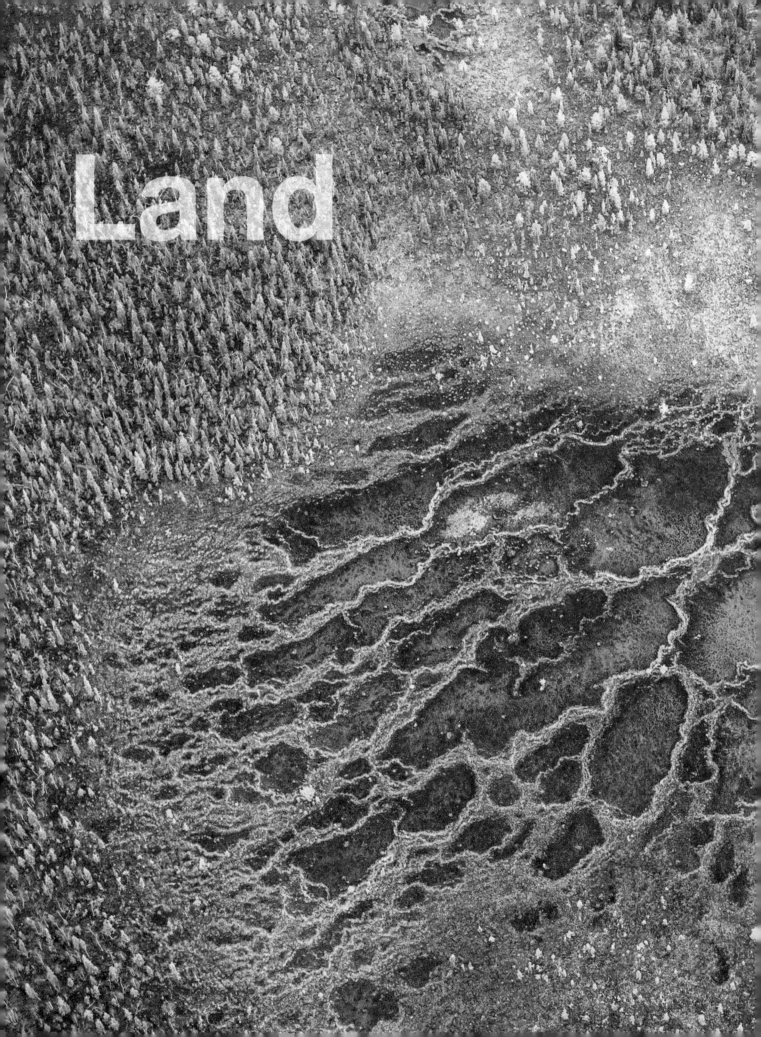

Land

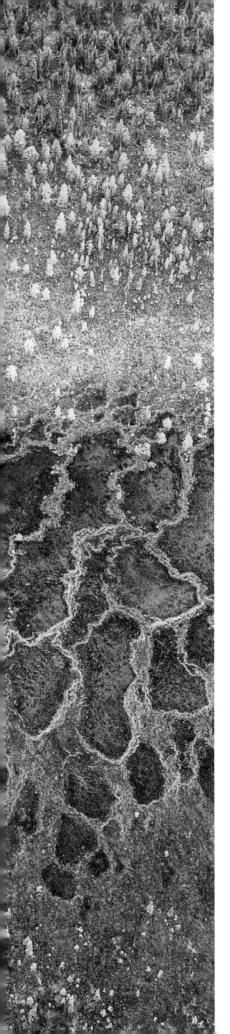

In English, the word *land* embodies many things: soil, ground, home, way of life, country, people.

The earliest known soil is 3.7 billion years old. That's only 800 million years after the planet came into existence. The soil was created by chemical weathering of rock under acid rains. It contains bits of carbon, hinting at life. Roughly three billion years later, algae moved from the sea to fresh water at the edge of land. The next step was momentous. Algae trap the light of the sun, photosynthesizing sugars from thin air. There were many types of algae at the time, but one group evolved into land plants. One hundred million years later, these plants had leaves and roots. Fungi were early and close allies. They collected nutrients from rocks and traded them to plants for carbon. It was an example of mutualism—a partnership in which different species interact and both benefit. As plants and microbes flourished and perished, they deposited organic sediment. It was the beginning of the soil we know today, the nutrient-laden medium that sustains 80 percent of all life on land.

Plant life evolved and became more complex. Fronds, needles, and bracts collected more energy from the sun. Dead roots created organic matter and passageways for water. Bacteria populated root zones. Sugars secreted by roots fed subsurface biota, providing energy for bacteria to increase the nutrient availability of minerals to the plant. Fungal filaments became mycorrhizal networks that collaborated with plants, shuttling nutrients, water, and information to the roots in exchange for energy. Greater soil life increased water-holding capacity, further expanding the range and diversity of plant species.

Photosynthetic light catchers transformed land around the planet, supporting a growing diversity of life. Terrestrial invertebrates, from worms to ants, spread out. Herbivores burrowed and intermixed the soils into finely aerated mediums. African dung beetles appeared at night, rolling balls of dung fifty times their weight into burrows. Tunneling and digesting organisms added fertility and structure. The manure and urine of large vertebrate species improved the fertility of the soils. Land became more permeable and held more rainwater, allowing it to sustain larger plants, which drew down ever more carbon from the atmosphere. Plants and trees transpired water vapor, creating mist, clouds, and rain that cooled the land, transforming the hydrology of the earth.

In the past two thousand years—a fraction of geological time—cities, agriculture, and deforestation have dramatically altered the natural carbon cycle. Forests were culled for fire, buildings, industry, and agriculture. With the advent of the plow, soil was exposed to the air and its carbon was oxidized back into carbon dioxide. Living ecosystems were transformed into lifeless dirt, easily eroded. Industrial methods of farming reduced soil carbon content from an average of 3 percent to 1 percent. Pesticides and herbicides killed off life in the soil, reducing lands around the world to deserts. Food for birds, butterflies, worms, and other creatures became scarce. The land suffered, as did the people who depended upon it. Herein lies both cause and cure. Land is resilient, like people. Regenerative agriculture and degraded land restoration reverse the loss of carbon and the drying out of our soils. We can move to methods of farming that capture more carbon and emit more oxygen. The benefits of regenerating our farms, forests, and grasslands are incalculable, helpful to all people and all species, everywhere on the planet. ●

String bogs and coniferous forests in the Laponia area of Lapland Province in the north of Sweden, now a UNESCO World Heritage Site. String bogs are mires with elevated ridges and islands formed where gradually sloping ground is frozen most of the year. They belong to a classification called patterned vegetation and contain on their edges wetland sedges and woody plants that eventually become peat.

Regenerative Agriculture

We know more about the surface of the moon than the surface of the earth. The moon is composed of known mineral fragments, whereas soil is its own ecosystem comprising trillions of living, diverse organisms, most of which remain unidentified. Hundreds of millions of years ago, biological systems began creating the pedosphere, the outer layer of the planet where rock, plants, atmosphere, and water meet and work together to make soil. Life has been evolving ever since. When you gather a teaspoon of healthy soil, you have at hand one of the most complex living systems on earth—one that in less than 150 years has been degraded by industrial agriculture. Roughly 35 percent of all carbon dioxide emissions generated by human activity since 1850 were caused by farming and deforestation.

The world over, 15 million smallholder farmers and tens of thousands of farmers and ranchers are employing agricultural methods to reverse the loss of soil health, restore land, and bring agriculture and food back to life. Regenerative systems include agroforestry, agroecology, silvopasture, pasture cropping, real organics, and advanced rotational grazing. Specific methods include no-till, complex cover crops, prairie strips, perennial crops, animal integration, and crop diversification. One of the outcomes is the sequestration of atmospheric carbon dioxide, resulting in significant increases in the carbon content of healthy soils. Regenerative agriculture has become the archetype and template for regeneration in general. It transforms farming into a method that promotes life, biological diversity, human and animal health, plant vigor, and pollinator viability.

The term *regenerative agriculture* was coined forty years ago by organic farming advocate Robert Rodale, yet its origin is in Indigenous agriculture. Lands have been continuously cropped for many thousands of years in the Americas, Africa, and Asia. One of the oldest forms was practiced by Mayan farmers using a forest garden system known as *milpa*. China, Japan, Korea, and India are supported by agricultural traditions that date back four thousand years. Regenerative practices were well in hand in

Western Africa, where the use of charcoal and green waste created durable "dark earths" that are three to four times more fertile than the thin, nutrient-poor rainforest soils. African Indigenous farming knowledge was used by slaves on White plantations in the American South. These practices caught the attention of George Washington Carver, who studied them scientifically and shared his knowledge, becoming the progenitor of regenerative agriculture in the United States in the early 1900s.

Today, there is an explosion of learning, experimentation, and collaboration around regenerative theory and practices, in no small part due to the damage done by industrial agriculture and the so-called green revolution, which promoted higher-yielding hybrid seeds (which don't reproduce themselves) and the increased use of chemical fertilizers and pesticides. Advances in crop diversity, microbiology, soil chemistry, and animal integration mean that regenerative agriculture can deservedly be called an emergent technology. It is more complex than a smartphone, involves more interactions than the internet, and has more moving parts than any machine or device—biological parts, that is, which is to say they are not parts at all but living elements within a multifaceted system. Though regenerative agriculture is complex and intricate, it is based on clearly defined principles and can be implemented by farmers and ranchers the world over. Importantly, its yields can be commensurate with or greater than conventional agriculture, with higher future returns due to enhanced soil resiliency and productivity.

The emphasis on carbon sequestration as a primary outcome of regenerative agriculture obscures its overall impact. If regenerative agricultural practices were phased in and implemented on one-fourth of the world's farms and grasslands, they would absorb and retain a much needed 55 billion tons of greenhouse gases in the next thirty years. That aspect of regenerative agriculture has become appealing to industrial agriculture companies, who see a way to profit from it by selling carbon credits. Anheuser-Busch, Bayer-Monsanto, Cargill, and other corporations are adopting the term as a preferred climate solution, which is causing confusion. Regenerative agriculture is a whole-systems approach to soil, farming, and crops, not a menu that can be cherry-picked for good publicity. For Bayer-Monsanto to promote carbon sequestration is contradictory, since they patented the largest-selling and most destructive agricultural chemical in history—glyphosate. The herbicide was patented as an antibiotic—for good reason. It kills microbial life and thus reduces carbon in the soil.

There is a simple way to understand the difference between industrial and regenerative agriculture. Industrial agriculture feeds the plants with chemical forms of nitrogen, phosphorus, and potassium. Regenerative agriculture

feeds the soil and its microbes, and the soil feeds the plants. Industrial agriculture takes a complex ecosystem of plants, soils, and insects and replaces it with monocultures, laser-guided tractors, sprayers, and chemical inputs to obtain its outputs (crops). As soil fertility declines, more inputs are required to maintain crop yields. As soils deteriorate, so does plant health, making crops more susceptible to insects. As more pesticides are used, natural predators vanish. Combinations of pesticides, herbicides, fertilizers, fungicides, and antibiotics have taken a toll on the soil, and on farmers, their families, their livestock, and wildlife. Rates of cancer, Parkinson's disease, and suicides among U.S. farmers are among the highest of any profession in the world. Industrial agriculture takes a toll on the climate as well. About one-sixth of global greenhouse gas emissions arise from the farm sector.

Scientists argue about how much carbon can be sequestered in soil over the coming decades, but it is a debate that does not mean much to a farmer. Farmers are transitioning to regenerative practices in order to receive and store more water in their soil, lower their costs, stop erosion, get out of debt, and produce healthier plants and animals to sustain their families. Here are the basic principles and techniques of regenerative agriculture.

Recarbonize the soil. The top six to seven inches of soil contain what is known as labile carbon, a type of carbon that cycles in and out with the seasons. Occluded carbon lies farther below and does not escape so readily into the atmosphere. Root sugars called exudates, packed with carbon, are released into the soil and devoured by microorganisms. The bacteria, fungi, protozoa, algae, mites, nematodes, worms, ants, grubs, insects, beetles, and voles in the soil eat one another, reproduce, metabolize waste, and solubilize minerals, making nutrients available to the plants above. Soil is a community, not a commodity. The average amount of carbon in depleted farmland and grassland averages 1 percent. How high can we build up carbon in the soil? We don't know. David Brandt's farm in Carroll, Ohio, started out at less than a 0.5 percent in 1978; today his regenerative farm is running at 8.5 percent carbon, greater than the 6 percent levels found in an adjoining woodlot.

Limit disturbance. Mechanical disturbance of the soil should be curtailed as much as possible, tillage especially. Plowing destroys soil structure, tearing apart aggregates of

A producer in Stanly County, North Carolina, rolls down a cover crop just minutes before planting corn. The "blanketlike" results of rolling or crimping provide season-long weed protection, moisture retention, and food for soil microbes.

mineral particles and tiny pores where air and water accumulate. It also destroys delicate assemblages of microbes, including beneficial fungi. Chemical disturbances, such as herbicides, pesticides, and synthetic fertilizers, can also badly damage biological life in the soil. No-till and reduced-till systems allow soil to evolve, build, and expand in breadth and complexity. When the soil is plowed repeatedly (mostly to prevent weeds), soil life is exposed to sun and weather, which causes it to dry out and die, creating additional greenhouse gases. Microbes emit carbon dioxide naturally as part of their digestive process, but repeated tillage causes decomposition and releases stored carbon into the sky.

Cover the soil. Cover crops protect the soil in winter and cool it in summer. The idea is to keep as much green on top of the soil for as long as possible in order to armor it against wind and water erosion while creating habitat for organisms. Photosynthesis is how the sun's energy gets in to the soil. Bare ground can only heat up, drying the soil and killing microbes. Except for deserts, beaches, and rocky slopes, you rarely see bare ground in nature. Cover crops also soften the impact of rainstorms, so that the water reaches the ground gently. They enrich the soil with their roots, feeding the microbiota with exudates. They fix nitrogen and convert minerals, including potassium and

phosphorus, so that they become available to plants.

A wide variety of cover crops are being employed by farmers today. It is a fiesta of plants that look more like a vegetable garden: cowpeas, buckwheat, plantain, chicory, sunflower, fescue, daikon radish, Kodiak mustard, Ethiopian cabbage, Dundale peas, black-eyed peas, Languedoc vetch, hairy vetch, common vetch, chickling vetch, crimson clover, alfalfa, winter rye, fava beans, flax, black oats, chickpeas, purple top turnip, mung beans. The original tallgrass prairies in America contained more than two hundred species of plants.

Hydrate the soil. What matters most to a farmer is not how much moisture falls from the sky, but how much soaks into the ground. As soils lose their microbial life, they lose structure; porosity is replaced by impermeability. The rate at which soils absorb rain or irrigation water is called the infiltration rate. A rich, crumbly soil structure is created by a sticky substance called glomalin, a carbon-rich protein created by fungi and humic acid that remains in the soil for centuries or longer. On conventionally farmed land, glomalin is rare and infiltration rates can be as low as a quarter inch per hour. More water than that pools and runs off, causing erosion. Regenerative farmers report ten- to thirtyfold increases in water infiltration, from being able to absorb a half inch per hour to fifteen inches. These carbon sponges make the soil an underground reservoir.

As a result of global warming, the atmosphere contains more water, and rainstorms are more severe. The capacity to capture as much water as possible is critical to preventing flooding and erosion. In Oklahoma, for every pound of wheat produced conventionally, three pounds of soil are lost to erosion. Improved water retention means greater plant growth, which creates greater photosynthetic capture of carbon. More carbon sequestration expands soil organic matter; more organic matter holds more water, and that means more food for people and animals. Increased soil moisture also provides farmers with more of the resilience they need to deal with more volatile patterns of rainfall and droughts. There is another benefit. Approximately 80 percent of the earth's surface temperature is determined by the hydrosphere, the sum of all the water in the atmosphere and on the earth. Over the past two centuries, the earth has been drying out. Deforestation, industrial agriculture, overgrazing, and increased heat have desiccated lands, raising surface temperatures. Regenerative agriculture cools its environs. Surface temperatures can be more than 1°C lower, which helps plants grow, and soil temperatures can be many degrees lower than bare soil.

Put creatures on the soil. Nature never "farms" without animals. However, one of the hallmarks of modern agri-

culture is the separation of crops and animals. For thousands of years, they had been paired—every farm, pasture, or paddy had horses, cows, sheep, goats, geese, chickens, ducks, or (particularly in Asia) fish. The ecological benefits of adaptive, rapid-rotational grazing on pastures has been known for years by practitioners. Grazing stimulates plant growth, pumping more carbon into the soil. This is equally true for crop fields and pastures. It's not just livestock. A healthy farm must also provide habitat for birds, pollinators, predator insects, earthworms, and microbes, integrated together.

Today, crops and animals are physically kept apart from each other, sometimes by many miles. Nutrient cycles suffer as a consequence. Fish can no longer survive in conventional rice paddies. Before the introduction of pesticides, they ate algae and insects and fertilized the crops. Farm animals grazed the land and were a source of manure and urea for the fields. In regenerative agriculture, the techniques of mob or rotational grazing engender increases in soil carbon and fertility. Traditional forms of herding can do this too. Careful grazing is effective because nothing stimulates exudate production better than grasses being chomped by a ruminant. It sends a powerful hormonal signal to the root to grow. If a farmer does not wish to be in the animal business, worms and vermiculture will also produce outstanding changes in soil fertility and carbon sequestration. Together, cover crops and animal grazing can replace most of the 460 billion pounds of nitrogen fertilizer consumed annually, much of which ends up polluting streams, rivers, and ocean dead zones.

Recognize that soil health is plant health is human health. There can be no plant health without soil health. And there can be no human health without plant health. Minerals, microflora, and phytonutrients are essential for human well-being. Deficiencies can cause chronic illness. Even if you are eating a nourishing diet, the plants you consume will not contain sufficient essential minerals if they are scarce in the soil. Plants depend on soil microorganisms to solubilize minerals and make them available to the plant. And plants use stress to become more fully nutritious. As with humans, stress pushes a plant to change and adapt—to make its roots go deeper in search of water, minerals, or nutrients; or to change the chemical composition of its leaves to resist pests. Conventional agriculture does the opposite. Everything a plant needs for rapid growth is provided close to the surface in the form of potash, superphosphate, and nitrogenous fertilizers. The soil is nearly sterile. It is a medium for holding plants upright. The plants have less stress and often look strong but are weak and vulnerable to insects, fungus, and rust. Competition is eliminated with herbicides, primarily glyphosate, a probable carcinogen. Conventionally farmed crops will grow faster, have shallower roots, and be less nutritious. The facts are simple: chemically dependent plants do not create healthy food; ultraprocessed foods do not create healthy children; antibiotics and pharmaceutical drugs do not create healthy animals. Our fruits and vegetables have significantly less nutrition than they did fifty years ago. Foods are "enriched" because they are impoverished. In 1965, 4 percent of the U.S. population had a chronic disease; it is now two-thirds. Today, 46 percent of our children have a chronic disease.

True regenerative farming improves plant and human nutrition without trying for a simple reason: it does not fight nature—it aligns with nature. Regenerative agriculture is at the heart of a regenerated society since it is the source of our food, nutrition, and well-being. A third of our total climate impact comes from our food and agricultural systems, as does a majority of human disease. The first principle of regeneration is to create more life. This is where we must start. ●

(*Left*) The rich, deep color of this soil indicates what healthy soil looks like. A diverse blend of crops, grasses, and cover crops creates a protective blanket that feeds and nurtures it.

(*Above*) The American pioneer in regenerative agricultural science is George Washington Carver of Tuskegee University in Alabama. Like regenerative farmers today, he created a science of farming "forward," devising methods to improve soil health that employed crop rotation, nitrogen-fixing legumes, and the regeneration of soil health on lands depleted by monocultures. He drew upon Afro-Indigenous agricultural knowledge and practices that go back more than two thousand years.

Animal
Integration

As he was driving home to his family farm in North Dakota a thought struck regenerative farmer Gabe Brown: What if he grazed his cattle on his crop fields *during* the growing season?

Normally, Brown kept his cattle herd on the farm's pastures and turned them into the crop fields only during the late fall, to graze on cover crops he had seeded after the harvest in order to keep the soil covered with living plants. The cover crops provided feed for his animals during the winter, and their manure provided nutrients for the soil. It was part of Brown's efforts to constantly improve the health of his soil. After a series of disasters in the mid-1990s, he decided to ditch conventional agriculture and take his inspiration from prairie ecology instead. He grew cover crops year-round on the farm fields, eliminating bare soil, and employed a diverse mixture of plants and animals, including chickens and pigs. To mimic the impact of a herd of bison and their graze-and-go behavior, Brown ran cattle on his pastures at a rate of 50,000 pounds per acre, moving them through the paddocks once a day. After his visit to Canada, however, he decided to raise the rate steeply, to 500,000 pounds per acre, and graze the cattle on part of his cropland during the summer. Instead of planting corn or wheat, he would turn his cover crop into a cash crop by using it to feed and grow his animals. He knew it was a highly unorthodox idea. But he gave it a try. It worked—the soil continued to improve.

That's when Brown realized that integrating animals into the entire farm operation was critical to creating true regeneration.

In a way, Brown was returning agriculture to its roots. Livestock species were domesticated thousands of years ago and globally over the subsequent centuries the cultivation of crops was intertwined with raising animals, including the use of manure as a fertilizer. In many traditional and Indigenous cultures today, livestock and crops are grown together in various combinations.

Historically, a farmer might raise a small herd of cattle or keep pigs in the barn. Before the advent of tractors, farmers used horses and oxen in the fields. In North America during the 1970s, agriculture began to change as national farm policies encouraged specialization, resulting in a segregation of livestock and crops as farmers chose one or the other. This split led to highly intensive crop monoculture systems such as corn and soybeans, on the one hand, and industrialized meat and dairy production, including feedlots and confined animal operations, on the other. Both systems were designed to maximize yield and deliver uniform commodities to the market as efficiently and cheaply as possible. Trouble soon followed. As evidence of environmental and human health damage from commodity agriculture began to mount, a countermovement sprang up, centered on organic and holistic practices that restored vitality to the land and provided healthy food. Taking their cue from nature, many organic farmers grew diverse crops and applied compost to their soils.

In the past two decades, there has been a resurgence of interest in reintegrating livestock and crops on farms, stimulated in part by important advances in our understanding of soil microbiology and led by a new generation of farmers and agricultural specialists. It is part of a shift to the soil-building and biology-enriching goals of regenerative agriculture, encapsulated by Gabe Brown's journey to restore the depleted land on his farm. Research has revealed the many links between aboveground plant and animal management and increased carbon storage and cycling belowground, deepening knowledge and broadening goals. Investigations by Christine Jones, an Australian soil scientist, into the benefits of regenerative agriculture for carbon sequestration and soil health directly inspired Brown to embrace a more active role for livestock on his farm and set his sights on improving the carbon cycle.

Examples of animal integration include chickens pecking in crop fields, cattle grazing in orchards, sheep in vineyards, and mixed herds of herbivores on pastures. A historical review of crop-livestock integration in the United States indicates a range of management practices, including animals and crops that occupy the same field at different times of the year; alternating annual and perennial forage in the field from year to year; and the intercropping of annual and perennial crops in the same field for grazing and crop production. Many types of animals can be employed in an integrated system, including turkeys, ducks, swine, rabbits, sheep, goats, horses, ostriches, llamas, and even fish. In China, the integration of fishponds with ducks, geese, and chickens nearly doubled fish production. In Southeast Asia, cattle and goats often graze under plantation trees, such as rubber, palm, and coconut, where they provide weed control while fertilizing the soil.

Animal integration happens in pastures as well. On Joel Salatin's Polyface Farm in Virginia, cattle, chickens, pigs, rabbits, and turkeys are rotated through the farm's pastures in a carefully choreographed sequence. Cattle are fed hay in the barn during the winter, and their manure is mixed with wood chips and corn, creating compost. When the cows are turned out to the pastures in the spring, pigs dig up the barn compost, aerating it. The carbon-rich compost is subsequently spread on the pastures. Chickens, turkeys, and even rabbits are guided into a pasture after the cattle have moved to the next field, sanitizing and recycling their manure. On Eric Harvey's Gilgai Farm in New South Wales, Australia, five thousand sheep and six hundred cattle are integrated together into a single unit called a *flerd*. Harvey uses their different forage preferences and grazing behaviors, as well as the composition of their manure, to benefit plant vigor, diversity, and density, expanding the number of plant species on Gilgai from seven to 136 in less than ten years.

Integrating animals into farm production has multiple benefits, including increasing the balance of nutrients across the entire enterprise. Poultry manure, for instance, is high in nitrogen, which plants need. Manure and compost applications can improve biological activity in the soil, which, in turn, can improve nutrient cycling, soil structure, and water-holding capacity. A five-year study of an integrated beef-cattle-and-cotton system in Texas found higher organic carbon, soil stability, and microbial activity in integrated forage/cotton plots than in continuous cotton plots. In Australia, integrating crops and livestock is helping farmers improve productivity, increase sustainability, and reduce risks associated with climate change and market fluctuations. One study showed that the dual use of cereal crops for forage and harvesting can significantly increase both livestock and crop productivity by 25 to 75 percent. Crop-livestock integration can also create diverse mosaics of landscapes, enhancing habitats for wildlife.

Despite its advantages, there are challenges to implementing animal-crop integration, including a persistent belief among conventional farmers that the two systems represent different types of agriculture and should not be comingled. Among climate-change activists, objections have been raised about the role livestock play in greenhouse gas production, particularly methane emissions from digestion, called enteric fermentation, though according to the EPA, direct emissions from livestock represented only 2 percent of total U.S. greenhouse gas production in 2016. Another obstacle is a cultural attitude among some consumers opposed to animal agriculture in principle. Lastly, many farmers do not have the experience or training needed to integrate crop and livestock proficiently.

One inducement to overcome these challenges is profitability. On Gabe Brown's farm, the rising amount of carbon in the soil, a result of improved soil health, has significantly increased the operation's bottom line. Animals are a key component of this success. As he travels, Brown often hears a lament from fellow farmers about the lack of profitability on their farms. "When I ask them about their model of production," he wrote, "I usually discover that they do not run any livestock on their land. I encourage all operators to take advantage of the many benefits that animals offer." ●

Greg Judy on his 1,620-acre Green Pastures Farm in Rucker, Missouri. He specializes in grass genetic South Poll cows, a British breed that is particularly suited to being grass-fed. Practicing rotational grazing, Judy does not use hormone implants and does not worm or employ fly ear tags or insecticides. Instead, the farm has 450 swallow birdhouses for natural fly control. Having gone nearly bankrupt employing conventional agricultural methods, Judy now teaches grazing classes to help other farmers create robust, healthy soil and animals.

Degraded
Land
Restoration

Walking up an eroded creek bed on a ranch in western New Mexico one day, Bill Zeedyk passed beneath a barbed-wire fence that stretched across the deep gully, fence posts dangling a few feet above his head. Inquiring, he learned from the rancher that the fence had been built sixty years earlier and the posts originally sat on the ground—a sign of how much erosion had occurred in a small amount of time.

This wasn't a surprise to Zeedyk, who knew first-hand that much of the Southwest existed in a similarly degraded condition. A decade earlier, after retiring from a career as a wildlife biologist with the U.S. Forest Service, he had embarked on a new mission: to find a way to restore damaged creeks to health. In his job with the Forest Service, Zeedyk had observed a litany of trouble in waterways—deep ditches that had once been meandering streams; wet meadows drained by head cuts (an abrupt vertical drop) advancing up the creek bed; twenty-foot head cuts and gullies where there had once been gently sloping drainages; and poorly designed dirt roads and stream crossings that had disrupted the natural movement of water across the land, causing extensive erosion. It added up to an unraveling of habitat for beaver, wild turkeys, and other animals as water tables dropped and vegetation changed. The erosion also affected livestock as creeks became steep gullies and water tanks silted up. But it wasn't just streams. Zeedyk saw degradation across the region's rangelands and forests as well.

According to the Intergovernmental Panel on Climate Change, about 25 percent of all ice-free land on earth is subject to human-induced degradation, affecting up to three billion people who depend on the land for their livelihood. Forty percent of the world's degraded lands are located in regions with high poverty rates. While millions of people have managed to migrate, an estimated one billion people are trapped in place by poverty.

Soil erosion is the most common type of land degradation. Natural landscapes are buffered against the erosive effects of wind, rain, and snowmelt by grasses, plants, shrubs, and trees. When this cover is disturbed or removed by agricultural tilling, overgrazing, logging, or land clearing, the soil is exposed to the elements. One storm can turn a farm furrow into a gully. Soil degradation is also caused by the application of chemical fertilizers and pesticides to crops, which damages the microbial life and nutrient cycling that bind soil particles together underground. As soil leaves a farm, it carries essential nutrients for plants along with it, such as phosphorus, nitrogen, and potassium. Soil loss in combination with climate change reduces crop yields, especially in vulnerable areas. The loss could contribute to human migration, increasing conflict, and political instability. Humans have been generating carbon emissions potentially for thousands of years through land use. Greenhouse gas emissions from agriculture, forestry, and other land use amount to 22 percent of global emissions as of 2018.

Land degradation is more severe in arid lands. Dry landscapes make up roughly 40 percent of all land and are home to nearly two billion people, the majority of whom live in poverty and are vulnerable to the effects of drought. Dry lands often have nutrient-poor soils and lack organic matter, which makes them prone to erosion if not managed properly. The primary causes of dry-land degradation are overgrazing by livestock, tree removal, tillage, and biofuel production. The steady loss of soil-holding vegetation in combination with decreasing amounts of rainfall under climate change mean these areas will continue to erode, putting them at risk of becoming deserts.

Most of New Mexico, like the greater Southwest, is dry country, characterized by drought-tolerant plants and low amounts of annual precipitation, about half of which comes as intense summer storms. Soils are thin and the ground cover is easily disturbed. The American conservationist Aldo Leopold, who began his career in the desert Southwest with the U.S. Forest Service, once described the ecology of the region as being "on a hair-trigger." By the 1950s, decades of hard use by cattle herds and sheep flocks in the area's rangelands, along with heavy logging in the uplands, had "pulled the trigger" on the land. The Rio Puerco, for example, a sizable stream that runs for two hundred miles through the center of the state, was once known as the breadbasket of New Mexico. Today it is an eroding, thirty-foot ditch along much of its length and carries one of the highest sediment loads of any watercourse in the region.

To counteract these trends and help heal damaged creeks, Bill Zeedyk developed a toolbox of restoration practices that utilize small, carefully designed structures placed in the channels. When waterways, big or small, erode, they often straighten and cut downward, losing their natural sinuosity. Floods caused by sudden deluges can be especially destructive. Zeedyk's structures, made of rock and wood, slow water down and let the creeks begin to heal. Many of his designs purposefully redirect streamflow from one side of the channel to the other, a technique he calls "induced meandering." His structures capture sediments from eroded uplands that wash into creeks. The sediment builds a new floodplain, adding stability to the creek. Sedges, rushes, and other plants take root in the water and on the banks, using the new sediment as a nutrient source. In all of his work, Zeedyk is guided by natural processes, which he calls "thinking like a creek." This approach has multiple regenerative benefits, including improved habitat for fish, more vegetation for wildlife and livestock, restored wet meadows for groundwater recharge, reduced wear and tear on roads, and a healthier watershed for everyone.

Zeedyk is part of a burgeoning corps of specialists, conservationists, scientists, and agriculturalists who are targeting degraded public and private land for restoration. Not many years ago, advocates for nature and wildlife believed in a hands-off approach when it came to confronting environmental damage caused by poor land management. Often, they preferred the curbing or removal of human activity altogether. Recently, as the scale of degradation has grown globally, many people have shifted their attention to restoration. Buoyed by a renewed interest in traditional Indigenous practices, on-the-ground innovators, often led by Indigenous people, farmers, and ranchers, and backed by scientific research, have dramatically expanded the toolbox of land-restoring practices, including the use of cover crops to shelter the soil, the integration of diverse plants and animals in agricultural settings, and reforestation.

The simplest step to restoring degraded land is removing constraints on natural regeneration. The cessation of overgrazing by livestock, for instance, can allow grasses and other vegetation to begin to grow back. Similarly, fish stocks in the ocean can improve once the pressure of overharvesting is reduced. The default mode of nature is regeneration. It is not the land that is broken, but our relationship to it. Nature has been recovering from disturbances for eons, including floods, fires, hurricanes, volcanic eruptions, and even an occasional asteroid strike. The natural world renews itself. Identifying human constraints on natural processes is often the most cost-effective step to restoring land. For places where natural recovery has been severely compromised or poses a risk of calamity, such as deeply gullied creeks or an overgrown forest, restoration may require human assistance or intervention.

Common to all land-renewal strategies is rebuilding soil carbon. Not only is it key to restoring natural processes by fostering the conditions for life, but it can also make a significant contribution to ending the climate crisis in a single generation. According to a major study, soil carbon represents a quarter of all regenerative climate solutions; of that quarter, 40 percent involves protecting existing soil carbon from degradation and 60 percent involves rebuilding depleted stocks. We can take a big step in this direction by ending behavior that causes land degradation in the first place and by removing the constraints on natural regeneration. ●

The Yarra Yarra Biodiversity Corridor Project in Western Australia aims to link existing nature reserves by restoring land to create a 200-kilometer corridor. Since 2008, more than 30 million trees and shrubs indigenous to the region have been planted on 14,000 hectares. Over 90 percent of the restored area was cleared in the 1900s and is no longer suitable for traditional agriculture. Pictured above is restoration in progress from one of their earliest plantings. With active management, shrubs and grasses will gradually return to join the overstory trees. Techniques to encourage concurrent seedling and understory growth are being implemented in newer sites, including more dense and close row spacing, curved and contoured row alignment, and full-time removal of sheep.

Compost

At age thirty-two, Sir Albert Howard was appointed imperial economic botanist to the government of India. His Cambridge and Rothamsted training in botany, mycology, and agriculture led him to postings throughout the British Empire, in which he had notably excelled. In Madhya Pradesh, Howard had a life-changing experience. He found that he was teaching "scientific" agricultural methods to farmers who practiced ancient Indigenous techniques of farming that were visibly and obviously superior to what he was espousing. Their core technique was based primarily on a method of composting that imitated the forest.

Compost is about decay and decomposition of organic matter. It is what you see on an intact forest floor. In nature, all waste becomes food. However, in cities, natural waste is not always allowed to. When buried in anaerobic landfills, organic waste becomes a potent greenhouse gas, methane. Forests do it differently. Trees shed their leaves, needles, cones, and seeds to become litter below. Animal waste from birds and scavenging mammals is added. Together they become nutrients for a community of worms, bacteria, protozoa, and fungi, decomposers that digest and break down the bark, pitch, cellulose, terpenes, and complex starches into humus, a rich black mixture that nourishes the soil below, creating further growth above. This cycle is the basis for all methods of sustainable agriculture, whether it is called organic, biodynamic, or regenerative (crops can be USDA Organic and not conform to these principles). Healthy agriculture requires death and decay in order to create growth and abundance. If a farm was on land that was once a forest, as is most of the farmland in the UK and Europe, Howard believed the farming system needed to be an analogue of the forest. He realized that modern agricultural science was an intrusion into understanding the "operations of nature" and referred to the people espousing industrialized agriculture as "laboratory hermits."

Anyone who has stepped into a primary forest, such as the California redwoods, the Cascades, or the Amazon, has experienced the extraordinary profusion of life that is self-organizing and sustaining. Howard saw how industrial agriculture did quite the opposite. Howard did not discover this alone. His wife, Gabrielle Matthaei Howard, became the second imperial economic botanist to India, and she and her husband did all of their research together, drawing the same conclusions. Farmers in Asia, Africa, and the Americas had known and practiced a regenerative and sustainable form of agriculture for millennia. The organizing principle was the same: return everything created by the land back to the land whence it came. Compost was the key, bridge, and pathway to complete this ancient cycle of agriculture.

Compost is made from what is mostly rejected. This includes food waste and scraps, a primary source, given that approximately one-third of all food produced is discarded at the farm, processor, or home, instead of consumed. Add to that food-processing by-products including seedpods, husks, and trimmings. On farms where there is animal agriculture, manure is a mainstay. Municipalities utilize leaves, grass clippings, straw, wood chips, and sawdust in addition to food waste. Paper can be included in the form of cardboard and printed material if it is shredded and printed with soy-based inks.

The composting process can be active or passive, as in the forest. Farm compost is usually made in trenches or long, narrow heaps called windrows. Passive techniques involve pits into which layers of natural waste are interlayered with animal manure, some moisture, and maybe some blood or bonemeal, then covered and allowed to cook for ten to twelve weeks. A variation of this method places compost material on the surface, where new organic waste, like urine-soaked animal bedding, can be added daily or weekly. Digested compost can be sprinkled on top, which inoculates the newer batch with microflora and microorganisms. Active methods involve mechanical means of turning and aerating compost to accelerate heat and oxidation. Municipal composting is all done with windrows, sometimes indoors in northern climes, to maintain sufficient warmth.

Farmers and gardeners don't command the majority of compostable material. Unlike a forest's contained ecosystem, centralized agriculture moves plant material away from its origin to far-flung suburban and urban areas. From there, almost all of the potential embodied energy and carbon in uneaten food exits the nutrient cycle. In the United States, 96 percent of food waste is landfilled or incinerated, leaving only 4 percent to be composted. In 2018, that meant that more than 50 million tons of compostable material was burned or buried in the U.S.

In 2009, San Francisco became the nation's first municipal service to require separation of all organic material, and today the city diverts more than 80 percent of trash from landfills and composts hundreds of thousands of tons of organic material every year. The composting success of San Francisco began when a group of primarily Italian immigrants formed the Scavengers Protective Association in 1921. Households would pay these scavengers to haul away their refuse on horse-drawn carts, while the scavengers would pay for the privilege of hauling away organic material from hotels and boardinghouses. The high quantity of food scraps and leftovers from these institutions could be sold for a good price to hog farmers on the outskirts of the city. As the city grew, this union of scavengers joined forces with competing scavengers and, after several iterations, eventually became Recology, a private company with exclusive rights to the city's garbage collection since 1932. This employee-owned company has become a billion-dollar enterprise and supplies compost to local farms and wineries in the Napa and Sonoma valleys, reconnecting the broken strands in the nutrient cycle.

In parts of the country without curbside pickup, citizen organizers are filling in the gaps with community compost piles and craft pickup services that partner with local farms and urban gardens. Where valuable nutrients are falling through the cracks, there's a business case to be made for finding them. Now equipped with vans and reusable plastic bins instead of horses and carts, a new wave of scavengers and compost activists could become citywide municipal services.

No matter how soil life is nurtured and enriched, the principle of feeding the soil is the core of agriculture and the bedrock of regeneration. When soil is not fed, the outcome is similar to what happens when humans are not nourished: disease. Disease in plants is experienced as sickly crops, poor yield, and infestation. Sir Albert Howard called disease a clue to be pursued, not an insect or organism to be poisoned.

Could spreading a thin layer of organic compost across the planet's vast grasslands address the climate crisis? According to scientists at the University of California, Berkeley, if a half-inch layer of compost were applied to only 5 percent of California's nearly 60 million acres of rangeland, it could offset greenhouse gas emissions produced by the state's agricultural and forestry sector for one year. They discovered that the compost significantly increased plant growth, improved the water-holding capacity of the soil, and boosted the sequestration of atmospheric carbon underground.

Although contributions to the soil are well known, compost has positive outcomes that cannot easily be measured. Human beings are now identified as a *supra-organism*, because our gut microbiome contains twice as many organisms as our somatic, human cells. The contributions of soil organisms were similarly undervalued. The complex array of fungi, bacteria, archaea, and protozoa in humans influence digestion and immunity, protect against disease, govern brain health, increase vital hormone production, and more. Soil is similar. The diversity, number, and functionality of microorganisms remain largely unknown. Just as pre- and probiotics feed and improve gut health, compost does the same for the soil. Compost represents an opportunity for nearly everyone to connect to a climate solution they can get their hands on. ●

Industrial compost made from composted vegetables and animal manure in the UK.

Vermiculture

When first-time farmers Molly and John Chester needed to bring their soil back to life, they turned to a humble partner: worms.

In 2011, the Chesters took over Apricot Lane Farms, 130 acres of orchards and pastureland an hour's drive northwest of Los Angeles in Ventura County, and began the process of transforming its conventional practices and worn-out soils into a model of regeneration. Their journey, chronicled in the documentary *The Biggest Little Farm*, began with the formidable challenge of restoring biological health to the farm's depleted soils as quickly as possible. A big step involved worms, specifically worm poop—called

castings—which they collected in a forty-foot-long compost bin located in a shed on the property. Into the bin went organic waste from the farm, including cattle manure, to feed a large population of red wiggler worms that happily chewed their way through the organic material, producing poop packed with nutrients and beneficial microorganisms. The Chesters then steeped the worm compost in water, adding a bit of sugar for the microbes, and blended the resulting "tea" into the farm's irrigation system, spreading the nutrient-rich mixture across the land.

What the Chesters did is an example of vermiculture—the use of worms to aid in the decomposition of

change public attitudes, which had previously considered the lowly invertebrate to be a pest.

Today, we know worms can help us in a variety of important ways. The mere presence of earthworms in the soil is a strong signal that conditions are right for regenerative farming, since they can't tolerate soils that are too hot, cold, dry, wet, acidic, or alkaline, and they need adequate amounts of oxygen and water, just as plants do. Pesticides, by contrast, wreak havoc on worm populations. For food, earthworms will eat anything organic, often consuming half of their body weight per day. They reproduce quickly—under the right conditions an earthworm population can double in sixty days. Castings are packed with nutrients that worms have unlocked from the organic material they've consumed, including phosphorus and nitrogen, which are essential to plant growth. The castings are water soluble, which makes these nutrients readily available to plant roots, especially since worms will burrow down through the soil, leaving their castings behind. One study revealed that on average the presence of earthworms in regenerative farming systems leads to a 25 percent increase in crop yield.

Through their burrowing, earthworms loosen compacted soil and create numerous pathways for water and oxygen. Soils with earthworms drain up to ten times faster than soils without, which helps prevent farm fields from flooding. They remove toxins, harmful bacteria, and heavy metals such as lead and cadmium from the soil, and their castings are metal-free. As Darwin learned, the constant churning by earthworms expands topsoil, including within pastures and forests, where they actively decompose animal manure, plant bits, and leaf litter. Earthworms are also food for predators, including birds, making them part of the great cycle of life.

In addition to boosting the health and productivity of regenerative gardens, farms, and ranches, vermiculture reduces the amount of organic matter that goes to landfills, an important source of greenhouse gas emissions. In the soil, the castings of some worm species bind tightly with carbon, making it more difficult for hungry microbes to break them down. Worms also consume microbes, which a recent study suggests could benefit the climate situation by reducing the amount of greenhouse gases produced in the decomposition process. This is significant because as soil temperatures rise under global warming, microbes will become more active and thereby emit additional carbon dioxide. Worms can help slow the rise in greenhouse gases while benefiting regeneration in the soil. ●

organic material. The creation of compost with the help of earthworms has a deep history in many cultures around the planet. The ability of worms to literally move earth has fascinated farmers and scientists, including their most famous admirer, Charles Darwin. In his last book, Darwin reported on the results of nearly forty years of study he conducted, mostly in his backyard gardens, on the amount of new soil worms build with their constant eating and movement. He watched a stone and bits of coal that he had placed on the ground slowly disappear as the soil rose up around them. He speculated that worm castings increased the soil's fertility as a result of their digestive process, a theory that helped

Aerial view of orchard plantings at Apricot Lane Farms in Moorpark, Ventura County, California.

Rainmakers

People can make rain, cool down the planet, rehydrate the land, and turn deserts green. It starts with imagination.

And microbes. Not the ones in the soil; the ones up in the sky. Bacteria—rivers of it—falling on our heads as rain, hail, sleet, and snow. It's called the water cycle, and it starts with evaporation on land and sea, which creates water vapor, a gas. As vapor rises into the atmosphere it cools. To make clouds, microscopic particles are needed. They act as seeds around which the water droplets and ice crystals form. For decades, scientists believed these particles were only inert minerals, such as dust. Although researchers knew that bacteria traveled through the atmosphere, riding air currents as high as four miles, they assumed the intense dryness, ultraviolet light, and bitter cold would render the microbes lifeless. However, bacteria are among the most resilient organisms on the planet. They can thrive in extreme environments, including lava vents, hydrothermal pools, and toxic waste. They may even exist on Venus. As it turns out, they are alive in raindrops and snowflakes.

Researchers determined that cloud-making bacteria contained proteins that allowed ice formation, similar to proteins used by ski resorts to make snow. Nature was doing the same thing in the sky. In fact, these bacteria condense and freeze water at warmer temperatures than dust particles do. Studies found cloud-making bacteria in every part of the world, including Korea, Montana, Louisiana, France, and even Antarctica. One-third of the ice crystals in clouds over Wyoming were built on organic matter. The clouds and mist above the Amazon forest are full of bioparticles. A single bacterial organism can be the source of one thousand ice crystals. Scientists studying the DNA of these microbes have discovered new strains, including species not previously linked to rain and ice formation. A study of cloud water directly revealed twenty-eight thousand different species of bacteria.

These discoveries are a potential breakthrough for cooling and rehydrating the planet. The reason: these rain-making bacteria originate on plants, indicating that local management of vegetation could stimulate cloudmaking. One of the first scientists to make the connection was Russell Schnell, who wanted to know why tea plantations in western Kenya kept experiencing record amounts of hailstorms. He discovered that microbes on the leaves of the tea plants were the same as the bacteria found at the center of the hail. Winds had apparently picked up the microbes from the plants and carried them aloft, where they became seed particles for the hail. It was a cycle. Restored to a host plant by precipitation, the bacteria quickly multiply, to be lofted again. It even takes place in the ocean, where bacteria from algae blooms are moved to the surface by upwelling currents, where they are churned into sea spray by storms and lifted into the atmosphere on winds.

The implications are profound. If plants contribute microbes that cause precipitation over land, then it follows that a lack of plants will cause a decrease in rain and snow locally. The removal of vegetation by overgrazing or the conversion of a plant-rich ecosystem to tilled and monocropped farms with little ground cover can create conditions for drought. Conversely, the restoration of plants could increase precipitation. In this scenario, the spread of deserts under hotter and drier conditions created by climate change suddenly looks less catastrophic. Cloud-seeding bacteria could be deliberately cultivated on plants so they could be lofted upward by winds, where they would condense water vapor. The ensuing rain and snow would green the land below.

This possibility is created by the small water cycle. The flow of moisture in nature is usually portrayed as a great circle: clouds raining or snowing over land; water flowing down the rivers to the sea; water vapor rising from the ocean surface to the atmosphere, where it condenses into clouds that drift over land to precipitate again. This flux between land and sea is the large water cycle. However, only 40 to 60 percent of precipitation that falls over land comes from oceans—the rest originates over the land itself. China gets 80 percent of its moisture from the continent to its west. Moisture evaporates from lakes, plants, trees, and the soil, making clouds overhead. Often, this moisture falls back to the ground within the area in which it originated. If the land is covered with green plants and the soil is a carbon-rich sponge, rain and snow will be absorbed. The water will evaporate back into the atmosphere to form clouds again.

Human disruptions to the large water cycle, such as climate-altering effects caused by the burning of fossil fuels, are serious and well-documented. But disruptions of the small water cycle can jeopardize ecosystems and day-to-day life in major ways as well. Deforestation, overgrazing, and industrial farming practices often cause the breakdown of localized water cycles. The consequences can include extended droughts, falling water tables, severe flooding, and excessive heat. In West Africa, for example, Dutch scientist Hubert Savenije discovered that the amount of rain that originated in the ocean was steadily replaced by moisture evaporating from the rainforests the farther one moved inland, reaching as much as 90 percent of total rain amounts. This put the land at risk. As forests have been cut down over the years, the interior country has become increasingly dry.

In these breakdowns is opportunity. The small water cycle can be restored using many of the practices that regenerate carbon stocks in soil, plants, and trees. Healthy, carbon-rich soils can store a great deal of water. Increased carbon levels, and the microbial communities that come with it, can promote plant growth, which, in turn, can increase evaporation, producing water vapor—and clouds. Rain and snow put water back into the soil and streams. Things cool down. Bare soil, on the other hand, produces little water vapor. It absorbs solar radiation directly, heating up. It makes dust but little rain.

The key to rainmaking is imagination. Visualize rehydrated land. A cool earth. Imagine abundance instead of scarcity. We have been trained to be scarcity-thinkers. We overlook abundance. There are huge amounts of sunlight, oxygen, nitrogen, carbon dioxide, soil, plants, and water available to us. Abundance is everywhere. We just need to shift the way we see. Evaporation, for example, is usually seen as a *loss*, something we want to minimize. But if we shift our perspective, evaporation can be seen as a *source* of precipitation. An abundant one. As much as two-thirds of the water that falls over land is generated by the small water cycle. Consider what that means for droughts. They are linked to global-scale weather patterns, amplified by global warming. But that's the large water cycle. Think small. Imagine yourself standing on a farm or grassland as cloud-seeding bacteria rise into the air on a burst of wind, floating high above the land. Vapor becomes water. Condensed into a cloud, it falls as rain.

Imagine regreening Egypt's Sinai desert. That's the dream of Ties van der Hoeven, a Dutch scientist and cofounder of Weather Makers, a company that plans to turn the upper half of the Sinai peninsula from brown to green, filled with farms, forests, animals, and plants. Centuries ago, the Sinai was green with life, but land-degrading activities by people took a heavy toll, ultimately creating a desert. The Weather Makers' plan involves dredging sediment from a lake on the Mediterranean coast, which will restore its vitality, and spreading the sediment across the land, encouraging renewed farming activity and building soil. Wetlands will be restored. Trees will grow and plants will spread. Fog nets will be constructed in the hills to capture moisture. Evaporation will create clouds and change the direction of sea breezes, bringing water. Temperatures will cool. The green will expand. It is possible.

Our future is not predetermined. It can be changed. It can be green. It can thrive, full of abundance and life. But we must imagine it first. ●

Rain showers over the sagebrush-steppe at the foot of the Sawtooth Mountains in Clark County, Idaho.

Biochar

In 2012, Australian farmer Doug Pow decided to test an unorthodox idea he had for reversing climate change: feeding biochar to his cattle.

Biochar is a type of supercharged charcoal. It is created when organic material, such as wood, grain stalks, grass plants, and even peanut shells, is roasted slowly at temperatures above 930 degrees Fahrenheit with very limited exposure to oxygen, usually in a specially designed oven. The process is called pyrolysis and it results in a light, stable, nearly crystalline substance rich in carbon and highly resistant to decomposition. Biochar can last thousands of years without breaking down biologically. It is used primarily as an additive to soils in order to improve their function, especially in degraded areas and places that are naturally low in fertility. In a classic example, archaeologists recently determined that Indigenous cultures in parts of the Amazon basin of South America infused the nutrient-thin soils of their land with massive amounts of biochar, for hundreds if not thousands of years. This dark, carbon-rich soil—called *terra preta*—boosted agricultural productivity and helped support a regional population estimated at between eight and ten million people. It remains so nutrient-dense today that it is often dug up and sold in Brazilian markets as potting soil.

Biochar is not a fertilizer, however. Instead, its solid structure and resistance to decay make it ideal long-term housing for essential players in regeneration, including microbes, minerals, and water molecules. As the organic material is pyrolized, empty chambers are created in the biochar from former cells and tunnellike pathways that carry water and nutrients to all parts of the plant. The sheer volume of these empty chambers means there is plenty of room for nutrients and microbes to move in, including the tiny filaments of beneficial fungi. Microbes like to congregate to support each other, which these chambers encourage. Plant health and growth depends on a complex "bartering" system among plant roots, microbes, and fungi, in which carbon from the roots, produced by photosynthesis aboveground, is bartered for essential micro-minerals delivered by fungi and microbes. Biochar gives them a place to congregate. As an added benefit, biochar has a negative electrical charge, which attracts positively charged minerals, such as potassium and calcium. It also readily stores water in its empty places, making this vital liquid available to plants and helping buffer them during periods of drought.

By being nearly indigestible to microbes and earthworms, biochar can help reverse climate change by "locking up" carbon for long periods that otherwise would have cycled naturally back into the atmosphere. Tree trimmings or lawn clippings, for example, if simply buried in a landfill, will decompose over time, releasing methane and other greenhouse gases. If this type of biowaste is transformed into biochar instead and added to the soil, the carbon won't make its way into the atmosphere for centuries. Widespread biochar production and addition to soils could reduce global greenhouse gas emissions by 2 percent per year. The pyrolysis process also creates gas and liquid by-products that can be turned into biofuels, a form of renewable energy. There is plenty of waste organic material available to use as feedstock for biochar, including vast quantities of animal manure from industrial dairies and feedlots. The biochar system is flexible and efficient and can be adjusted according to use. Different feedstocks produce different types of biochar, each with its own properties, and there are a variety of ways to generate the substance, from traditional pits in the ground to high-tech ovens. To speed up the effects of biochar in the soil, it is best to mix it with additional organic matter, such as compost, before application and moisten it with water or a nutrient-rich liquid in order to attract microbes and fungi.

No matter what feedstock is used or how the biochar is produced and prepared, the next challenge is the same: to get it into the ground efficiently and economically. At the scale of gardens and small farms, incorporating biochar into topsoil is usually done by hand, using a shovel, hoe, or other implement. On larger areas of land, it is most often applied with a manure spreader or broadcast seeder, or by tractor. However, if it is not mixed into the soil quickly, up to 30 percent of the biochar can be lost to wind and water erosion, especially if applied to bare soil. After the mechanical spreading, incorporation of biochar is best achieved by shallow plowing, though the resulting disturbance to the soil will likely result in a negative impact to the microbial world underfoot. Alternatively, biochar can be incorporated into the ground via a seed planter attached to a no-till or keyline plow, which cuts a narrow slice through the soil, depositing the biochar as it goes and then covering the slice with a trailing disc. The disadvantage of this method is that it applies less biochar than broadcast spreading does. There are financial costs associated with each of these methods, including fuel, labor, and resource expenses, which could limit their utility at the farm or ranch scale.

Which is where Doug Pow's cattle come in.

Desiring to help reverse climate change from his small farm in southwestern Australia, Pow decided to apply biochar to his pastures, but he lacked the necessary machinery to work the material into his soil. So he came up with a novel strategy: feed the biochar to his cattle. Pow knew that small amounts of charcoal had been fed to livestock since Roman times as a way to combat illness (it can absorb toxins), so he mixed the biochar in tubs with molasses and placed the tubs in front of his cattle—who ate it eagerly. Apparently unaffected by the digestive tracts of the animals, the biochar was subsequently distributed efficiently across the pastures as manure—essentially for free. Dung beetles finished the job. Dung beetles are not native to Australia, so Pow had to employ a species of dung beetle native to Africa, which had been introduced decades earlier. Working in pairs, the beetles made fast work of the manure, burying it belowground—along with all the biochar.

Pow repeated this novel strategy over a three-year period, in collaboration with agricultural researchers. He witnessed a significant improvement in his farm's soil during this time, an observation supported by the scientists. "The preliminary investigation results suggested that this strategy was effective in improving soil properties and increasing returns to the farmer," they wrote in a research article. The biochar picked up nutrients from the cow's gut and the dung, showing little evidence of breakdown during digestion, and the biochar-dung mixture was transported into the soil by the beetles to a depth of fifteen inches. They also conducted a financial analysis of the costs and benefits of this approach, concluding that it "provided very positive and initial evidence that biochar practice could be commercialized very quickly." Pow noted that the quick action of the dung beetles in addition to biochar prevents the nitrogen in the manure from becoming nitrous oxide, another contributor to global warming.

There may be an additional important benefit to feeding cattle biochar: a reduction in the amount of methane, a potent greenhouse gas, emitted by the animals. There are around one billion cattle in the world, and collectively they generate 70 percent of all greenhouse gas emissions originating in the livestock sector. A majority of this contribution is methane, which is produced by enteric fermentation and released through eructation (burping) in cattle. Various strategies have been explored to reduce this type of methane emission, including feeding cattle oilseeds, distillers' grain, and seaweed. In 2012, a research group in Vietnam tried biochar, adding a small amount of it to cattle feed. They discovered that it reduced methane emissions by more than 10 percent. Further studies are underway to determine biochar's potential in this area. Meanwhile, the authors of an extensive review of the scientific literature in 2019 concluded that biochar as a feed additive has the potential to improve animal health, reduce nutrient losses and greenhouse gas emissions, and increase soil organic-matter content and fertility. "In combination with other good practices," they wrote, "co-feeding of biochar may thus have the potential to improve the sustainability of animal husbandry."

Institute for Rural Entrepreneurship and Economic Development member Joseph Kapp (left), with poultry farmer Josh Frye at Frye's poultry farm in Wardensville, West Virginia. Kapp is helping Frye to market his biochar product, which he produces from chicken waste.

Pow's use of biochar on his farm—he also grows avocados with it—demonstrates the regenerative power of this "black gold." He was able to (1) increase agricultural productivity, (2) decrease farm costs, (3) improve soil health and resilience, (4) transform a waste product into a resource, (5) contribute to resolving the climate crisis by sequestering carbon, (6) potentially reduce methane and other greenhouse gas emissions generated by livestock, and (7) produce healthy food in cooperation with nature. "Mimicking a natural system as much as is possible," Pow told an interviewer, "does not need to be difficult and it can certainly be economic. In the light of the recent disastrous fires in our country's forests and to reduce forest floor fuels, maybe we need to transfer or divert the massive amounts of available combustible material into agriculture where it will quickly rebuild our precious soil carbon."

Recently, biochar's utility beyond agriculture has been expanding quickly. It is being added to concrete, for example, to reduce cracking, increase resistance to erosion, and create a more flexible material. After water, concrete is the second most commercially used substance in the world. Concrete is made by baking limestone, clay, and other materials at very high temperatures, consuming a great deal of energy. Over 20 billion tons of it are generated every year, accounting for 5 percent of global greenhouse gas emissions. This means concrete has the potential to be a significant destination for biochar—the product is sometimes called "charcrete"—especially for feedstocks that may not be acceptable for use in agriculture for regulatory reasons (i.e., food, sludge, human waste). According to biochar advocates Albert Bates and Kathleen Draper, "switching ten percent of our cements and mortars from silicates to carbon would cancel out another one percent of annual greenhouse gas emissions."

Other industrial uses for biochar include highways and paved roads, internal building material, roofing tiles, insulation and humidity control, plaster, batteries, recyclable plastics, sports clothing, paper, packaging, tires, housewares, air fresheners, "gray" water filters, electromagnetic radiation absorption, 3D printer ink, cosmetics, paints, medicine, and many others. If practical after its initial use, the biochar can then be recycled into another useful product or turned into an amendment for soil, from which it originated, closing the circle.

With its focus on biological principles and the promotion of life, biochar is as much an attitude as it is a practice. As char specialist David Yarrow describes it, our challenge in the twenty-first century is to transform agriculture—and our lives generally—from antibiotic to probiotic. "This reversal of relationship respects microbes as allies, not enemies," he wrote. "Rather than eradicate them, enlightened growers encourage population explosions of beneficial organisms, starting with microorganisms." The soil regeneration that biochar helps to set in motion can accelerate carbon capture and sequestration in ways that support and expand life on earth. ●

Biochar made of chicken waste and wood chips from Josh Frye's farm in Wardensville, West Virginia.

Call of the Reed Warbler
Charles Massy

Charles Massy combines three unusual qualities: he is a farmer, a scholar, and a sage. Growing up in southeastern Australia, he was schooled in conventional agriculture as a young man—or, in his words, "inducted" into the dominant industrial technologies employed by his father, friends, and neighbors. After decades of diligent practice, he realized he knew little about nature itself—biosystems, ecology, soil biology, energy flows, and the networks of living organisms that suffuse the soil and the living world. Moreover, he had not stopped to inquire about what the neighboring Indigenous Ngarigo people, who had lived there for more than twelve thousand years, might know about the lands he tilled and grazed. Although a highly successful merino sheep breeder on his 4,500-acre grazing property in the Snowy Mountains, he decided, after thirty-five years of farming, to go back to school in Canberra and received his doctorate in human ecology from Australian National University in 2012. He began to work with Aboriginal elders to learn how they had been regenerating "country" for millennia, including the use of fire ecology. Massy is a gifted writer and poet, authoring one of the most literate and treasured books on regenerative land practices, Call of the Reed Warbler: A New Agriculture, A New Earth, *published in 2017. As it is said in Australia, the book is fair dinkum, as honest and genuine as the man who wrote it. —P.H.*

It is late August, 4:00 a.m. Hard frost embosses the earth, trees, the bleached-straw grass. As I walk to my farm shed, the Milky Way curves above me, each star pulsing sharply in the clear air, the Southern Cross slowly turning on its elbow. I can see the dark shape of what some Aboriginal nationals call the "emu in the sky," the emu's head nestling below the Southern Cross. As my feet crunch the grass, a male magpie somewhere in the canopy of a candlebark gum maintains a gentle, melodious night-warbling.

I ponder the clarity of this night-world. My story is the story about country. About my own country: country on which my family have lived for five generations and I for most of my life; country on which I grew up, running barefoot the long day, gamboling with poddy lambs, tadpoling, and exploring thick bush. It is country on which I discovered the natural world and a lifelong interest in birds, mammals, reptiles, and other creatures, along with their vegetative and subterranean abodes, on which I trapped and banded birds in study, collected butterflies, and spot-lit owls and gliding marsupials, and on which I learnt to track wallabies and kangaroos, and to flush secretive nightjars from peppermint gum hollows.

But also country on which I learnt to hunt with gun and rifle, to fish for trout in nearby streams (first with worms, then fly), to milk a cow, to cut and split wood, to kill and butcher a sheep or steer, to muster livestock (sheep, cattle, goats), to ride and jump a horse, to drive farm trucks, tractors, and motorbikes, to start diesel engines and pumps, and to use a shearing handpiece (badly).

And yet, though intimately my country, I came to realize that for a long time I didn't fully understand it. Consequently, at times, I caused immense damage to this country—in some paddocks, perhaps at least a few thousand years' worth. I now know if we want to profitably manage, nurture, and regenerate country, then we need to fathom where it came from and how, what it is made of, how it works and functions, how it was managed before us, what organisms and vegetation reside on it, and how they in turn function and play a role. In a journey of ecological enlightenment, I had a remarkable awakening to a new regenerative agriculture in Australia.

Regenerative agriculture is about landscape management, about healthy and unhealthy food and how modern industrial agriculture and the connected food system is not just poisoning us but is also confoundingly making us

obese while starving us at the same time; about the potentialities of rural and urban societies and human well-being; and about the actions of seemingly ordinary farmers' extraordinarily regenerating their landscapes. Groups of farmers and others are providing answers to this provocative question: "What would the world feel like that created solutions from the ground up?" I realized this is about the future survival of humanity in the most critical of all moments for our species in its entire time on Earth.

Humanity has shifted from an organic, ever-renewing economy to one that is overwhelmingly extractive. This is because, as the late, brilliant Jesuit ecological thinker Thomas Berry pointed out, we now continue to "see ourselves as a transcendent mode of being. We don't really belong here anymore."

Agriculture, in occupying 38 percent of the Earth's terrestrial surface, is both the largest user of land on the planet and humankind's largest engineered ecosystem. Because it is based on plants, which take carbon out of the atmosphere to make and store sugar through photosynthesis, and because these plants have roots growing in the ground, a healthy agriculture has the potential to bury huge amounts of carbon for long periods. Moreover, when a healthy agriculture puts more long-lasting carbon into the soil while minimizing the loss of such carbon, this in turn has a major impact on the water cycle and its crucial role in thermoregulation of our planet (85 percent of the Earth's surface temperature is controlled by the hydrosphere, all the liquid on the planet including groundwater, fog, lakes, clouds, rivers, and the ocean). The problem is that traditional industrial agriculture—through practices such as burning vegetation for land clearing, using fossil fuels (in fertilizers and chemicals, and to power farm machinery), overgrazing, plowing, and fallowing—emits, rather than stores, carbon.

It need not be this way. An ecologically and socially enhancing agriculture—regenerative agriculture—can reverse this harmful carbon-emitting signature of industrial agriculture. It can do this via various methods, but all are based around revegetation and inculcating healthy, living soils (that is, soil containing plants, insects, bacteria, fungi, and other organisms). The practice of regenerative agriculture across vast swaths of the world's farmlands, grasslands, marginal savannas, and arid regions can provide a major solution to climate change.

The other day I was driving to town with my grandson Hamish to watch him play soccer. On the way, we passed a farmer on his tractor, boom-spraying a paddock with glyphosate. Hamish, all of nine years old, turned to me with a puzzled look on his face. "Grandpa," he asked, "why do people have to kill things to grow things?" I was speechless for a moment. It was a profound question, but a question with a simple answer: "You don't have to kill things to grow things."

My journey in coming to understand this and the importance of regenerative agriculture really began when I was twenty-two. My father, forty-six years older than me, suffered a sudden and massive heart attack. I immediately left university and came home to take over management of the farm. While I finished my degree part time, my education on the farm began from day one.

Growing up on a farm does not equip one for management. It takes years to become hardened to physical work, and an immature youth is not ideal material for absorbing important lessons. I assiduously set about learning management skills. I sought my father's counsel; I co-opted Department of Agriculture officers; I read the scientific and departmental literature; and I perused my father's books on managing lucerne, "improved" pastures, and ruminant animals. I also sought advice from those regarded as the district's best and most progressive farmers. Within a decade, I had become what I thought was a competent livestock and pasture manager for the Monaro region. In those first few years, I was being inducted into the dominant Western industrial-agriculture approach.

I find it sobering now that, despite my "education" over the next two decades, I actually understood very little. Irrespective of my university training in plant and animal physiology and in soils, plus my own interest and training in the natural world and holistic human ecology, I was somehow oblivious to the existence of an entirely different management approach and its accompanying body of knowledge. In time this blindness meant the escalation of a debilitating debt, before, finally, I was able to open my mind to another way of seeing and thinking.

This alternative view held that soils were not inanimate chemical boxes, that our farm was instead a complex living entity of dynamic cycles, energy flows, and networks of self-organizing functions and coevolved systems beyond imagining. Later still, I would discover that such a parallel universe paradoxically comprised both the most

ancient Indigenous and yet also newest scientific knowledge, and that it related profoundly to human health. Moreover, this approach could be just as (if not more) profitable in both a capitalistic and ecological sense. Most certainly, it could feed the world without destroying the environment at the same time. I began to journey far and wide to seek out farmers who had gone before, who were both transforming agriculture and transformed. I realized that these transformative farmers were, and are, at the forefront of an underground agriculture insurgency.

Recently I visited one of these regenerative farmers: a friend, David, who had once been one of Australia's leading economists. We drove out through his farm to visit a creek and paddocks he had regenerated, using the methods of "Natural Sequence Farming." It was a drought year and the neighbors' farms on either side were grazed into the dirt; there was no grass or shrubs, or biodiversity of any kind, let alone any green color, and the creek that came from the upstream neighbor into David's farm was badly eroded. However, through the fence, the same creek in David's paddock was in stark contrast. The erosion had healed; the creek was trickling through rocks and into large pools; there were large patches of *Phragmites* reeds, and his paddock either side of the creek had green feed extending hundreds of meters outward. As we stood near the creek discussing this transformation, there suddenly erupted out of a large patch of reeds a beautiful sequence of birdsong. It was a reed warbler, invisible yet present. And the gorgeous sound cut deeply into my psyche, for I realized that this would have been the first time in over 150 years of European mismanagement that a reed warbler had returned to that valley. It was all because David had restored the land once more to health and regeneration.

The reed warbler's call for me became a powerful metaphor. For I realized that in their Earth-sympathetic thinking, and in their connections to like-minded urban sisters and brothers (who are equally passionate about healthy food, human and societal health, and about the Earth and its natural systems), regenerative farmers like David are part of a powerful vanguard that is rapidly gathering momentum both in Australia and around the world. This movement is returning landscapes, humans, and their societies to the state of health that our evolutionary history has designed us for and can turn around our destruction of Mother Earth and human societies as we enter this potentially cataclysmic era. Collectively, these farmers provide a template for regenerating Earth. ●

Great reed warbler (*Acrocephalus arundinaceus*) diving for fish in the water.

People

Things are good for many people, yet if we look more broadly, a greater portion of humanity is anxious, hurting, or living in fear. People face loss of rights, land, livelihood, income, food security, or opportunity. These result in cascading hardships that can worsen fragile lives, including migration or poverty. For some people, daily life can be a constant stream of obstacles, even insults. Indignities include marginalization, racism, exclusion, exploitation, disrespect, and scorn. Children's lives are cut short or stunted by lack of nutrition, health, and schooling. Women the world over live in fear of what may happen to them, or fear the consequences of reporting what did happen to them. Men are humiliated by backbreaking, low-paying work that provides no purpose or future. Indigenous people are being arrested and sometimes killed when they return to lands they have occupied, hunted upon, and foraged for thousands of years.

There is a Māori custom known as *manaakitanga*, which is taught in all of the public schools in Aotearoa (New Zealand). It translates as showing kindness, generosity, and care for others. It includes looking out for someone, protecting them, providing hospitality and support. At the core of the word is the care of all people, whether great or small. It makes every person you meet significant and valued. It places the collective above the individual. Guests, strangers, and others have importance equal to if not greater than one's own. *Manaakitanga* points to what is missing in our lives at this tim: a sense of connectedness to others that provides respect and honor, a universal need, a quality that is essential if we are to come together and end the climate crisis.

The words and subjects in this section are women, Indigenous people, people of color, and children. At this time in human history, we would be wise to listen to the people who are enduring the greatest harm caused by climate change, who endure hunger and poverty, people who experience firsthand what needs to change and how we might do so. We hear a lot about technical solutions to climate. However, there are people who understand that comprehensive solutions to climate change encompass a far broader landscape of understanding and practices, yet their voices are seldom heard because they can be drowned out by the opinions of the privileged. These essays reflect some of those voices. Their words are principled, discerning, and universal. ●

Indigeneity

If sustainability is the highest science, then we should look to those for answers that have a proven track record of living in one place for thousands of years without destroying its life-bearing capacity. Those people, by definition, are Indigenous Peoples.

    —Patricia McCabe, Diné Thought Leader

We were told by the Creator, "This is your land. Keep it for me until I come back."

    —Thomas Banyaca, Hopi Elder

It might help if we non-Aboriginal Australians imagined ourselves dispossessed of the land we lived on for 50,000 years, and then imagined ourselves told that it had never been ours. Imagine if ours was the oldest culture in the world and we were told that it was worthless. Imagine if we had resisted this settlement, suffered and died in the defense of our land, and then were told in history books that we had given it up without a fight. Imagine if non-Aboriginal Australians had served their country in peace and war and were then ignored in history books. Imagine if our feats on the sporting field had inspired admiration and patriotism and yet did nothing to diminish

prejudice. Imagine if our spiritual life was denied and ridiculed. Imagine if we had suffered the injustice and then were blamed for it.

    —Paul Keating, prime minister of Australia

Indigeneity means "originating or occurring naturally in a specific land or region. When applied to human beings, it connotes human cultures that are innate, as intrinsic to a specific bioregion as are native plants and animals. That being said, Indigenous Peoples determine what *indigenous* means and define it through their language and identity. There were supposedly 590 distinct nations when colonists first arrived in the territories that are now the United States; however, the concept of "nations" as understood by colonists is not in accord with Indigenous understanding.

(Left to right, top to bottom) Waorani woman, Bameno Community, Yasuni National Park, Ecuador. Naro San man, Kalahari, Ghanzi region, Botswana. Afar woman, Danakil Depression, Ethiopia. Tamang woman, Thuman, Langtang region, Nepal. Nenet reindeer herder, Yar-Sale district, Yamal, northwest Siberia. Chang Naga woman, Tuensang district, Nagaland, Northeast India. Záparo man, Llanchamacocha, Ecuadorian Amazon. Hamer woman, Omo Valley, Ethiopia. Arbore man, Lower Omo Valley, Ethiopia. Waorani man, Bameno Community, Yasuni National Park, Ecuadorian Amazon. Fulani woman, north Senegal, West Africa. Waorani child, Bameno Community, Yasuni National Park, Ecuadorian Amazon.

The word contributes to the idea of Native people living as isolated nations, rather than having vast, overlapping, intercultural, and fluid governance formations. Today, nearly 400 million Indigenous people live among more than five thousand Indigenous cultures that speak the oldest languages on earth.

The brutal conquest of Indigenous cultures was based on fifteenth-century papal decrees that became known as the Doctrine of Discovery. In Asia, Alkebulan (Africa), Turtle Island (North America), Abya Yala (the Americas), Australia, and Aotearoa (New Zealand), lands could be claimed in the name of a Christian monarch when a flag was planted on the soil. Counterclaims by Indigenous peoples were deemed insufficient if their culture did not meet European standards, and since none were Christian, none did. Compounding the claims to sovereign title by the invaders was the concept of ownership. The vast majority of Indigenous nations did not (and still do not) see the land as a belonging, but saw themselves as belonging to the land. This difference in understanding—an individual who is separate from the land versus a biological community dwelling within the land—was the excuse used by colonists to steal 1.5 billion acres from Indigenous people on Turtle Island. These are known as the stolen lands.

The Doctrine of Discovery was embedded into American law in an 1823 Supreme Court case, *Johnson v. M'Intosh*. The ruling dispossessing Native people of their lands was written by Justice John Marshall, who claimed that "the principle of discovery gave European nations an absolute right to New World lands." This "New World" was anything

but new to Indigenous peoples, who had stewarded it since time immemorial. Thomas Jefferson claimed that the Doctrine of Discovery was international law. Although Justice Marshall owned lands that fell under the jurisdiction of the case and stood to benefit from the ruling, he did not recuse himself. Both men were later echoed by Justice Ruth Bader Ginsburg, who ruled against land title for the Oneida Nation in 2005 under the Doctrine of Discovery, writing that title to lands "occupied" by Indigenous peoples before colonization belonged to the discovering nation, which would be the English and then the United States. In June 2015, Pope Francis apologized to all Indigenous people for the "grave sins" of the church; still, the fifteenth-century papal bull authorizing the Doctrine of Discovery has never been rescinded by the Catholic Church.

Indigenous people thrived on land and sea for thousands of years because of intimate knowledge of weather, botany, animals, migrations, medicines, forests, food, and the oceans. They practiced observational science, discoveries about the natural world gleaned over millennia, etched into metaphor and conveyed through stories passed down in unbroken oral traditions. Indigenous nations had experienced intense war and conflict long before Columbus arrived, and from these crucibles developed advanced mechanisms of peace and coexistence, as exemplified by the Peacemaker who assisted the Five Nations of the Haudenosaunee in the creation of democracy.

Some of the Indigenous nations existed in places where most people would quickly perish—the Yup'ik along the Bering Sea, the Tlingit in the Yukon, the Inuit of Greenland. To enable their survival, the Yup'ik learned to foretell the weather two years in advance. Integrating timely and minute observations of seawater, ice, moss, seals, fog, fur, fish, gulls, caribou, and more, they learned to recognize early climatological signs and indicators because their lives depended on it.

Today, Indigenous peoples steward about one-fourth of the world's landmass and inhabit around 85 percent of areas designated for biodiversity conservation globally. It has not gone unnoticed that the areas of greatest language diversity are also the areas of greatest biodiversity. Native languages instruct and guide users to a complex understanding of their homeland. This is a different way of seeing, knowing, and being. In the West, students are steeped in a scientific method of learning that separates and isolates people, plants, animals, species—life itself—into distinct

---

Fixico Akicita, member of the Muscogee-Creek Nation, brushes away tears as he speaks about the land as his mother. "This is our home. By choice because we choose to live this life this way. . . . I made a commitment to the future generations. For the ones who aren't even here yet. If it means my life is in harm's way, it's going to be okay."

parts disconnected from habitat, river, grassland, or forest floor, as if one plant or species could exist without everything else. This method of science was proudly called the "disenchantment of nature," which allowed for manipulation, study, control, and precise prediction. It made sense. How could you perform an experiment unless you controlled or eliminated all the variables? How else could you discover a pesticide, or synthetic fertilizers, or make a drug? That ability is brilliant, incomplete, and dangerous. The need to control is devoid of foresight. It exchanges long-term sustainability for short-term objectives. Its ability to perceive and measure accurately occludes relationships of mutual interdependency that permeate all of existence. Relationships cannot be measured. The biological and scientific knowledge of Indigenous peoples was complex and sophisticated. It was an ontology imbued with respect for the sacredness of creation, informed by detailed observations and insight accrued over thousands of years, encoded into language and practices, honored in prayer, remembered through ritual, celebrated in ceremony, and absolutely critical if we are to restore a relationship to the biosphere that can calm the atmosphere.

Indigenous knowledge persisted in the face of determined efforts to erase it. Colonization instigated centuries of rape, violence, genocide, and dispossession. At certain times and places, when such violence was not considered "acceptable," colonists tried to strip away Indigenous cultures instead. Children were taken from their parents, forcibly moved to distant boarding schools, put in uniforms, forbidden to speak their native language, controlled like cadets at a boot camp, and taught a way of thinking that demeaned their culture and history. On top of the effort to obliterate their culture, they were subject to physical and sexual abuse by the boarding schools and some of the churches that entered into Indigenous homelands. How the Diné and Tsalagi of Turtle Island, the Gunggari of Australia, the Aguaruna of Abya Yala, the Beothuk of Newfoundland, and thousands of other cultures survived wars, disease, deracination, and massacres is extraordinary.

The efforts to marginalize, criminalize, and displace Indigenous cultures is still present in the twenty-first century. The Sengwer and Ogiek peoples in the Cherangani Hills and Mau Forests of Kenya are facing human rights violations, being forcibly evicted, and getting arrested for "poaching" and even killed by eco-guards. At one time, foreigners were issued licenses to track and kill San people. The San's presence here has been traced back 140,000 years, and yet they are being evicted from their lands in Botswana. In India, Adivasi people are being evicted from tiger reserves, and face prosecution for engaging in the traditional practice of collecting wild honey. The Baka people of Cameroon are forbidden from entering their traditional hunting areas and are often arrested and harassed.

Worldwide, Indigenous people are struggling to save their traditional lands from exploitation and species extinction. "If you are going to save only the insects and animals and not Indigenous people, there is a big contradiction," noted José Gregorio Diaz Mirabal, who serves as the coordinator of COICA, the Coordinator of the Indigenous Organizations of the Amazon Basin. First Nations are creating parks to block mining in Alberta, the Diné shut down a coal-fired power plant on their reservation, though it meant losing high-paying jobs, and in the Amazon the Kayapo, Waorani Nation, Uru-Eu-Wau-Wau, and other peoples are dealing with armed intruders who try to log, hunt, mine, and raze their lands for agriculture.

Their vigilance has gained an ally, a worldwide movement known as 30 by 30, which aims to protect 30 percent of all land and waters on earth before 2030. This effort has been taken up by virtually every conservation organization and by fifty-seven countries that have formed the High Ambition Coalition for Nature and People. Its success will depend on stopping attacks on Indigenous lands and people. It means supporting efforts to restore cultures that are being subjected to racism, that have suffered trauma, that wish to revive their language and reclaim their sovereignty where it has been violated. There is an extraordinary teaching about the earth that is needed, a way of knowing that erases the separation between people and nature, a disconnection that has caused the climate crisis. That knowledge is here. ●

Debra Anne Haaland is an enrolled member of the Kawakia tribe of the Native American Pueblo, who have lived in what is now New Mexico since the 1200s. She and Sharice David were one of two Indigenous women elected to the U.S. Congress, and Haaland is now the fifty-fourth United States secretary of the interior, the first Native American to serve as a cabinet secretary.

Hindou Oumarou Ibrahim

Traditional knowledge and climate science are both critically important for building resilience of rural communities to cope with climate change, and Indigenous Peoples are ready to share their knowledge to help to mitigate and adapt.

—Hindou Oumarou Ibrahim

A member of the Wodaabe, a pastoral people in the Sahel region of sub-Saharan Africa, Hindou Oumarou Ibrahim is a cofounder of the Association of Indigenous Women and Peoples of Chad, which advocates for Indigenous rights and environmental integrity. In 2016, she was chosen to represent civil society organizations at the signing of the Paris Agreement, committing the world to act on climate change. Unusual for a rural woman growing up in Chad, Ibrahim received a formal education in the nation's capital, learning in the process about the many ways women are excluded from important roles in society. She also learned about environmental issues, including climate change. Today, she is a leader in the global movement to elevate the role of Indigenous Women and marginalized communities in shaping policy and practices that affect the future of the planet.

The Wodaabe are nomadic cattle herders, part of the larger Fulani people. They move from one place to another in search of water and grass for their animals, sometimes traveling as much as a thousand kilometers a year. Pastoralism allows her people to live in harmony with the natural world. "We understand each other," Ibrahim says about the close relationship. "Nature is our supermarket, where we can collect our food, our water. It's our pharmacy where we can collect our medicinal plants. It's our school, where we can learn better how to protect it and how it can give back what we need." Her grandmother, Ibrahim notes, can predict not only the day's weather but if the rainy season will be a good one or not by closely observing her environment, including wind direction, cloud patterns, bird migrations, the size of fruits, the timing of plant flowers, and the behavior of her cattle. It's the sort of deep knowledge one can only gain by living on the land—knowledge that Ibrahim believes needs to be shared with researchers who study climate change.

She tells a story about a scientist she invited to her community who was surprised one day when Ibrahim said it was about to rain. She quickly began packing up her belongings. The scientist protested that the sky was clear. Ibrahim shook her head. An old woman, she told the researcher, had noticed that the insects were taking their eggs inside their homes for protection. It was a sign. Soon, it was raining heavily and the scientist had to hide under a tree. After the storm, Ibrahim and the scientist began a serious discussion about how to combine traditional knowledge with analytical weather forecasting. "That's how I started working with meteorological scientists and my communities," Ibrahim said, "to give better information to get people adapted to climate change."

She has experienced the effects of global warming firsthand. When her mother was born, Lake Chad—one of the most important freshwater lakes in Africa—held about 25,000 square kilometers of water. At Ibrahim's birth, the lake had shrunk to 10,000 square kilometers. Currently, it is roughly 1,500 square kilometers. Ninety percent of the water has vanished. More than 40 million people depend on the lake for survival, including pastoralists, fishermen, and farmers. Water has become scarce in other places as well. As a result, conflicts are on the rise in the region as the climate crisis deepens. Ibrahim says it has become increasingly difficult for her people to survive. The rainy season has grown shorter while droughts grow longer. The distances the Wodaabe must travel to find food and water are longer, the stays briefer. When the rains do come, they are more variable. Flooding is more frequent. Ibrahim has seen the effect on the cattle as well, which are giving less milk. She remembers milking four liters a day, two in the morning and two toward evening. Now, during the dry season a cow will produce only one liter of milk every other day. In the rainy season, only a single liter per day. The falloff has happened almost entirely during Ibrahim's lifetime.

Climate change is having a big impact on the social fabric of her people. Traditionally, a man is supposed to feed his family and take care of his community, Ibrahim says, and if he cannot do that his dignity is under threat. As a result, some men now migrate to cities to look for work. They can be gone as long as twelve months. If they don't get a job, they migrate to Europe. For the women who are left behind, the stress of these changes has been enormous. In addition to their customary roles, including finding enough food to ensure the health of their families, many women must assume responsibilities of the men, such as providing security. Ibrahim says these changes have galvanized women to become innovators and solution makers, transforming scarce resources into regenerative assets for the entire community. Ibrahim calls these women her heroes.

She has become an advocate for combining science, technology, and traditional knowledge together in order to protect Indigenous people, the planet, and restore faltering ecosystems. She has personal experience here too. In 2013, she led a project in her community that brought together hundreds of people to catalog the natural resources in the area using a process called 3D Participatory Mapping. It gave further voice to women, who knew the precise spot of certain resources and what time of year they cold be utilized. Often, the mapping process meant challenging the men in her community. "After the men figure out all the knowledge to put on the map," Ibrahim recalled, "they say to the women, 'Come and have a look.' When the women come, they look at the map. They say, 'Mmm, no. This is wrong. Here's where I collect the medicine. Here's where I collect the food.' We changed the knowledge in the map." Ultimately, men, women, youth, and elders documented mountains, sacred forests, water sources, migration corridors, and other culturally and environmentally significant locations. The project caught the attention of government officials, who decided it would be useful in their efforts to mitigate conflicts over natural resources.

The project gave Ibrahim both motivation and an opportunity to speak up. "People gradually accepted me as a leader," she says. "I have been changing the way women are seen and treated in our communities."

As a leader on the global stage, Ibrahim's message is clear: the knowledge held by Indigenous peoples is crucial to the future of life on earth. She points out that while scientific knowledge has been around for two hundred years and technology has provided vast amounts of data recently, Indigenous knowledge is thousands of years old. It needs to be respected. Our goal must be to combine all of these knowledges together to help each other in the climate crisis, especially Indigenous peoples, who often find themselves on the frontlines of global warming. There is strength in sharing. Ibrahim points out that developed nations are seeing the effects of climate change firsthand as well, including fires, floods, and stronger hurricanes. All our knowledge needs to come together, she insists, and we need Indigenous people at the center. Decision makers must change what they do, and to encourage them we must share and educate them with our collective knowledge. Time is short. ●

Wodaabe pastoralists, the last nomadic peoples in these areas, migrate the Sahel from northern Cameroon to Chad and Niger. Hindou Oumarou Ibrahim and her fellow pastoralists walk for many miles a day balancing water, a precious resource before the rainy season in the parched lands, on their heads. Ibrahim introduced geospatial imagery to improve Wodaabe success on their thousand-kilometer journeys and is one of the founders of the Association of Indigenous Women and Peoples of Chad.

Letter to Nine Leaders
Nemonte Nenquimo

Nemonte Nenquimo is a member of the traditional Waorani community, who dwell in the Pastaza Province in the Oriente lowlands of eastern Ecuador located between the Curaray and Napo rivers. It comprises one of the largest areas of unexploited Amazonian rainforest, parts of which have seen virtually no human presence save for the Waorani, one of the last nations to be found and contacted. Five village communities continue to spurn contact with outsiders and have relocated to ever more remote parts of the rainforest. Within and around their eleven-thousand-square-mile territory reside hundreds of species of mammals, some 800 species of fish, 1,600 species of birds, and 350 species of reptiles. The extraordinary concentration of biodiversity includes pink river dolphins, anacondas, marmosets, monk sakis, sloths, silky anteaters, spear-nosed bats, kinkajous, and jaguars. Their ecological and botanical knowledge is vast and likely unfathomed. The approximately five-thousand-member community speak the WaoTededo language, which is unrelated to any other, a linguistic isolate with no ancestors. Their material and spiritual life are inseparably linked to and integrated by the trees and forests. In the WaoTededo language, the word for "forest" is the same as the word for "world." Starting in the 1990s, their land started to be exploited by oil companies and illegally logged. As incursions increased, the Waorani found themselves pulling back into more remote and

smaller areas of their ancestral lands. That changed in 2019. Nemonte Nenquimo, who was initially educated in a missionary school, rebelled and became an activist, cofounding the Indigenous-led Ceibo Alliance. The alliance consists of the A'I Kofan, Siekopai, Siona, and Waorani nations. Nenquimo was the lead plaintiff in a lawsuit against the Ecuadoran government that called for the protection of half a million acres of Amazon rainforest from oil exploration and illegal logging. In 2019, a three-judge panel in Ecuador ruled in her favor, and for the first time in the history of the Amazon, a national government was required to conform to standards of international law and engage in an informed and open consent process before any lands could or would be ceded to oil corporations. It is a legal precedent that has inspired Indigenous people throughout the Amazon. In 2020, she was awarded the prestigious Goldman Prize and was named one of the most influential people in the world by Time. Below is her letter to the leaders of nine Amazonian countries, which began, "My message to the western world—your civilization is killing life on Earth." —P.H.

Dear presidents of the nine Amazonian countries and to all world leaders that share responsibility of plundering the planet. We Indigenous people are fighting to save the Amazon, but the whole planet is in trouble because you do not respect it.

My name is Nemonte Nenquimo. I am a Waorani woman, a mother, and a leader of my people. The Amazon rainforest is my home. I am writing this letter because the fires are raging still. Because the corporations are spilling oil in our rivers. Because the miners are stealing gold (as they have been for 500 years), and leaving behind open pits and toxins. Because the land grabbers are cutting down primary forest so that the cattle can graze, plantations can be grown and the White man can eat. Because our elders are dying from coronavirus, while you are planning your next moves to cut up our lands to stimulate an economy that has never benefited us. Because, as Indigenous peoples, we are fighting to protect what we love— our way of life, our rivers, the animals, our forests, life on Earth—and it's time that you listened to us.

In each of our many hundreds of different languages across the Amazon, we have a word for you—the outsider, the stranger. In my language, WaoTededo, that word is "cowori." And it doesn't need to be a bad word. But you have made it so. For us, the word has come to mean (and in a terrible way, your society has come to represent): the White man that knows too little for the power that he wields, and the damage that he causes.

You are probably not used to an Indigenous woman calling you ignorant and, less so, on a platform such as this. But for Indigenous peoples it is clear: the less you know about something, the less value it has to you, and the easier it is to destroy. And by easy, I mean: guiltlessly, remorselessly, foolishly, even righteously. And this is exactly what you are doing to us as Indigenous peoples, to our rainforest territories, and ultimately to our planet's climate.

It took us thousands of years to get to know the Amazon rainforest. To understand her ways, her secrets, to learn how to survive and thrive with her. And for my people, the Waorani, we have only known you for 70 years (we were "contacted" in the 1950s by American evangelical missionaries), but we are fast learners, and you are not as complex as the rainforest.

When you say that the oil companies have marvelous new technologies that can sip the oil from beneath our lands like hummingbirds sip nectar from a flower, we know that you are lying because we live downriver from the spills. When you say that the Amazon is not burning, we do not need satellite images to prove you wrong; we are choking on the smoke of the fruit orchards that our ancestors planted centuries ago. When you say that you are urgently looking for climate solutions, yet continue to build a world economy based on extraction and pollution, we know you are lying because we are the closest to the land, and the first to hear her cries.

I never had the chance to go to university, and become a doctor, or a lawyer, a politician, or a scientist. My elders are my teachers. The forest is my teacher. And I have learned enough (and I speak shoulder to shoulder with my Indigenous brothers and sisters across the world) to know that you have lost your way, and that you are in trouble (though you don't fully understand it yet) and that your trouble is a threat to every form of life on Earth.

You forced your civilization upon us and now look where we are: global pandemic, climate crisis, species extinction and, driving it all, widespread spiritual poverty. In all these years of taking, taking, taking from our lands, you have not had the courage, or the curiosity, or the respect to get to know us. To understand how we see, and think, and feel, and what we know about life on this Earth.

I won't be able to teach you in this letter, either. But what I can say is that it has to do with thousands and thousands of years of love for this forest, for this place. Love in the deepest sense, as reverence. This forest has taught us how to walk lightly, and because we have listened, learned and defended her, she has given us everything: water, clean air, nourishment, shelter, medicines, happiness, meaning. And you are taking all this away, not just from us, but from everyone on the planet, and from future generations.

It is the early morning in the Amazon, just before first light: a time that is meant for us to share our dreams, our most potent thoughts. And so I say to all of you: the Earth does not expect you to save her, she expects you to respect her. And we, as Indigenous peoples, expect the same. ●

The Forest as a Farm
Lyla June Johnston

Lyla June Johnston is a poet, performance artist, and scholar of the Diné (Navajo) and Tsétséhéstáhese (Cheyenne) lineage. She graduated with honors from Stanford University with a degree in environmental anthropology, with an emphasis on studying the cycles of violence that eventually gave rise to the Native American Holocaust. It is estimated that 90 percent of the Indigenous populations in Abya Yala (the Americas) perished from enslavement, violence, and disease. She is a cofounder of the Taos Peace and Reconciliation Council, which works to heal intergenerational trauma and ethnic division in northern New Mexico. She is a walker within the Nihigaal Bee Iiná Movement, a thousand-mile prayer walk through Dinétah (the Navajo homeland) that is exposing the exploitation of Diné land and people by the uranium, coal, oil, and gas industries. She is the lead organizer of the Black Hills Unity Concert, which gathers Native and non-Native musicians to pray for the return of guardianship of the Black Hills to the Lakota, Nakota, and Dakota nations. She is also the founder of the Regeneration Festival, an annual celebration of children that takes place in thirteen countries around the world every September. She spends her free time learning her endangered mother tongue, planting corn, beans, and squash, and devoting time to be with elders who retain traditional spiritual and ecological knowledge. —P.H.

Growing up, I was not connected to my food, even though my ancestors were. I ate from grocery stores, restaurants, a little bit of fast food … I ate your normal Bureau of Indian Affairs school lunch menu—and it was trash. They had all us little Native kids drink milk at school even though we were genetically lactose intolerant. Like most, I survived on the colonized American diet.

When I was about twenty-seven, an elder came and told me that it was time to plant the seeds. It was time to burn around the oak groves again. It was time to transplant the kelp gardens to make room for herring roe and herring spawning grounds. It was time to replant the chestnut forests in the east; space them apart so disease would not wipe them out. It was time to clear the understory of the forests to make room for the deer again. It was time to cultivate corn that might be smaller in size but more nutrient dense than what we eat today. It was time to harvest the saguaro seeds again from the cacti. It was time to pick berries again. It was time to propagate those bushes again so future generations would have more to harvest. It was time to regenerate the highly sophisticated food systems that we as

Indigenous Peoples had cultivated and, through trial and error over many thousands of years, perfected. Food is not a noun in our language. It is a verb, because food is not a thing. It is a dynamic, living process that is constantly in flux. It was time to jump into those actions again.

There's a long-standing myth that Native people in North America were simpletons—primitive, half-naked nomads running around the forest, eating hand-to-mouth whatever they could find. That's how Europeans portrayed and continue to portray us. It's been that way for so long that even Native people are beginning to believe it. The reality is that Indigenous nations on Turtle Island were highly organized. They densely populated the land and managed it extensively. And this had a lot to do with food, because the motivation to prune the land, to burn the land, to reseed the land, and to sculpt the land was about feeding our nations. Not only our nations, but other animal nations as well.

If you want to know what was going on in the land thousands of years ago, you drill a soil core down into the earth. These soil columns can be up to thirty feet deep, and with them you can analyze the fossilized pollen of a specific place. From the bottom to the top, you can date each layer and determine when the pollen was deposited. They show evidence of fossilized charcoal, revealing how people would burn the land routinely and extensively. There's a soil core from what is now called Kentucky that goes back ten thousand years. It shows that from ten thousand years ago up to about three thousand years ago, it was mainly a cedar and hemlock forest. Then, around three thousand years ago, in a relatively short time, the composition of the whole forest changed to a black walnut, hickory, chestnut, and oak forest. Additionally, the pollen evidenced edible species like lamb's-quarters and sumpweed. People who moved in three thousand years ago radically changed the way the land looked and tasted.

These are anthropogenic or human-made foodscapes, where inhabitants would shape the land in a non-dominating and gentle way. Similarly, in the Amazon there were food forests where soil cores reveal numerous varieties of fruit and nuts trees. Studies show that human beings co-created the Amazon rainforest as we know it. They utilized *terra preta*, a soil-modifying technology found near former human settlements, known today as biochar. *Terra preta* generates deep, highly fertile soils that endured for thousands of years. Our best soil scientists are only now beginning to understand how it works.

Another example is in Bella Bella, Canada, on the Central Coast of British Columbia. Here, kelp gardens are planted and cultivated by the Haíɫzaqv (Heiltsuk) Nation.

Black oak trees (*Quercus kelloggii*) in their fall foliage at Cook Meadow, Yosemite National Park, California.

Their kelp groves provide spawning ground for the herring, where they lay their eggs. Herring roe is crucial to the web of life in that ecosystem. The humans eat the roe, the wolves eat the roe, and the salmon that eat the roe feed the killer whales in turn. Everyone eats the roe, and everyone eats the things that eat the roe. Without this human touch along the coastline, the entire ecosystem would diminish.

What we're finding, and what European scientists may be realizing, is that human beings are meant to be a keystone species. A keystone species creates habitat and living conditions for other species. If you remove or extirpate a keystone species, the ecosystem degrades and unravels. Wolves, beavers, sea otters, and grizzly bears are all keystone species because of the ecological role they play.

We are here for a reason. Every being is here for a reason—every rock, every deer, every star, every person. Creator does not make things without purpose or function, or that aren't a piece of the larger puzzle. Indigenous people are trying to bring the human being back into the role of the keystone species, where our presence on the land nourishes the land and its inhabitants. We can't just sustain ourselves; that's a low standard. We are going for enhanceability, as my friend Vina Brown calls it. This is the ability to amplify ecological health wherever I walk, the capacity to make the land better than I found it.

Where you live, which biome or ecosystem, will determine how we are meant to work with the land. For example, the Amah Mutsun Nation, which is indigenous to what is now called Santa Cruz, California, has a ceremony that they do with the oak trees. If you look at their oak trees, the bark is hard and fire-resistant. This is because they have been coevolving with human fire for thousands of years. The Amah Mutsun had a rule of thumb: only fourteen trees per acre. Today in California you might see two to four hundred trees per acre. The land can't handle that. Those trees are stressed, if not starving, because there are limited nutrients and water in the soil. The Amah Mutsun people would create savannas with larger, more abundant, appropriately spaced oak forests to provide rich, green pastures in between for deer, elk, and other hoofed beings.

According to elders, the Amah Mutsun cut down the low-hanging branches annually because they could catch fire. In the fall they would gather the fallen leaves and burn a circle around the oak trees. They blessed the trees with the smoke, which would go into the leaves and inhibit or prevent tree disease. The bugs would fall into the fire, ensuring a healthier acorn crop. The competing saplings would be killed off so that only the hardiest and strongest plants would survive. Native tribes would do this throughout California. A gentle pressure that held order and health for all beings of the forest. The forest needs us. We are here for a reason. Our big brains aren't an accident and can be leveraged to heal and enhance the land for all our relations. Because the tribes have been prohibited from traditional burning practices, there are now catastrophic fires throughout the state.

When European explorers first landed on the Eastern Seaboard, they marveled at the forests, and wrote how they looked like parks. There was space between the trees. There were deer walking through. They called this "wilderness" beautiful. *Wilderness* is a word that we need to examine and reconsider. The land is not necessarily wild. And if we call it wilderness, we separate ourselves from it, as if I am over here in the non-wilderness and real nature is over there in the real wilderness. And it may not be as "wild" as you think, as healthy ecosystems need the careful tending of human hands.

The Great Plains, with their tens of millions of buffalo, were also anthropogenic. Meaning made by humans. People used to call the late summer "Indian Summer" because the sky would go darker due to the Native-set fires. Without fire, the famed tallgrass prairies would have turned to scrub, forest, and inedible browse. And yes, we hunted buffalo—but we hunted them in the lush grasslands we made for them. We did not follow them. They followed us.

There is a term used for what we practiced in the Buffalo Commons: successive regrowth. If you burn an area it will grow back in stages. One year after the burn, a specific set of flora and fauna will emerge. Two years later, another set of flora and fauna emerge. Three years later it will change again, and four years later it will have further evolved. There were always areas throughout the Great Plains in different stages of annual and perennial regrowth with a diverse set of flora and fauna. Thus, a mosaic of varying stages of regrowth quilted the Plains and the overall biodiversity of that region was enhanced. This is the kind of genius our ancestors created through patient observation of land and place.

Sometimes, even our own elders tell us Native people are not as smart as other races. However, we need to understand that 90 percent of our people were wiped out before colonists started writing stories about us, before they started taking pictures. Every picture you see—the black-and-white daguerreotypes and tintypes—were taken after the Native North American population was decimated by disease or massacre. Wisdom and knowledge were also lost and obliterated. All of those photos are false representations of great civilizations.

The Native nations we know of today—Cherokee, Seminole, Cheyenne, Sioux—are survival bands. These are the small percent of Native tribes that survived and got together to make things work. They do not reflect the original composition of the people. It doesn't belittle the descendants; it is simply an invitation, a beckoning for the world to look deeper at the stories we have been told. The story that unfolded on this continent is far greater than what any of us now know. The original composition of the people upon North America was vast, and it was highly organized. Archaeologists have a supposition that if there were vast populations, there would be marks on the earth that we could see today. But we did not leave marks on the earth that you could see hundreds of years later, because we knew if you did that, you had done something disrespectful. You could say we were the original leave-no-trace society. What we did leave in our wake, however, is biodiverse biomes. Much of that biodiversity survives and supports the earth today. Much of it is being depleted. There is almost no record of our enormous populations other than oral histories, biodiverse food systems that the world eats today, and fossilized records found in things like soil cores.

It's not what you do with food, but why you do it. What you do will change from biome to biome; but why you do it should remain the same. You do it to honor what Creator has made. You do it to enhance the land you live on. You do it to diversify genes at every opportunity. You do it to honor the natural flow of water. You do it in the spirit of selflessness, in the spirit of service, in the spirit of community. And as long as you're doing that, the technical skills will follow.

It's hard to know what happened on this planet, and how it flourished—the civilizations that were once here. There were over eighty different languages spoken in California alone before European arrival. Extraordinary things happened within that diverse base of knowledge. Expanding your imagination about what happened will help us set the record straight about who was primitive and who was civilized—and will help us regenerate the world again. Maybe by thinking about which seeds we plant. Maybe by trying to have twelve different kinds of squash, and twelve different kinds of corn in your garden. Maybe instead of cutting down a forest to make room for a farm, realize that the forest is already a farm. If you know how to take care of it, it will make food for you, better than any monocrop. It is time for us to remember that a forest is a farm. And if it's not a farm when you find it, then delicately, respectfully, and carefully turn it into a farm. Don't cut it down. ●

———

Wild roe deer (*apreolus capreolus*) eating acorns.

Women
and Food

A key climate action pathway is where two major solutions overlap: the transformation of global food systems and the empowerment of girls and women. Achieving gender equity at household, community, and policymaking levels makes for improved agricultural yields and social outcomes. Agriculture is responsible for a significant share of the world's greenhouse gas emissions: almost one-quarter when accounting for land clearing. Climatic stresses on agriculture present significant food security challenges to large sectors of the population, especially rural women. In coming decades, adverse environmental and climate factors are expected to boost world food prices up to 30 percent and to increase price volatility. Amplifying the agency of rural female farmers—among the most marginalized groups in society and particularly vulnerable to food insecurity—is essential to building community resilience in the face of climate change. Reaching parity in training, education, credit, and property rights is critical: women own less land than men, while they do around 40 percent of labor related to food production. Just as important is recognizing the value of women's traditional knowledge of land, farming, and culinary practices and drawing this wisdom into the center of agriculture policy.

Women are the backbone of food systems in many parts of the world, deeply involved in every step of the process, from planting and harvesting crops to the planning and preparation of meals. Yet they receive a small percent of agricultural advisory and support services globally, and their involvement at the production level does not translate to increased food security or financial benefits for them. Nine out of ten nations have at least one law impeding women's economic opportunities, including access to credit and the ability to own land. The Food and Agriculture Organization of the United Nations reported that if women farmers were given the same access to resources that male farmers have, they could increase their crop yields by 20 to 30 percent and reduce malnutrition 12 to 17 percent globally. Boosting the yields of women farmers helps keep forests standing, as farmers are less inclined to expand their crops into nearby forests when their existing land is productive. This solution could reduce emissions by 2 gigatons of carbon dioxide by 2050. It highlights the deep connections between forests and food systems—integrating the two through agroforestry and ecological farming practices has been at the heart of a number of successful regenerative movements led by rural women. These initiatives, aimed at improving food and water security, result in restored ecosystems and demonstrate a powerful, multifaceted response to the climate crisis.

The effects of deforestation and industrial agriculture, compounded by warming temperatures, have resulted in severe drought conditions and food insecurity in many parts of the world. Wangari Maathai's Green Belt Movement organized women to plant trees on a large scale, restoring land and water resources and galvanizing a resurgence in traditional and organic farming in Kenya. The potency of women's potential to innovate the food system is rooted in a deep understanding of the interconnectedness of nature and agriculture and a resourcefulness born out of their daily labor to support their families. As Maathai pointed out, women who walk miles to fetch water daily are keenly aware when sources run dry. They are often the first to be involved in assessing changes in the availability and quality of natural resources, and adaptively managing those resources to build resilience in the food chain.

Since 1977, the Green Belt Movement has planted more than 51 million trees and trained tens of thousands of women in trades such as agroforestry and beekeeping, serving as a model for women-led approaches to the transformation of food systems. The Women's Earth Alliance Seeds of Resilience Project, in Karnataka, India, partners with a women-run seed-saving collective, Vanastree, which means "Women of the Forest." In a region where chemical agriculture and climate instability are destroying forests, biodiversity, and people's age-old control over their food sources and medicinal plant traditions, Seeds of Resilience supports women farmers who are promoting forest-based agriculture and small-scale food systems through conservation of traditional seeds. After participating in the year-long training, farmers went on to launch seven community seed banks, which increased seed biodiversity by 43 percent in the region. Women farmers also learned to become successful seed entrepreneurs who now grow and sell seed. These farmers train others to utilize drought- and flood-resistant native seeds that secure healthy food and generate income. Earnings are reinvested into families and into training more women entrepreneurs. These seed banks act as a safeguard for preserving and storing critical seed varieties, alongside the landscape, which acts as a seed sanctuary itself.

These movements demonstrate the fierceness with which women are defenders and protectors of local resources, and the exponential effect of women's leadership on food systems. Encouraging women's agency at the farm level involves ensuring that education and training are equally accessible to them. This takes the support of both women and men, who must also recognize the benefits of inclusion for their families and communities.

In India, three-quarters of rural women work in agriculture, a sector that has been hit hard by decades of economic liberalization as the amount of arable land has dwindled as a result of government policies that support industrialization and corporatization of farming. In early 2021, women occupied the forefront of one of the biggest and longest-lasting protests in the country's history, when farmers demanded that the government withdraw legislation that supported corporations at the expense of smallholders. In spite of the patriarchal traditions that have prevented women from achieving equality in agriculture in India, these protests were a rare grassroots uprising in which women and men stood arm in arm, despite an increasingly authoritarian response from government.

In the United States, there is a significant movement of women into agriculture. Family farms always involved women as part and parcel of agriculture, but women are now taking over farm operations or are farming on their own in increasing numbers. Between 1997 and 2017, women as principal agricultural producers increased from 209,800 to 766,500, one of the greatest demographic shifts in agricultural history. Because of resistance, barriers, and sexism in the traditional farming community, women are forming networks and organizations that provide them with safe spaces free from the pressures of "farming like a man." The agricultural challenges they face are the same for all farmers: commoditized markets, toxic pesticides, depressed prices, and small or negligible profits. They encounter greater difficulties in obtaining loans and working with equipment designed for men's bodies. What women bring to farming are qualities that emphasize personal relationships, a greater focus on sustainability for oneself and the land, regenerative techniques, networking, and collaborative learning. Around the world women carry forward Native and Indigenous knowledge of the land, climate, and plants, passing it on from generation to generation. For many reasons, women more easily recognize that our well-being is inseparable from the well-being of the soil. The restoration and renewal of farmlands means changing from a male-dominated form of extractive agriculture to a community-led form of regenerative agriculture that includes everyone. ●

Coodad is a women-led cacao growers cooperative established in 2017 on lands adjoining the Virunga National Park, a UNESCO World Heritage site in the Democratic Republic of Congo, celebrated for its spectacular scenery and the world's last mountain gorillas. The cooperative emerged after regenerative chocolate company Original Beans focused expansion of production on women. Recognizing their role in forest management for firewood and in healing their war-torn communities, the company organized leadership and artisanal enterprise training for hundreds of women in the remote villages around Virunga Park. The "Femmes de Virunga" cooperative shares know-how and income from cacao crop sales with more and more women. In 2020, each participating woman planted more than fifty new trees on average, more than one hundred thousand in total. Communal forest conservation areas have grown to the size of thirteen thousand soccer fields.

Soul Fire Farm
Leah Penniman

The life and work of Leah Penniman is an extraordinary story of regeneration. She connects food sovereignty, racism, agriculture, incarceration, topsoil, nourishment, and people of color into a seamless expression of bountiful goodness at Soul Fire Farm, near Albany, New York. Growing up in a racist society as a person of color, she realized that "to feed ourselves is to free ourselves." The deracination of Black agriculture in the South was an orchestrated, destructive scheme that nearly destroyed the wisdom, intelligence, and foodways of a people who were expert agronomists. Imagine being a Ghanaian Krobo woman: you were kidnapped in 1740, enslaved, placed in chains on a ship, and sent to an unknown land by unknown people for unknown reasons—and yet you had the wisdom to hide seeds in your hair to hopefully be planted somewhere, someday … if you survived. The history, knowledge, and innate understanding of regenerative agriculture that arrived with the Atlantic slave trade was almost wiped out. Leah went back to the land. She learned from the land. She retraced the steps of her ancestors, created rich, dark loam from hardpan, built a living sanctuary for urban youth, and began to grow pure, healthy food. She taught and inspired young people of color, many of whom had never seen a farm or garden, to love the land, to grow food, to cultivate a crop, to know how life is created. Leah put the soul back into food, and created a magical farmland of celebration and renewal. Soul Fire demonstrates, practically and visibly, how regeneration connects health, nourishment, soil, society, education, and a renewed sense of dignity and self. —P.H.

As a young person, and one of three mixed-race Black children raised in the rural North mostly by our White father, I found it very difficult to understand who I was. Some of the children in our conservative, almost all-white public school taunted, bullied, and assaulted us, and I was confused and terrified by their malice. But while school was often terrifying, I found solace in the forest. When human beings were too much to bear, the earth consistently held firm under my feet, and the solid, sticky trunk of the majestic white pine offered me something stable to grasp. I imagined that I was alone in identifying with Earth as Sacred Mother, having no idea that my African ancestors were transmitting their cosmology to me, whispering across time, "Hold on daughter—we won't let you fall."

I never imagined that I would become a farmer. In my teenage years, as my race consciousness evolved, I got the message loud and clear that Black activists were concerned with gun violence, housing discrimination, and education reform, while White folks were concerned with organic farming and environmental conservation. I felt that I had to choose between "my people" and the Earth, that my dual loyalties were pulling me apart and negating my inherent right to belong. Fortunately, my ancestors had other plans. I passed by a flyer advertising a summer job at The Food Project, in Boston, Massachusetts, that promised applicants the opportunity to grow food and serve the urban community. I was blessed to be accepted into the program, and from the first day, when the scent of freshly harvested cilantro nestled into my finger creases and dirty sweat stung my eyes, I was hooked on farming. Something profound and magical happened to me as I learned to plant, tend, and harvest, and later to prepare and serve that produce in Boston's toughest neighborhoods. I found an anchor in the elegant simplicity of working the earth and sharing her bounty. What I was doing was good, right, and unconfused. Shoulder-to-shoulder with my peers of all hues, feet planted firmly in the earth, stewarding life-giving crops for Black community—I was home.

Through BUGS [the National Black Farmers and Urban Gardeners Conference] and my growing network of Black farmers, I began to see how miseducated I had been regarding sustainable agriculture. I learned that "organic farming" was an African-indigenous system developed over millennia and first revived in the United States by a Black farmer, Dr. George Washington Carver, of Tuskegee University in the early 1900s. Carver conducted extensive research and codified the use of crop rotation in combination with the planting of nitrogen-fixing legumes, and detailed how to regenerate soil biology. His system was known as regenerative agriculture and helped move many southern farmers away from monoculture and toward diversified horticultural operations.

Dr. Booker T. Whatley, another Tuskegee professor, was one of the inventors of community-supported agriculture (CSA), which he called a Clientele Membership Club. He advocated for diversified pick-your-own operations that produced an assortment of crops year-round. He developed a system that allowed consumer members to access produce at 40 percent of the supermarket pricing.

Further, I learned that community land trusts were first started in 1969 by Black farmers, with the New Communities movement leading the way in Georgia. Land trusts are nonprofit organizations that achieve conservation and affordable housing goals through cooperative

ownership of land and restrictive covenants on land use and sale. In addition to catalyzing the community land trusts, Black farmers also demonstrated how cooperatives could provide for the material needs of their members, such as housing, farm equipment, student scholarships and loans, as well as organize for structural change. The 1886 Colored Farmers' National Alliance and Cooperative Union and Fannie Lou Hamer's 1972 Freedom Farm were salient examples of Black leadership in the cooperative farming movement.

Learning about Carver, Hamer, Whatley, and New Communities, I realized that during all those years of seeing images of only White people as the stewards of the land, only White people as organic farmers, only White people in conversations about sustainability, the only consistent story I'd seen or been told about Black people and the land was about slavery and sharecropping, about coercion and brutality and misery and sorrow. And for good reason. Brutal racism—maiming, lynching, burning, deportation, economic violence, legal violence—ensured that our roots would not spread deeply and securely. In 1910, at the height of Black landownership, 16 million acres of farmland—14 percent of the total—was owned and cultivated by Black families.

Now less than 1 percent of farms are Black-owned. Our Black ancestors were forced, tricked, and scared off land until 6.5 million of them migrated to the urban North in the largest migration in U.S. history. This was no accident. Just as the U.S. government sanctioned the slaughter of buffalo to drive Native Americans off their land, so did the United States Department of Agriculture and the Federal Housing Administration deny access to farm credit and other resources to any Black person who joined the NAACP, registered to vote, or signed any petition pertaining to civil rights. When Carver's methods helped Black farmers be successful enough to pay off their debts, their White landlords responded by beating them almost to death, burning down their houses, and driving them off their land.

And yet here was an entire history, blooming into our present, in which Black people's expertise and love of the land and one another was evident. When we as Black people are bombarded with messages that our only place of belonging on land is as slaves, performing dangerous and backbreaking menial labor, to learn of our true and noble history as farmers and ecological stewards is deeply healing.

Fortified by a more accurate picture of my people's belonging on land, I knew I was ready to create a mission-driven farm centering on the needs of the Black community. At the time, I was living with my Jewish husband, Jonah, and our two young children, Neshima and Emet, in the South End of Albany, New York, a neighborhood classified as a "food desert" by the federal government. On a personal level this meant that despite our deep commitment to feeding our young children fresh food and despite our extensive farming skills, structural barriers to accessing good food stood in our way. The corner store specialized in Doritos and Coke. We would have needed a car or taxi to get to the nearest grocery store, which served up artificially inflated prices and wrinkled vegetables. There were no available lots where we could garden. Desperate, we signed up for a CSA share, and walked 2.2 miles to the pickup point with the newborn in the backpack and the toddler in the stroller. We paid more than we could afford for these vegetables and literally had to pile them on top of the resting toddler for the long walk back to our apartment.

When our South End neighbors learned that Jonah and I both had many years of experience working on farms, from Many Hands Organic Farm, in Barre, Massachusetts, to Live Power Farm, in Covelo, California, they began to ask whether we planned to start a farm to feed this community. At first we hesitated. I was a full-time public school science teacher, Jonah had his natural building business, and we were parenting two young children. But we were firmly rooted in our love for our people and for the land, and this passion for justice won out. We cobbled together our modest savings, loans from friends and family, and 40 percent of my teaching salary every year in order to capitalize the project. The land that chose us was relatively affordable, just over $2,000 an acre, but the necessary investments in electricity, septic, water, and dwelling spaces tripled that cost. With the tireless support of hundreds of volunteers, and after four years of building infrastructure and soil, we opened Soul Fire Farm, a project committed to ending racism and injustice in the food system, providing life-giving food to people living in food deserts, and transferring skills and knowledge to the next generation of farmer-activists.

Our first order of business was feeding our community back in the South End of Albany. While the government labels this neighborhood a food desert, I prefer the term *food apartheid*, because it makes clear that we have a *human-created* system of segregation that relegates certain groups to food opulence and prevents others from accessing life-giving nourishment. About 24 million Americans live under food apartheid, in which it's difficult to impossible to access affordable, healthy food. This trend is not race-neutral. White neighborhoods have an average of four times as many supermarkets as predominantly Black communities. This lack of access to nutritious food has dire consequences for our communities. Incidences of diabetes, obesity, and heart disease are on the rise in all populations, but the greatest increases have occurred among people of color, especially African Americans and Native

Americans. These diet-related illnesses are fueled by diets high in unhealthy fats, cholesterol, and refined sugars, and low in fresh fruits, vegetables, and legumes. In our communities, children are being raised on processed foods, and now over one-third of children are overweight or obese, a fourfold increase over the past thirty years. This puts the next generation at risk for lifelong chronic health conditions, including several types of cancer.

At Soul Fire Farm we had to invest in the soil in order for her to yield the food our community needed. Working hard to build up the marginal, rocky, sloped soils using no-till methods, we managed to create about a foot of topsoil. Into this rich, young earth, we were finally ready to plant over eighty varieties of mostly heirloom vegetables and small fruits, centering crops with cultural significance to our peoples. Once per week we harvested the bounty and boxed it up into even shares that contained eight to twelve vegetables each, plus a dozen eggs, sprouts, and/or poultry for the members of South End Community.

In early spring members signed up for the program and committed to spend whatever they could afford on our farm's bounty. We used a sliding-scale model where people contributed depending on their level of income and wealth. In turn we committed to providing members with a weekly delivery of bountiful, high-quality food throughout the harvest season, which lasts twenty to twenty-two weeks in our climate. We delivered the boxes directly to the doorsteps of people living under food apartheid and accepted government benefits as payment, such as the federal Supplemental Nutrition Assistance Program (SNAP). This reduced the two most pressing barriers to food access: transportation and cost. Using the farm-share model, we can now feed 80 to 100 families, many of whom would not otherwise have access to life-giving food. One member told us that their family "would be eating only boiled pasta if it were not for this veggie box."

Even as we continued to provide nourishing food to our people living under food apartheid in six Capital District neighborhoods, we knew we needed to do more. So we started to organize. We expanded our work to include youth empowerment and organizing, specifically working with court-adjudicated, institutionalized, and state-targeted youth. Arguably, the seminal civil rights issue of our time is the systemic racism permeating the criminal "justice" system. The Black Lives Matter movement has brought to national attention the fact that people of color are disproportionately targeted by police stops, arrests, and police violence. And once they're in the system, they tend to receive subpar legal representation and longer sentences, and are less likely to receive parole. The 2014 police killings of Eric Garner and Michael Brown were not isolated incidents, but part of a larger story of state violence toward people of color.

Black youth are well aware that the system does not value their lives. "Look, you're going to die from the gun or you are going to die from bad food," one young man said while visiting Soul Fire Farm. "So, there is really no point." This fatalism, a form of internalized racism, is common among Black youth. It's a clear sign that this country needs a united social movement to rip out racism at its roots and dismantle the caste system that makes these young people unable to see that their beautiful Black lives do matter. We started the Youth Food Justice program in our third year, aiming to liberate our young people from the criminal punishment system.

Through an agreement with the Albany County courts, young people could choose to complete our on-farm training program in lieu of punitive sentencing. It was imperative that we interrupt the school-to-prison pipeline that demonizes and criminalizes our youth. We felt that young people instead needed mentorship from adults with similar backgrounds, connection to land, and full respect for their humanity.

According to the think tank Race Forward, even today, Blacks, Latinx, and Indigenous people working in the food system are more likely than Whites to earn lower wages, receive fewer benefits, and live without access to healthy food. Our Black ancestors and contemporaries have always been leaders in the sustainable agriculture and food justice movements, and continue to lead. It is time for us all to listen. Owning our own land, growing our own food, educating our own youth, participating in our own healthcare and justice systems—this is the source of real power and dignity.

As Toni Morrison wrote in her 1977 novel *Song of Solomon*: "'See? See what you can do? Never mind you can't tell one letter from another, never mind you born a slave, never mind you lose your name, never mind your daddy dead, never mind nothing. Here, this here, is what a man can do if he puts his mind to it and his back in it. Stop sniveling,' [the land] said. 'Stop picking around the edges of the world. Take advantage, and if you can't take advantage, take disadvantage. We live here. On this planet, in this nation, in this county right here. *Nowhere* else! We got a home in this rock, don't you see! Nobody starving in my home; nobody crying in my home, and if I got a home you got one too! Grab it. Grab this land! Take it, hold it, my brothers, make it, my brothers, shake it, squeeze it, turn it, twist it, beat it, kick it, kiss it, whip it, stomp it, dig it, plow it, seed it, reap it, rent it, buy it, sell it, own it, build it, multiply it, and pass it on—can you hear me? Pass it on!'" ●

Clean Cookstoves

While all clean cookstoves aim to reduce harmful particulate emissions and increase fuel efficiency, there are many types, each utilizing different technologies and fuel sources. Some cookstoves use wood pellets from wood that would otherwise be discarded, displacing wood harvested from intact forests and reducing forest degradation. A closed-loop system such as a biodigester turns methane from animal waste into fuel for cooking, compounding the reduction of greenhouse gases and bypassing the need for fossil fuels. Other models are powered by solar, electricity, or fossil fuels, including liquefied petroleum gas, or natural gas.

Even when stoves rely on fossil fuels, the reduction in black carbon particulate emissions, as well as the efficiency gains, ultimately reduce greenhouse gas emissions associated with cooking. Replacing wood with low-emissions fuel sources has the potential to remove approximately 0.4 gigatons of carbon-dioxide-equivalent emissions per year. One study found that replacing traditional stoves with improved forced-draft stoves in a household can achieve a 40 percent reduction in the average concentration of black carbon particles. Another study observed pollutant reductions greater than 90 percent with the use of pellet stoves.

Black carbon particles created by traditional cookstoves influence cloud formation, changing rainfall patterns, which impacts plants, animals, and people. When black carbon particles settle on snow and ice, they increase surface temperatures by absorbing radiation and reducing the reflection of sunlight back into the atmosphere. Black carbon isn't the only worrisome by-product of traditional cookstoves. The grayish-white smoke produced as a fire cools down contains carbon particles that can absorb as much radiation from sunlight as sooty smoke, and its health effects may be worse. This form of carbon is known to be carcinogenic.

In Haitian households, women spend twice as much time as men on domestic chores, including firewood collection and food preparation. When women and girls collect wood and other fuels for cookstoves, they are at increased risk of gender-based violence. Access to clean cooking solutions reduces the time it takes to cook meals and reduces the lengthy task of wood collection, both of which allow women and girls time to pursue other activities, such as education, earning income, expanding family interests, or simply resting.

All clean cookstoves have a similar purpose: to reduce the amount of fuel used to prepare food, reduce the amount of time involved in cooking, and be functional, flexible, and safe while meeting local needs and cultural considerations. Cuisines and cooking practices are often deeply embedded in a country's history and traditions. In Africa, many nations have set targets for widespread liquid petroleum gas (LPG) stove adoption but have run into barriers at the community level, including the need for training, safety concerns, and the high cost associated with LPG, especially for remote locations. Studies show that while LPG stoves are highly effective at reducing black carbon emissions, they also have the lowest long-term use among households. Though forced-draft stoves that use organic fuels have higher emissions of black carbon, they have a higher usage rate because they are the most similar to current cooking practices.

Despite large-scale initiatives launched in 2010 by the United Nations Foundation, between 2015 and 2017 only a small fraction of households in low-income countries were using clean fuels and technologies for cooking. Recently, researchers have identified mistakes that have slowed the adoption of clean cookstoves. Initially, the smoke problem was treated solely as an engineering chal-

lenge, overlooking the needs of the people actually doing the cooking. Also, advocates were frequently outsiders from abroad. New strategies have emerged, with more promise. Local women's voices need to be heard. The support of opinion leaders is crucial to the adoption of new technology, as are educational efforts to inform people about the health risks associated with sooty smoke. The ability of communities to repair broken stoves is also critical. Affordability and accessibility are likewise key and need to work in tandem. Clean stoves and their fuels cannot cost too much or be too hard to replace, otherwise people will simply return to using traditional stoves. Discounted prices, subsidies, rebates, home delivery, and effective support systems have all been demonstrated to work.

In the long term, electrification may hold the greatest promise of all for bringing cleaner cookstoves to large populations. As electrical grids spread into rural areas and as the energy they provide becomes renewable and less expensive, opportunities for electric stoves to make a difference will expand. When integrated into communities through culturally appropriate and practical means, clean cookstoves are at the heart—and hearth—of regeneration within homes and societies. ●

The Gyapa cookstove is made in Ghana, which has one of the highest deforestation rates in the world. It is the cocreation of ClimateCare and Relief International. The liners and claddings are made by local ceramicists and metal workers, providing local employment. Over 4.1 million Gyapa cookstoves have been made, saving users more than $75 million thus far. It reduces smoke and energy use by 50 to 60 percent. Richard and Gladys Eken make ceramic liners that are designed to create more complete combustion of charcoal or biomass.

Education
of Girls

Every girl, no matter where she lives, no matter what her circumstance, has a right to learn. Every leader, no matter who he or she is or the resources available to him or her, has a duty to fulfill and protect this right. —Malala Yousafzai

The universal education of girls is the essential precursor of full gender equity and the empowerment of women. Unto itself, realizing the potential of women is the single most important pathway to planetary regeneration. A universal principle of any system, whether social systems, ecosystems, or immune systems, is that the way to create a healthy and regenerative system is to connect more of it to itself.

Out of custom, beliefs, and ignorance, cultures continue to treat girls as secondary. How this came about and continues to propagate is an ancient story of dominance, fear, and ignorance. Erasing it is a matter of survival.

The first definitive study on the impact and benefits of girls' education was written in 2004 by Barbara Herz and Gene Sperling for the Center for Universal Education. That work was expanded by Sperling and coauthored with Rebecca Winthrop in 2016. It was entitled *What Works in Girls' Education*, and the short answer to the question in the title is simple: Everything works better when girls are fully educated and empowered.

As Sperling and Winthrop state, there should not be a need for a book on girls' education. Respect and inclusion are common sense, deeply human, and obvious. The impact girls' education would have is extraordinary. However, the value of women's empowerment cannot be reduced to metrics.

One of the reasons that girls' education continues to lag in many areas of the world is its expense. It costs money, and in many countries that is a limiting factor. Not providing separate latrines and menstrual products is also a limiting factor. The greatest expense the world will face in the coming years is global warming and its impacts on land, forests, commerce, food, migration, water, and cities. Ignorance is not a cost-cutting measure. There is overwhelming evidence that an educated populace of women radically reduces infant mortality, child marriage, family size, malaria, and HIV/AIDS. Conversely, it increases health, economic well-being, agricultural yields, and social stability. When girls are forced to be child brides as if they were chattel, they have an average of about five children, they usually lack access to family planning, and their children have poorer health outcomes. Girls who complete the equivalent of a high school education give birth to an average of two children, and because of their education they are able to earn more income, which they use to ensure the health and educational opportunities of their children. Such virtuous circles start with girls being supported to continue their education, without the pressure of having to go to work to put their brothers through school or to be married to relieve their family of one more mouth to feed. Poverty creates poverty. Transforming a vicious cycle to a virtuous one begins with a classroom.

But there is something more profound at stake: the minds and intelligence of women. It is hard to invent, imagine, and innovate if daily life is a never-ending struggle. Human well-being is determined by connectedness, cooperation, and community, yet the full promise of all three is suppressed by exclusion of girls and women. When more women join the workforce, wages rise—for women and for men. There is a direct correlation between the number of women in elected offices and fairness, justice, and economic well-being. There would not be a global health system were it not for the fact that 76 percent of essential workers are women—educated women. The advent of quality circles in manufacturing arose directly from observations of women munition workers in World War II, who were segregated from the men but produced far better outcomes. Women bring different skills, thinking, perspectives, and attitudes about collaboration to institutions, governance, and business. Productivity expands, innovation increases, and revenues grow. That this comes as news is emblematic of how the world can still ignore what has been missed and overlooked in virtually all fields of endeavor. As with girls' education, the message needs to be repeated over and over again.

Malala Yousafzai is a Pakistani activist and perhaps the most influential leader in the global struggle for girls' education. Her Malala Fund operates in eight countries with among the lowest secondary education enrollment rates and strives for a simple goal—"twelve years of free, safe, quality education for every girl." The Malala Fund's approach cuts directly to the most pervasive and deeply entrenched barriers holding women back. They invest in local activists and educators, advocate for better policy in national and multinational arenas, and amplify the voices of women and girls on the front lines of education advocacy through their digital publication, *Assembly*. In these ways, they show the rest of us where we should concentrate our energy to advance this crucial cause—cultural attitudes, policy, and representation.

In the meanwhile, the impetus and drive to go to school is overwhelming. Forty-one percent of all schools in Afghanistan do not have buildings. Girls study in tents, stairwells, and homes. There are discriminatory and violent attacks by the Taliban. And yet despite sexual harassment, acid attacks, lawlessness, poverty, gender norms, lack of teachers, and many other barriers, there is an irrepressible demand for education. There is a custom in Afghanistan of short poems called *landai* that are passed on orally from woman to woman. Fifteen-year-old Lima Niazi wrote this one:

> *You won't allow me to go to school.*
> *I won't become a doctor.*
> *Remember this: One day you will be sick.*

That is precisely where we are now globally. We need to be doctors, to heal the wounds that divide us and harm the only home we have. We need one another; we need everyone involved and engaged in the regeneration of the earth, its people, places, and inhabitants. It doesn't always work to describe the climate crisis as a crisis or, even more pressing, to say it is a climate emergency. However, it is both. If unattended to, the coming changes in climate will break the bones of communities, shatter people's lives, and break our hearts. Solving the climate crisis means solving many other problems, but this makes solving pressing issues easier, not more difficult. What we know about systems dynamics and the work on intersectionality is that systemic change encompasses and enlarges the potential for human involvement, participation, and engagement. In a sense, our climate radar has been pointing in the wrong direction—at coal, cars, and carbon. Of course, these are crucial causes, and they are being addressed brilliantly by many. However, the radar needs to point the other way too, to the true cause, which is what we believe and how we treat one another. ●

Sisters returning home from the Ewaso Primary School in Ewaso, Laikipia, in Northern Kenya. This and other local schools are funded by ecotourism revenues from the Loisaba Wilderness Conservancy.

Acts of Restorative Kindness
Mary Reynolds

Mary Reynolds is a renowned landscaper and garden designer from Wexford, Ireland. She is an author, a graduate of University College Dublin in landscape horticulture, and a garden philosopher. She recounts becoming lost on her family farm as a young girl and feeling how the grasses and plants were letting her know that she was with family. That experience has never left her. Reynolds made her mark in the studied world of traditional British horticulture with her submission to the famed Chelsea Flower Show in 2002, at age twenty-eight. Her proposal to the Royal Horticultural Society was wrapped in mint leaves and read: "People travel the world over to visit untouched places of natural beauty, yet modern gardens pay little heed to the simplicity and beauty of these environments." Her entry, "Tearmann si—A Celtic Sanctuary," was as different as anything ever seen at Chelsea. It had druid thrones, a moon gate, and a fire bowl placed over a pond, and was surrounded by a profusion of native plants from Ireland. It was a sacred place within a wild setting. It was awarded the Gold Medal and became hugely influential in the garden world for showing us how landscapes can be created that heal our relationship to all that is precious and wild. In 2015, a biopic was made about her Chelsea exploits, Dare to Be Wild, *featuring actors Emma Greenwell and Tom Hughes. It has been viewed in theaters and streamed all over the world. —P.H.*

For too long we have barely acknowledged that we share the earth with millions of other life forms, rooted and unrooted, above and below the ground. Until we recognize this, wild creatures will have fewer and fewer places left to go. Agricultural land, public land, and private gardens are treated with potent chemicals that make it impossible for most life to exist. Our water systems are poisoned, our soils are being degraded to dust, blown or washed away by wind and rain, and native habitats are being depleted at a rapid rate. The steadily warming climate is forcing wild creatures to migrate in order to adapt and survive, but urban and agricultural sprawl has left them with no safe corridors to travel through. They are trapped in small island sanctuaries and they are running out of options. The Web of Life is being shredded and we two-legged creatures are inextricably tethered to that web. They go, we go.

It was a winter's morning in Ireland when this all came home to roost, leading me to form an idea that could help anybody lucky enough to have a garden or even a window box to become part of the solution. It wasn't the startled fox that grabbed my attention from the drawing board where I was daydreaming out the window at home that day. It was the pair of hares that were unusually chasing the fox across the garden. Soon afterward, I spotted a hedgehog scurrying along, following the hare's path, but well tucked under the protection of the thick hawthorn hedge that edged the lawn in front of me. They all disappeared into the scrubby wildness that was one half of the land I was minding. Seeing as it was early winter and a bright midmorning, I figured something must be up for these normally hibernating and nocturnal creatures, so I went outside to investigate. I followed the direction they were coming from, and wandered to the end of my laneway, onto the quiet country road where I lived.

Not so quiet today, however. Across the road, there was once an acre of thick, impenetrable, self-willed land, dense with prickly gorse, thorny brambles, spiky hawthorn, blackthorn, and lush fern and bracken. Today, a big yellow monster of destruction had landed. My neighbors had finally gotten planning permission to build a house on the field and so they did what everyone does, they sent in a digger to clear out "the rubbish" and make a garden, without any thought for the multiple families that already called it home.

I stood there in horror, forgetting to breathe. I had done this myself so many times in so many places. For over twenty years I worked internationally as a garden designer, carrying out similar clearances everywhere I worked. It was only in that moment that I realized what I had been doing, and my career as a landscape designer, in the traditional sense, was instantly over.

I went back inside and started researching the collapse of nature. I learnt quickly that the biodiversity crisis was as insidious and dangerous as the looming threat of climate collapse. It has not been given as much attention and yet it is rapidly reducing the ability of the earth to maintain clean air and water, and to provide food and habitat for all her creatures—including us. It is happening at incredible speed, principally within the last fifty years. Multiple species are falling prey to extinction every day. They are never coming back. Some of our most beloved creatures are on the Red List of endangered species, or were. Most everyone is suffering from a phenomenon called "shifting baseline syndrome." Within a couple of generations, we forget how abundant and alive the earth, seas, and skies used to be. How oceans were kept crystal clear by

the vast fields of oysters on the seabed and how the waters were literally hopping with life. How the flocks of migrating birds and butterflies could block out the sun with their sheer numbers. At this rate, we will be living on an almost barren planet that we will accept as "natural" because it will be all we know. This is the Great Forgetting.

There is no sanctuary for wildlife in gardens filled with pretty, purchased "garden plants" that have not evolved within the local food web. Gardens are controlled and sprayed to the point of being a still life, with no room at the inn for anything other than our own creative visions. Unwittingly we were at war with nature; thus, at war with ourselves.

So, sitting with a cup of tea and my thoughts at the kitchen table, I came up with a plan. Great changes in history come from the ground up, from small movements of passionate and focused people. After the parade of animal refugees I had witnessed from my window that day, I started writing about a seed of an idea called "We Are the Ark" (Ark standing for Acts of Restorative Kindness to the

Earth). It is a simple concept that asks people to give as much of their garden as possible back to wildness, to share it with as many other creatures as possible. To restore lands so that they are true to nature. To mimic the natural, successive processes and rebuild a network of native plants and wildlife in the island nations we call our gardens. These would be supportive corridors of hope for our nonhuman kin that are being cut off and herded to the edge of life.

It seemed quite a leap, to get people to think so differently about their gardens. To allow grasses to grow long and let the native weed-seed bank to emerge from the earth to reboot the ecosystem. To tackle nonnative invasive plants at a local level and add native plant diversity back where nature needs a hand. Native plants are the foundation stones of nature, the earth's protective clothing. They

Mary Reynolds in her patch of gorse, hawthorn, brambles, and bracken declaring it an ARK zone.

give us oxygen to breathe, filter our water, clean our air, feed us, and yet we barely give them any heed unless they look pretty or taste good. Insects have intricate, specialized relationships with native plants that has a knock-on effect on everything else. We ask people to replace outdoor lights with amber-toned bulbs (or better still—darkness), to build wildlife ponds and log piles, and to consider wildlife access tunnels in the boundaries. Our Arks are about what we can do for the web of life, not what our gardens can do for us.

The key to making a movement is a simple thing that captures people's imaginations. I ask people to put up homemade signs in their wild patches exclaiming THIS IS AN ARK so that they can remove the stigma of people seeing a messy garden. Instead, they can be proud of being part of a new and kinder world. A world where we are embracing our roles as caretakers of the earth and as many wild plants and creatures we can fit into our Arks as possible. We need to learn how to share, and where better to start than at home.

We have an informative, homemade website, a resource for people to reference and turn their garden, farm, window box, park, or school into an Ark. It tells how they can add more creature support systems depending on their abilities and the size and scope of their land. We ask that people try to step outside of the destructive food system as much as they can. To grow their own food if possible and support local, regenerative, organic producers. Nature needs as much of this earth returned to her as possible. We need to become guardians, not only gardeners.

This is something different, an action anybody with land can undertake. Hope is gold dust these days and it is astonishing to observe the speed at which nature recovers. Within months we had thousands of "Arkevists" around the world, all active members of an online group. Some Arks were as big as fifteen acres in the U.S., and one was as small as a window box full of local weed seeds in Norway. All were precious symbols of hope.

People were finding meaning and joy in their Arks. Their families had expanded to include each hedgehog, dragonfly, dung beetle, and wild weed that came to stay. Having watched how many creatures of all types came to live in the simple Arks they had restored, people were suddenly looking over the neighbor's fences and seeing the wasted opportunities that nonnative gardens are. Unused green spaces in universities, schools, parks, and industrial

estates were suddenly obvious places for Arks. People began campaigning at their local councils and schools to give over those spaces to nature, a win-win for everyone, as it required much less maintenance, if any.

In our relatively small Arks we can't re-create wild, balanced landscapes such as those being restored on a large scale through rewilding processes. Apex predators and large herbivores are ecosystem engineers that ensure balance in the web of life. There is constant destruction and re-creation in all the stages of an ecosystem. The earth, like us, is in a constant state of death and renewal. We obviously cannot introduce larger, wild creatures into our Arks, as they cannot survive in fragmented habitats. Therefore, we must become the wolf, the deer, the beaver, and carry out the ecosystem services these creatures would normally assume. Depending on your part of the world you inhabit, it helps to learn the attributes of your local ecological systems. If you don't have that information, a simple way of structuring your role is to create as much diversity as possible in the small Arks.

Everything you can possibly imagine already exists on this magical planet we call home. The beautiful, weird and wonderful creatures that we can and can't see are our reason to be here now—to mind them, provide for them so that they may thrive, and that we may survive.

It's time for human beings to step up and learn to become the weavers of the web of life, to restitch the threads we have broken. Set your land free. Build an Ark for life and your heart. ●

(*Left*) As seen here, yellowhammers (*Emberiza citrinella*) love to sit on top of hedges and sing their hearts out. They are now Red-Listed as endangered in the UK. Their population has plummeted in the past decades, in no small part due to the EU's intensive agriculture policies that caused the elimination of insect populations and their preferred habitats of woods and hedgerows. Its characteristic song influenced the motifs of Beethoven's Fifth Symphony. Having developed different dialects throughout the UK and the EU, female yellowhammers prefer to mate with birds that sing their same dialect.

(*Below*) The British hedgehog (*Erinaceus europaeus*) is a familiar creature weaving its way through gardens. Their nests are built in hedges, bushes, or shrubs, and the "hog" part comes from their characteristic snorting sounds. They can also make their homes in urban habitats, including cemeteries, drain pipes, and abandoned railway yards, but do prefer gardens. They mostly eat beetles, slugs, and caterpillars, which makes them invaluable to gardeners, and one hedgehog may have a "territory" of a dozen different gardens.

Who's Really Trampling Out the Vintage?
Una nota de gracias

Mimi Casteel

Food and agriculture systems are inextricably entwined. They are extractive industries that exploit people and land. In her essay, regenerative viticulturist Mimi Casteel speaks to the people part—what regenerative agriculture means when it comes to social justice, workers' health, job security, compensation, and dignity. Brown and Black farmworkers are the backbone of the food system in America. Without them, the system would collapse. Yet they are paid the least of all classes of worker, with no benefits or security. From a monetary point of view, workers are not people; they are inputs on the financial statement, and those "costs" are minimized by using contract labor. When workers are paid piecemeal, by the pound, box, or crate, they often make less than minimum wage. This practice of paying the least to those who create what is invaluable (food) has its root in slavery and has not changed with respect to its inhumanity to this day. As climate changes, as crops fail, as temperatures on the field exceed ninety and a hundred degrees, suffering is magnified. Our food system depends on immiserating fellow human beings. At Mimi Casteel's farm, Hope Well Vineyard, in western Oregon, it is a different story. —P.H.

About a year back, when I was still working with my dear friends at Lingua Franca Wines, I found myself navigating the darkest hour between their night harvest and my dawn harvest at Hope Well. Not enough time to go home and get a nap, too tired to try to even bother getting warm. I was sitting in my car drinking coffee from a thermos in the 7-Eleven parking lot, as I did not want to wake my family by going home. Some of the people from the night pick were milling about the parking lot, eating the only warm, cheap food that can be had at that hour of the night/morning. Why on earth wouldn't they already be home in their beds, you ask?

Because of course they were all going on to other jobs. Maybe their day job, maybe another pick in the daytime. We were all cold, sweaty, dirty, and tired. I looked over to the car next to me and saw a woman sleeping with two little babies on her chest. Her hair indicated she was the *abuelita*, and the babies slept soundly, but she was clearly stirring. I watched her transition the babies to their mother, who had finished her snack and was taking her place in the car with the children while *abuelita* put on her gear and boarded the van to go with the pickers to the next job.

I watched them go as I tried to thaw my fingers on my thermos. I thought of *abuelita* and the three generations of her family working on farms across the valley. Mama was strapping her babies into their car seats and looking so tired that I tapped on her window and offered to follow her home. We delivered her babies to their older siblings. Mama quickly changed into the clothes she would wear while she cleaned houses the rest of the day.

As dawn drew close, I asked her to be careful that day, please sleep soon, please thank her family for the beautiful work they were doing, and left to meet my crew at Hope Well. It was an incredibly emotional drive, which helped keep me awake, but I still struggle to convey what was burning in my chest.

At the gates of Hope Well I was greeted by the shouts and smiles of my well-rested compadres. This group of remarkable people I've known for so long, the crew that has worked with my family through the decades, also works with me at Hope Well. The work I am able to do because they belong here means everything to the mission of this place.

Since I was a child, I have watched the same group of men and women work the vines, first at Bethel Heights Vineyards, and now here as well, and how differently they work than any of the very talented crews of contract labor I have had the opportunity to work with in other vineyards.

The way I work is, to put it lightly, different. I am asking these people to take a journey with me and this place. To take that walk I cannot just ask them to go out and do a particular job. I need them to understand the reason everything looks so different here. I need them to understand that our work is always toward the goal of repairing a relationship with this land. I need them to see how everything fits together.

There is real levity in the vines here. We do serious work with great joy in our hearts. As I crossed the gate that morning, into the arms of my crew, I've never felt more grateful for the system that my family has fought to maintain, even as the world turns away from any permanent labor model, outsourcing every job that can be systematized, placing the responsibility and liability on someone else's business. Bethel Heights formed an alliance with Temperance Hill more than twenty years ago to provide full-time work, benefits, and proper working conditions.

I bring this up because that parking lot at the 7-Eleven was full of the very same talent and skill that was waiting for me back at Hope Well. They are all a part of an unstable economy, built on their backs, that won't take responsibility for the most important component of the work we do: the human component.

We have marginalized the most important work being done in this country. We have painted labor in terms of drudgery, mindless, unskilled tasks that anyone with education or talent would be a fool to tarry with. I would be the last person to point the finger at labor contractors or even the businesses that hire them for starting this epidemic. At the heart of this is the systemic death of a type of working landscape that understands how beholden we are to that work, and how our society is failing as that uniting force is vilified, alienated, and persecuted.

To work the land in the growing of food and fiber for your country is, I submit, the very highest calling. To put your body and mind into bringing food to the first responders, the soldiers, the doctors and the lawyers, lauded all, cannot the people who feed and clothe them all be justly honored? To be very clear, the work we do, the synergy between bodies moving through space, the way the naked and healthy environment inspires a free mind to think about the work of the hand, the tool, the effort, there is no cubicle or desk that offers the same promise.

The people in that early-morning parking lot, the people behind my gate, the people who have picked me up when I have fallen and suffered every loss with me as viscerally as I, we are you. We are building your children's future. Those hands that we classify as "unskilled" and the direction of their work could change the fate of our world. Can we please stand up and give the greatest thanks of this season to the backs that brought that bounty to your table?

They are we. Our land turns on their affection. Our futures are their futures.

A mi gente: José Luís, Victor, Chapo, Tito, Joaquín, Blanca, Catalína, Asunción, Jesús, Nicolasa, Francisco, Maria, Isabel, Cente, *gracias muchísimas por todo, cada día, cada año. Y a todas las familias trabajando en las granjas, ranchos, escuelas, casas, hospitales, ustedes tienen mi gratitud tambien.* ●

Philanthropy Must Declare a Climate Emergency
Ellen Dorsey

Ellen Dorsey is the executive director of the Wallace Global Fund, founded by Henry A. Wallace and his son Robert B. Wallace in 1996. In 2008, Time magazine named Henry Wallace one of the ten best cabinet ministers of the twentieth century, as secretary of agriculture in Franklin D. Roosevelt's administration. There he created the food-stamp and school-lunch programs that endure to this day. Wallace was a third-generation agriculturist who was keenly concerned with human rights and the "common man," influenced no doubt by witnessing the plight of farmers during the Dust Bowl. His concern went beyond the fate of rural populations. He lectured in the South about endemic racism, at great risk, writing in 1947 that "our greatest weaknesses as a progressive democracy are racial segregation, racial discrimination, racial prejudice, and racial fear." That awareness and purpose is carried on to this day by the Wallace Global Fund, with its emphasis on threats to human rights and democracy, climate, corporate power, and movement building. It was a meeting at the Wallace Global Fund in 2011 that accelerated the divest/invest movement, a call for cities, pension funds, universities, faith groups, and foundations to divest their portfolios of all fossil fuel holdings and invest in climate solutions. Many institutions, including Brown and Harvard universities, resisted the call, citing portfolio theory and the need for growth and income. Divesting was not just the morally responsible thing to do; it turned out to be far and away the most fiscally responsible thing to do. By 2020 the value of Exxon stock had dropped 50 percent in value. If in 2011 a pension fund had sold its Exxon shares and invested in the world's largest wind turbine company, Vestas, their funds would have increased twenty times over. In 2020, the most valuable energy company in the United States for several months was NextEra Energy, a wind turbine company that was valued more than Chevron or Exxon. Regardless of portfolio size, Ellen Dorsey calls for a radical spend-down of assets so that there will be a future. Environmental and climate causes receive less than 2 percent of all philanthropic dollars. Money does not need to be saved at this point in history. It is needed to regenerate the planet, its people, and all of its creatures, and that in turn depends on equity, social justice, and honoring the "common human." Philanthropists are listening. Laurene Powell Jobs, the widow of Apple cofounder Steve Jobs, has committed to giving away her $28 billion in assets during her lifetime or shortly after her death, becoming the largest climate funder in the world. —P.H.

The scale and pace of the climate crisis require philanthropy to act with an urgency commensurate with the science. We have one decade to change the trajectory of emissions and the fundamental practices that are driving them up. To effectively do so will require radical changes, and we are the last generation of philanthropy to comprehend the scale of the problem that may still be able to do something in time.

Philanthropy must declare a climate emergency. It matters how much we spend and what we fund. Donors must be willing to put climate at the center of everything we do, including funding climate-based advocacy and movements working for systemic change alongside more traditional research, policy, and technology. Given the time frame, foundations should also consider significantly increasing how much they distribute entirely, or whether to give all their money away over the next decade. And, of course, how we invest that money matters. If we don't take these actions, we will not achieve the necessary solutions in time.

Today, the majority of foundations are not focused on climate, are not acting with the urgency required, and are not coming together to devise a unified response. That has to change. With one decade left to turn the tide, business as usual cannot be our response. Our financial resources can power the advocacy needed to move governments, corporations, and financial institutions to act in time. We can fund communities to be resilient and adapt to a changing economy, participate in the design of the energy transition, and generate economic opportunities in a new energy economy that benefit everyone, not just the wealthy few. We can fund pivotal research and new technological solutions with grants and investments, while helping develop new economic models that place justice and sustainability at the front of our decision making, fundamentally replacing the extractive economy that drove this crisis in the first place.

To keep our planet within livable limits of warming, philanthropy must be guided by clear analysis of what we are up against—economic vested interests that have fought for decades to maintain their profit and power, corrupting governments and co-opting officials to ensure inaction. If we want lasting change, our strategies need to build power to confront power. We cannot succeed unless we provide sufficient resources to build powerful movements that pressure governments to regulate the industries and financial players who are refusing to act, while building a new energy economy that is inclusive for all.

To do so, we need to sharply expand how much we spend to fight climate change. We will need to invest in audacious and bold ideas for systemic change that give the planet and future generations a fighting chance. Achieving transformative change requires supporting advocacy, pushing many foundations past their comfort levels, and getting money to frontline leaders and communities. We will need to provide sufficient resources so that movements have the scale to confront inaction; that requires a different approach than simply funding traditional nonprofits.

In a climate emergency, we must also call into question our endowments—both how much we are paying out and how the money is invested. Philanthropy can no longer justify distributing little more than the 5 percent of annual returns required by law. That is particularly impossible to justify when our endowments are growing. Nor can we justify being invested in the fossil fuel companies. We cannot fund grantees to fight climate with a small percentage of our assets if the vast majority of our endowment assets are invested in companies driving the crisis in the first place. Such a position is no longer acceptable, and future generations will consider it unforgivable.

Here is what declaring a climate emergency should entail:

Make climate central to the mission. The warming of the climate affects everything. It determines the short- and long-term well-being of the people foundations serve, regardless of whether the focus is on social services, human and civil rights, the arts, or social and economic justice. Making climate a priority will inform how we define our missions, advance change, build our strategies, and identify which grantees are most able to advance an urgent agenda. Creating small environmental grants programs that are marginal to a foundation's overarching strategic priorities is not sufficient. We must make climate an immediate cross-cutting priority in order to ensure the long-term survival of humanity.

Spend more, spend quickly, spend it all. Governments are failing to meet the challenge at the scale required. As a result, philanthropy needs to play a key role in funding any emergency combat response. We need both more money and smart money to address climate at all levels and create the transformative systemic change that will save us. If philanthropy is serious about fighting climate change, we need to spend more immediately. At a recent philanthropic gathering, former U.S. secretary of labor and University of California, Berkeley professor Robert Reich implored the audience to fund the movements that are our greatest hope, including urging us to spend more of the corpus to build their strength and ensure success.

Images on pages 144, 146, and 147 represent direct or related outcomes created by grantees of the Wallace Global Fund.

(*Left*) Julieth Mollel, 61, prepares to cook dinner on her clean cookstove at her home near Arusha, Tanzania. She is a Solar Sister entrepreneur, a nonprofit that is working to eradicate energy poverty by empowering women to be entrepreneurs. It is a women-run direct sales network that distributes clean energy technology, such as solar lights, mobile phone chargers, and clean cookstoves to communities across rural Africa.

It is immoral to continue increasing our endowments and paying out the lowest possible percentage required by law when faced with such a monumental existential threat. Each foundation and donor will need to assess its own legacy to determine at what level it should spend, but all must give more. Foundations must raise the question at the board level: Consider paying out half of the endowment over the next decade or spend down the endowment entirely over the remaining ten years left to prevent a global catastrophe.

Drive systemic change. Climate change is no accident—it is the result of economic practices and political choices that placed profit and power over the common good. Climate change and endemic inequality are driven by our current global order, where finance is far more powerful than governments. We cannot replace one extractive economy with another; the obligations of corporations and financial systems must be tied to climate risk. Real transformative change also necessitates making racial, economic, gender, and environmental justice a priority in our thinking, planning, and funding of action on climate.

We will need to build a government that works for the people instead of for its biggest donors, and every sector of the economy will have to change. Agriculture, transportation, water systems, and sourcing of chemicals will all need to be reengineered. Infrastructure must be rebuilt, and government investment in expanding new energy systems is required. Each of these challenges is an opportunity to invest in communities, create high-quality jobs, and usher in a just transition for workers and communities that depend on the extractive industries of the past.

The need for innovative and system-level response presents a special moment for philanthropy to support bold and audacious thinking that places justice at the top of the agenda, challenges inequality, and invests in the power of communities to lead. We've done it before; today's proposals for a Green New Deal, along with innovative models of agroecology to replace industrial agriculture, are a continuation of American progress in the face of disaster that led to the New Deal and continued through the Great Society. These systemwide, bold programs tackled tremendous economic and political crises with innovation that truly benefited society.

Collaborate with movements. History has taught us real change has never happened without mass public mobilization. Today, all around the world, the climate movement is increasingly being led by young people. Indigenous communities, women, communities of color, and LGBTQIA+ people are rising together. They are providing the clearheaded analysis and viable proposals that we need in order to shake the entrenched interests. Philanthropy cannot patronize them or dismiss their protests as merely the energy or naiveté of youth. We need to work with these movements and help bring the best of the non-profit world behind them to ensure their success, instead of trivializing their power. We will not solve the climate crisis with status quo solutions designed by consultants in a foundation boardroom.

Many foundations and donors shy away from advocacy or funding movements, but when powerful interests block government action at the scale we need, advocacy is crucial. Funding scientific research or policy recommendations alone will not be sufficient to fight the power of those who are profiting from the current energy system while demanding governments advance the common good, not the profit of a few. Once we have done that, we must stand alongside movements and use our collective resources to help build power in step with—not removed from—the grass roots.

Philanthropy can support courageous social movements that protest injustice, such as the protests at Standing Rock, Fire Drill Fridays, Sunrise Movement, and others. We can help expand their leadership skills, learn from their strategies, support their organizing efforts, and bring the research and advocacy talent from traditional nonprofits to help out when economic interests block government action. Our climate strategies must include working to help advance their priorities. Philanthropy can find appropriate ways to fund these political movements while remaining nonpartisan.

Deploy every bit of the endowment for good. In 2021, there can be no wall between grants and investments. Declaring a climate emergency means using all our tools. Getting the world to 100 percent renewable energy in time will require every institutional investor to put at least 5 percent of our assets into renewables, efficiency, clean tech, and energy access. Foundations must also use our voice as shareholders to put pressure on every industry to curb carbon use.

Imagine what philanthropists could accomplish by investing in projects that advance both a green world and a just one. Imagine that a foundation invested in a grid-scale wind farm owned by native and indigenous communities that brought substantial economic benefits to them and fair returns to you. Imagine that your investments in women-led small businesses generated significant income while bringing energy—clean energy—to communities for the very first time. The opportunities are endless and financially viable.

Foundations must also divest from fossil fuels. Any assets held in fossil fuels drive climate change. Investing in such fuels and those who finance them ultimately undercuts the grantee work that we fund. More than twelve hundred global investors with management assets of over $14 trillion have divested, along with two hundred foundations worldwide. Divestment turns out to be a very smart economic choice, too, as fossil fuels have been delivering terrible returns for more than six years. Foundations that began divesting more than five years ago saw better returns than those that did not. An invest-divest commitment will bolster returns, meet fiduciary duty, align pro-

gram and investments, and make a foundation's portfolio consistent with the demands of a climate emergency.

With a global systemic crisis of this magnitude, the only rational response for philanthropy is to declare a climate emergency and act in radically different ways commensurate with the challenge. We are accountable to no one but our boards, we have enormous privilege in the level of resources we expend and consume, and we have deep ties to the very system and economic players that produced this mess. We seldom challenge ourselves to the types of actions we demand of our grantees, governments, even businesses. Now is the time to do so.

We are all climate funders now—but if we act quickly, we may not have to be forever. ●

(Left) Kayapo leader Cacique Raoni Metuktire watches various tribes perform ceremonial dances in Piaracu village, near Sao Jose do Xingu, Mato Grosso State, Brazil. Dozens of Amazon indigenous leaders have gathered to form an alliance against Brazilian president Jair Bolsonaro's environmental policies. The Kayapo have enabled protection of nine million hectares of intact Amazon forest in their contiguous ratified territories, so far one of the most effective "conservation organizations" in the world.

(Above) The elephant represents WILD foundation which was founded in South Africa in 1974 and is headed by Vance Martin. It has become one of the premiere organizations protecting wild nature around the world. They have convened wilderness conferences on six continents and literally wrote the book on wilderness guidance and international wilderness law and policy.

The City

Cities are tapestries of culture, science, art, cuisine, music, scholarship, theater, diversity, and innovation. And they are also depleting the world's resources at breakneck speed. The needs, wants, and demands of cities ultimately degrade life in all its forms, on all continents and oceans, largely unseen by its inhabitants. What becomes of civilization in this century will be determined by what happens in urban and suburban environments.

In cities, the waste stream of paper, plastic, clothing, electronics, cars, and food is never-ending. For every human on the planet, there are two hundred fifty pieces of plastic in the Great Pacific Garbage Patch, between Hawaii and California. Seventy percent of global greenhouse gas emissions come from consumption in cities, including electricity, manufacturing, heating and cooling of buildings, transport, and waste. To achieve the goals set forth by the Intergovernmental Panel on Climate Change and adhered to in this book, greenhouse gas emissions must be cut in half by 2030. That puts cities and the 4.3 billion people who occupy them at the center of the problem and the heart of the solution.

Cities contain high concentrations of air pollution, causing millions of untimely deaths. The emissions from the combustion of fossil fuels contain poisons—carbon monoxide, nitrous oxides, particulates, benzene, toluene, ethylbenzene, xylene, and polycyclic aromatic compounds. Ninety-two percent of the world's population lives in areas where air quality does not meet the standards of the World Health Organization. This can cause irreparable damage to the near- and long-term health outcomes of children and lead to premature death of the elderly.

Fortunately, the mayors of the world's cities lead the way with respect to air pollution, noise, waste, mobility, green housing, and climate. Mayors are more effective than national or provincial leaders, because what unites people in communities, towns, and cities is more important than what divides them. For the world to thrive, the city will need to be the principle driver of regenerative practices, methods, and outcomes.

The question is how quickly can renewable energy replace fossil fuels used in transport, electricity, heating, and cooling? Can urban air be as clean as forest air? How soon can child health be the norm and asthma eliminated? How many dignified, living-wage jobs can be created by regenerating natural and human systems? How rich and productive can regenerated agricultural soils become? How much water can we return and store in grasslands, wetlands, and farmlands? We do not know what the limits are to regenerating life. If there is to be a regenerated earth, if one day greenhouse gases are reduced and drawn down from the atmosphere, it will happen because of the practices, policies, and leadership of citizens and cities.

What might a regenerative city look like? Buildings and transport are electrified, and the energy source is renewable. Mobility is barely audible, affordable, nonpolluting, and largely free. Polluted air is a distant memory. It is quiet day and night since only electric buses, cars, and trucks are allowed. The burning of natural gas, diesel, oil, and gasoline are banned inside city limits and nearby. Along with a renewable energy system, there is now a renewable food system. Fresh local food is available within walking distance at affordable prices. The food is grown and supplied by diverse urban farming communities within and surrounding the city.

Housing will be dense in order to create open spaces and the city will begin to look more like a forest. Buildings will be clad in vines; trees and shrubs will permeate rights of way, and rooftops will will be farms and gardens. Unused industrial spaces will become vertical farms. Landscaped terraces and medians are sanctuaries for birds and pollinators. Local and regionally made products, especially food and clothing, will displace remote manufacturing where practical. Because the city is designed for people, not cars, roadways and parking lots transform into housing, parks, gardens, and recreation. The city begs to be walked in because of its distinct neighborhoods, safety, peacefulness, and greenery.

Paved-over creeks are restored. Many species long absent return to dwell in the parks, brooks, and forests. The city is designed to be traversed by humans, animals, and streams. People cannot imagine a city that is not zero waste. Every municipality has facilities for food and industrial composting. Community infrastructure is more resilient, more diverse, and more decentralized in order to deal with more disruptive weather events.

If this sounds like a fantasy, bear in mind that cities were the birth of civilization. These initiatives are being created and implemented in cities the world over, cities that can become the needed forerunners of a new civilization. ●

View through the twin towers of Chaoyang Park Plaza, Beijing, China, designed by MAD Architects. Chaoyang Park began construction in 1984 and is now Bejing's largest park. Although it consists primarily of green space, water surfaces cover over 170 acres.

Net Zero Cities

More than twenty-five hundred years ago, Greek civilization faced an energy crisis. The growth of city-states and their colonies required ever larger amounts of fuel to heat homes and public buildings. The main source was wood, and forests were being destroyed. In response, the Greeks turned to an inexhaustible supply of warmth: the sun. Excavations in Priene, on the coast of modern Turkey, and other cities reveal the implementation of a sophisticated checkerboard pattern of streets that enabled every building to utilize passive solar energy by facing south. Socrates himself noted "sunshine during winter will steal in under the verandah, but in summer the roof will afford an agreeable shade . . . we should build the south side loftier to get the winter sun and the north side lower to keep out the cold winds."

Ancient Chinese and Romans also incorporated solar design into their urban planning, but by the medieval period circumstances had changed. Implementation of forest protection laws, new agricultural practices, and the development of coal as a fuel source eased wood shortages. Cities abandoned solar design in favor of densely packed centers. By World War II, cities in the developed world were almost exclusively using fossil fuels—coal, oil, and natural gas—for power generation, heating, and transportation. Cooling was soon added to the mix. A rapid expansion of air-conditioning use took place as home construction boomed in the postwar years. Architecturally popular but thermally inefficient "glass box" office buildings required large amounts of air-conditioning to keep cool. It was a sign of the times. Architects no longer paid attention to where south was.

In 1973, a series of disruptions to oil supplies exposed urban dependence on fossil fuels. A global effort to develop renewable energy took off, including solar, wind, biomass, geothermal, and hydropower. By 1990, non–fossil fuel sources accounted for more than a third of global electricity production, with renewables providing 19 percent and nuclear power 17 percent. Fossil fuels made up the other two-thirds. Thirty years later, however, the proportion of renewables to fossil fuels remains the same: one-third to two-thirds. The only real change has been a drop in nuclear power's share, to 10 percent.

Enter Reykjavík, San Francisco, Jeju, Barcelona, Hamburg, Sydney, Munich, Vancouver, San Diego, and Basel. All have set targets to operate on 100 percent renewable energy or have already achieved it. More than one hundred cities worldwide already get 70 percent or more of their power from renewable energy, double the

total in 2015. Many are in Latin America, where hydropower generates a great deal of electricity. In the United States, more than one hundred fifty cities, representing nearly 100 million people, have formalized commitments to achieve 100 percent renewable energy goals in the electricity, heating and cooling, or transportation sectors. Los Angeles plans to shut down all oil and gas production within city limits, including the approximately one thousand active wells; shutter all natural gas plants; require new buildings to be certified zero-emission for greenhouse gases; phase out gas and diesel engines; required all construction equipment to run on electricity; and install public infrastructure for electric cars, including twenty-five thousand new charging stations.

Copenhagen is almost there. In 2010, the bicycle-friendly Danish capital, where there are six times as many bikes than cars, gave itself fifteen years to become the world's first carbon-neutral city. Seven years later, it had cut its production of greenhouse gases nearly in half from 2000 levels. The city's buildings are almost entirely warmed by waste heat created by power plants that incinerate wood pellets and refuse. Copenhagen relies equally on onshore and offshore wind energy. Denmark set a formal goal of 100 percent renewable energy by 2050 and is now halfway there. The main strategy is to use less energy, allowing the nation to meet its electricity needs solely with wind and solar power. Denmark is a pioneer in wind power generation, dating back to the 1970s, and remains a global leader in turbine innovation and manufacturing.

Can renewables provide reliable power to meet demand all day and all night, even during periods of peak use? The quick answer is yes. It will involve a combination of reduced overall electrical demand, increased battery storage capability, grid upgrades, coordination between power plants, and the employment of a diverse array of renewable energy sources, including an expansion of base-load power sources that operate continuously, such as ground-source geothermal. More than half of all homes and buildings in the United States rely on natural gas or other fossil fuels for furnaces, kitchen ranges, and hot-water heaters, generating one-tenth of the nation's total greenhouse gas emissions. A number of city councils have passed ordinances requiring new buildings to be entirely electric (sometimes called "gas bans"). A traditional option for buildings is rooftop solar.

Many of these developments are beyond the reach of low-income residents and disadvantaged communities, requiring policies that create a more equitable pathway to achieve net zero. Fortunately, the cost of solar, wind, and other renewable sources has fallen dramatically, easing the pressure to raise utility rates for consumers.

Electrification has important collateral benefits for cities and citizens. It would eliminate the greatest sources of air pollution, improving the health and safety of many people, particularly children, the elderly, and communities of color. In 2018, nearly nine million people died as a result of exposure to air pollution caused by fossil fuels, accounting for one in five of all deaths around the world that year. Electrification reduces noise pollution. Traffic noise is a chronic source of stress that can lead to anxiety, insomnia, high blood pressure, and hearing damage. Electric cars and motorcycles don't need mufflers, as their engines are nearly silent. By reducing our overall demand for energy, an electrified city encourages prosperity, not privation, in a warming world. ●

Vancouver, Canada, has committed to reducing carbon emissions by 50 percent by 2030, and to be a net-zero city by 2050. The areas of focus will include natural gas use in buildings, gas and diesel in vehicles, walkability, and overcoming historic discriminatory legacies of social injustice. By 2030, 90 percent of citizens will be within walking or rolling (bike, scooter) distance of their daily needs. Two-thirds of all trips will be by public transit or nonmotorized active transportation. Half of all mileage will occur in zero-emission vehicles. Carbon pollution limits for existing buildings will be set to transition from fossil fuels to renewable energy. All replacement heating and water systems will be zero emissions (geothermal heat pumps). Embodied carbon in new building construction will be reduced by 40 percent.

Buildings

Buildings are where we spend most of our daily lives, and yet so many buildings are not designed for either human or planetary well-being. Building materials are full of toxins; condos are designed for construction efficiency but not for supporting positive social dynamics; and developers do not have the knowledge base, experience, or incentive to look at the long-term performance of the buildings they create. These problems are intensely magnified in the poorest parts of cities and rural areas. Regenerative construction means looking at the ecological, social, and personal health impacts of every building.

Today, new building construction is expanding at the fastest rate in history. Over half of the world's population lives in cities, and 1.5 million new residents are added weekly to the global urban population. By 2050, it is predicted that 6.7 billion will live in cities. To accommodate this unprecedented growth, it is estimated that 2.5 trillion square feet of new and renovated buildings will need to be constructed. This is twice the existing global building stock, tantamount to building one New York City every thirty days for the next forty years.

Foreseeing this growth, architect Edward Mazria created the 2030 Challenge in 2006. Its goal was for all new buildings and major renovations to be carbon neutral by 2030. It also aimed for 50 to 65 percent reduction in carbon emissions in the existing built environment. Mazria defined carbon-neutral buildings as ones that use no more energy than they create on-site, and/or renewable energy they purchase off-site, not to exceed 20 percent. Globally, 12 percent of building emissions come from the embodied carbon created by the materials and the construction process, and 28 percent is from operational emissions. Creating zero-emission buildings addresses both. Mazria and his organization have collaboratively mapped the pathways, financing, standard, and principles that can create zero-emission cities. They include:

Achieving Zero—an integrated policy framework that can eliminate all carbon dioxide emissions from the built environment. It focuses on three areas: construction of new zero-carbon buildings, retrofitting and upgrading existing building stock, and creation of policies and recommendations on how to reduce embodied carbon in materials and construction methodology. The framework creates new financial models to pay for upgrades, significant energy reductions, local jobs, and healthy communities, including new pathways for ownership.

Zero Code—a model international building code that enables cities to upgrade their existing codes to the highest standards. Cities have relied on the International Building Code, which came into existence in order to have one uniform building code, ensuring that safety standards were upheld throughout any country. However, the code has been undermined by corporate interests who wanted lower standards, which prevented efforts to address climate change. Mazria's Zero Code provides an efficiency standard for upgraded building codes that favor the future instead of the interests of corporations.

In 2006 we also saw the founding of the Living Building Challenge (LBC), a design philosophy that views buildings as regenerative structures that are fully integrated into their local ecosystems. Developed by architect Jason McLennan, LBC pushes design objectives, whether they are carbon footprints, energy use, or water harvesting, beyond neutrality to *net positivity*. LBC transforms buildings and their inhabitants from passive users of resources to active stewards of all elements involved in a structure's construction, maintenance, and purpose. LBC certification, for example, requires a building to generate more fresh water and clean energy than it consumes. All materials and substances must be nontoxic to the greatest degree possible and sourced sustainably, right down to the raw material.

LBC certification isn't final until the building has been continuously occupied for twelve months and has met one or more of twenty separate standards. The building's impacts on the surrounding community, both ecological and human, must protect and restore healthful living conditions. Water must be collected on-site and used in a way that mimics hydrological cycles in nature. Buildings must produce 105 percent of the energy they need each year. Indoor air quality, lighting, cooling and heating systems, room layout, and access must all be designed and integrated with positive emotional, physical, and equity impacts in mind. The building should be beautiful and inspiring to all who interact with it.

The first large structure to meet the LBC challenge was the Bullitt Center, a six-story office building in Seattle. Other buildings conforming to some of the LBC requirements include Etsy's two hundred thousand square-foot headquarters in Brooklyn, New York; and Google's new seven-story office in Chicago. Since 2013, more than one hundred buildings have been certified LBC, with five hundred more in process, spread across fourteen nations.

In densely populated, established cities, the main issue is upgrading existing buildings. Global emissions goals cannot be attained unless the more than one billion older buildings in the world are upgraded to new energy standards. Not all of them need retrofitting, owing to local climate, construction, and use. However, most structures that house people, manufacturing, offices, schools, and places of worship use far more energy than is required to operate, heat, and cool them.

A massive worldwide building upgrade may seem both onerous and impossible. However, if looked at in another way, the potential for retrofitting every building in the world to net-zero emissions over the next thirty years would create more well-paid jobs for a longer period than any other known initiative. We associate energy jobs with workers who mine coal, extract oil, or frack for gas. Creating energy-efficient buildings is an energy job, too. The value of energy efficiency is money not spent over time. That is the return on investment. You can be paid for every barrel of oil you pump—or for every barrel of oil you save. The total savings amounts to trillions of dollars over the life of existing and new buildings.

In 2021, the Lawrence Berkeley National Laboratory calculated that it would cost approximately one dollar per person per day to completely rid the U.S. economy of fossil fuels and operate it on renewable energy. Those dollars would be spent at home, not exported to other nations for oil. They would spur economic activity that would create jobs and cascading economic benefits. The one dollar will come largely from investments made by the private sector, or from homeowners' pockets in a way that provides a return on investment. It can come from a mortgage premium charged to homeowners and businesses that are paid back through energy cost savings.

In New York City, 70 percent of total carbon emissions come from buildings. If you calculate the total

---

The Atlanta headquarters for ASHRAE (American Society of Heating, Refrigerating, and Air-Conditioning Engineers) is a deep-energy retrofit that transformed an energy inefficient building from the 1970s into a daylit, industry-leading, efficient facility with advanced HVAC systems, LED lighting, and building automation systems. It is a fossil fuel-free and net-zero energy facility designed by McLennan Design.

annual energy expenditures in the five boroughs and Westchester County, they exceed $12 billion per year. If you could save 80 percent of that expenditure annually, $9.6 billion, what would you invest? A $100 billion investment/loan would provide a 9.6 percent return, assuming energy prices do not go up. The cost of retrofitting New York City (and the rest of the world) could come from savings.

This is already occurring. New Yorker Donnel Baird grew up with two passions: climate change and civil rights. He combined them into a company called BlocPower. His company goes after midsize buildings and apartments in low-income neighborhoods in Brooklyn and other cities, generally older structures that are energy hogs burning large amounts of natural gas to heat home and water. BlocPower places raised solar panels on the roof (for shade and energy), but the key to his success is the heat pumps below that replace the gas boilers. BlocPower finances the retrofit and is paid back by savings from the residents' utility bills. If the co-op or building owner chooses to purchase renewable power, it becomes a zero-carbon building within months, at no cost to its owners. What Baird and BlocPower can do is make the changes that reluctant

building owners do not know how to do. Baird has converted more than a thousand buildings, and his next goal is a hundred thousand.

Building retrofits may be the most labor-intensive solution to climate change, which is a good thing. They provide an opportunity to upskill capabilities of currently unemployed or underemployed people in every city in the world, training that provides expertise and knowledge that create future opportunities. There are five important ways to upgrade our homes and buildings: heating and cooling, insulation, windows, lighting, and converting to renewable energy. The sixth is to supply renewable electricity to the 800 million people who have no power and the two billion without consistent, reliable power.

Heating and cooling. Most buildings rely on age-old technologies to heat and cool. These include boilers using heating oil or natural gas, massive rooftop air-conditioning units in big buildings, and window-based appliances for smaller buildings. Many cities around the world employ district heating and cooling, networked systems that redistribute hot and cold water to districts or neighborhoods in underground insulated pipes. They can significantly reduce energy—by up to 50 percent—but rely largely on conventional heating methods, including capturing waste heat from power stations, incinerators, or sewage treatment. The answer is heat pumps, whether for single-family homes, condominiums, apartments, or commercial buildings. Think of a heat pump as an air conditioner that can run backward. Instead of converting hot air to cool air, it converts cool air to hot. In the summer, it can do the opposite and cool the air. Like an air conditioner, it uses external power to compress a refrigerant (the noise you hear from a refrigerator is the compressor), which can transfer the latent heat in cold air into greater levels of heat. The source of thermal energy for a heat pump can be the air or the ground.

Because they are far more efficient, heat pumps reduce overall energy demand for the same thermal output by 50 percent. When analyzing the cost of buildings transitioning to renewable energy, it is important to register the social costs of fossil fuels. Those include 8.7 million premature deaths each year caused by air pollution in cities across the world. There are the costs of seemingly endless wars in the oil-rich Middle East, which have exceeded $5 trillion for the United States alone in the past fifteen years. These dollar expenses do not count the enormous suffering endured by women, children, and soldiers. They do not include the environmental damage caused by fracking, oil wells, the BP *Deepwater Horizon* oil spill, or the Alberta Tar Sands. And finally, nowhere in cost analyses are the impacts of global warming included. When those costs are totaled, heat pumps are not only 40 percent less costly, but easily three times less energy intensive.

Retrofits. Retrofitting includes installing efficient windows; preventing air leaks; insulating roofs, attics, and walls; and swapping out existing lighting for LEDs. If retrofitting buildings would save hundreds of billions of dollars every year for home and building owners, why is it not being done? There is a joke about a $1,000 bill lying on the floor of an inefficient factory. A guest is being escorted by a manager, who walks right over it. The guest later asks the host why he did not pick it up, and the manager replies that "if it was real, someone would have already picked it up." We have known for decades that saving energy is cheaper than creating fossil fuel energy. Energy efficiency puts money into the pockets of home and building owners. Fossil fuel investments concentrate profits and are returned to fewer hands. Since the Paris Agreement was signed in 2015, the banking industry has loaned and invested more than $3.8 trillion in the oil and gas industry, more than enough money to retrofit every building in America to be a zero-waste structure. The energy saved would exceed the amount gained by oil and gas exploration by a factor of ten.

The jobs created by retrofits would be local and primarily with small companies; more than 70 percent of companies that do retrofits are small companies. A nationwide retrofitting program would mobilize more than a million people if the goal was to complete the retrofits by 2035. There are 79 million single-family homes and 45 million more households in apartments, condos, and duplexes. Americans spend an average of $1,380 per year on energy bills. Nearly half of residential buildings were constructed before 1973, when building standards for efficiency were nonexistent. With at least 40 percent reductions in energy use, total savings for all households would be $102 billion per year. Worldwide, there are 1.46 billion more households, generally not as large as American homes. However, their potential savings approach half a trillion dollars annually. The facts and figures here are instructive when you consider what those big numbers replace. In other words, if you save $1,380 per year in household energy costs, what was it being spent on? Overwhelmingly on coal, gas, and oil. If buildings are electrified and the source is renewable by 2050, one-sixth of the world's total greenhouse gas emissions would be eliminated. ●

The Edge in Amsterdam is an advanced net-zero building. Constructed in 2015, it uses thirty thousand sensors to continuously measure occupancy, movement, and temperature. It automatically adjusts settings to maximize efficiency and can even inform facility managers about anticipated food needs. Using a mix of solar panels, innovative design, and an aquifer thermal-energy storage system, the Edge generates more energy than it uses. Its designer and builder, Edge Technologies, is credited with creating the smartest buildings in the world, including Edge Olympic, also in Amsterdam.

Urban Farming

Urban-grown food is consumed where it is produced. It can happen anyplace in the city where there is light, vacant lots, parks, rooftops, landfills, brownfields, medians, warehouses, community gardens, home gardens, and soil-filled containers of various sizes and shapes. From the inner city to the more spacious urban perimeters, urban farming produces more than vegetables. It connects people to fresh, vibrant, nutrient-dense food and to a variety of flavors and textures. The gardens and farms attract birds, butterflies, bees, and other pollinators, elements often absent from the lives of children growing up in cities. The minifarms are classrooms, and just like regular farmers, people working the crops learn more every year about seeds, timing, water, light, soil, tilth, and taste. They offer training, reentry programs for the formerly incarcerated, and education for all. Fresh and diverse fruits, vegetables, and herbs can serve specific culinary traditions, sustaining traditional foodways for the many world cultures that tend to cluster in urban environments. The food can be shared by churches, schools, and soup kitchens or sold to restaurants. There grows the neighborhood.

Will urban farming make a dent in climate change? Yes, but not much. However, it does something that can make a difference. It reawakens people's understanding of food and its total impact on human health, happiness, and well-being. People living in food apartheid are surrounded by ultra-processed foods laden with fat, starch, and sugar. Farmers' markets, community gardens, and urban farms help people literally walk back their obesity and ill health to vitality and a sense of well-being. People begin to vote with their fork; they purchase better food and avoid the junk. Ultimately, this is the only way the larger food system will change, a system that has the greatest impact on global warming, with its attendant and accelerating

influence on fires, floods, and storms. Conversely, those same fires, floods, and storms make supply chains more fragile and questionable. Although food security is only slightly impacted by urban farming, it raises awareness of the long distances a head of lettuce or tomato travels to the table. Change in demand for fresh, local food stimulates localization of food through increased numbers of small farms in the region surrounding a city. Collateral benefits include a decrease in the urban heat island effect, improved habitat for pollinators and other wildlife, the recycling of organic matter into compost, reduced emissions from fossil fuels, and increased resiliency to interruptions in food procurement and storage.

Can cities feed themselves? It might depend on the city. A study of Cleveland, Ohio, which has many vacant lots, determined that converting 80 percent of its empty spaces to vegetable plots and adding chickens and bees could supply the city with up to half of its fresh produce, 25 percent of its poultry and eggs, and all of its honey. When rooftop gardens are added, the amount of potential produce jumps to all of Cleveland's needs. New York City, on the other hand, has far more people and far less vacant land. A study by Columbia University in 2013 determined that between 162,000 and 232,000 acres of farmland are required to meet New York's annual fruit and vegetable needs, not counting tropical food, which cannot be grown locally. However, there were fewer than five thousand acres of vacant land available for agriculture within the city (whose economic value for development far outstripped its value as food-producing land). Potentially, between 50 and 75 percent of the 400,000 acres of open land in the counties surrounding New York could supply the fruit and vegetable needs of the city, but only if the land were dedicated to food production.

One idea that has blossomed in recent years is an old one: rooftop gardens. There is recorded evidence dating back thousands of years of plants being grown on the tops of buildings, possibly including the famous Hanging Gardens of Babylon. In the modern era, the trend of installing rooftop gardens began in Italy in the fifteenth century and became popular in New York City in the 1890s, when a massive garden was built atop the building then known as Madison Square Garden. Recently, rooftop gardens have expanded into an exciting model of urban agriculture. In 2010, a group of young farmers formed the Brooklyn Grange and opened what was then the world's largest rooftop vegetable farm, a two-and-a-half-acre installation on the Standard Motor Products Building in Queens. Their goal was to grow healthy, tasty food in the "unused spaces of New York City." The farm includes a commercial apiary, a training program for interns, an educational center, and weekly markets and open houses. In 2012, the Grange opened its second farm on the sixty-five-thousand-square-foot roof of the historic Brooklyn Navy Yard building (twelve stories above the street). They opened their third and largest farm in 2019, on the sprawling roof of a waterfront structure in the Sunset Park section of Brooklyn, including a large greenhouse, an events hall, and a kitchen. Altogether, the farms total nearly six acres and produce more than one hundred thousand pounds of organic vegetables annually, all sold locally.

The Michigan Urban Farming Initiative runs an urban agriculture campus in Detroit's North End neighborhood to increase food security and promote education, sustainability, and community. Over ten thousand volunteers have grown and distributed over 120,000 pounds of organically grown produce to over 2,500 local households. One hundred percent of the produce is available free of charge, using a pay-what-you-can model.

The rooftop movement has expanded and diversified. In Washington, D.C., the nonprofit Rooftop Roots supports social and economic justice through projects that help community members convert roofs, balconies, patios, and other spaces into food-producing ventures. In Hong Kong, Rooftop Republic is finding creative ways for residents of the densely packed city to realize a rooftop farm. In 2020, Lufa Farms, in Montreal, opened its fourth rooftop greenhouse—the largest in the world—enlarging the company's capacity to grow fresh tomatoes, eggplants, and other vegetables to more than three hundred thousand square feet. The new greenhouse produces twenty-five thousand pounds of food each week. It came into existence quickly, in response to a doubling of demand from customers during the early months of the COVID-19 crisis. Lufa also inaugurated seven-days-a-week service, tripled its home delivery runs, signed up thirty-five new local farmers and food makers, and added thirty thousand new members. "Growing food where people live is our mission, and this greenhouse is an acceleration of that mission," said Lauren Rathmell, a Lufa Farms cofounder, "the timing of which couldn't be any better as we respond to an ever-growing demand for fresh, local, and responsible foods."

Another urban agriculture idea being explored is vertical farming. Vegetables and other crops in this system are grown inside city buildings, including warehouses, offices, and restaurants, using a stacking arrangement. Components range from simple trays of plants to complex arrays of artificial lights, heaters, pumps, rows of bins, and computer-controlled timers. One type of vertical farming grows its food hydroponically—without soil—often under LED lights that are fine-tuned for optimal photosynthesis conditions in plants. Another type is an aeroponic system, in which plants are suspended in the air and their roots sprayed with a nutrient-dense mist by specialized high-tech appliances. These "aerofarms" can be customized to fit almost any space, including schools, restaurants, government buildings, community centers, and apartment complexes. A third type of vertical farming is an aquaponic system, which grows edible fish along with vegetables. In this self-contained system, water from a fish tank, carrying nitrogen-rich fish waste, is filtered through a hydroponic plant arrangement and recirculated back to the fish tank. The plants love the natural fertilizer, and the fish love the clean water that comes back. Aquaponic systems produce little waste, can be adapted to fit almost any indoor space, and can grow a wide variety of fish and plants.

Advanced technology is also enabling vertical farming to be done in standardized modules. In the case of Square Roots, an innovative start-up company based in New York City, that means shipping containers. Square Roots was founded in 2016 by entrepreneurs Kimbal Musk and Tobias Peggs, with the goal of creating a technologically sophisticated, climate-controlled food system that grows fresh, local, and great-tasting food. To do that, the company repurposes ordinary shipping containers—each enclosing 320 square feet—to grow food. Each container can produce as much as one hundred pounds of food per week. Each one is managed by a local farmer who grows the plants from seedlings, oversees the hydroponically delivered nutrients, coordinates the full-spectrum lighting system, harvests the crop, and takes it to market. In the Square Roots model, energy use—often the bane of energy-intensive vertical farming systems—is carefully measured and tightly controlled by technology to maximize efficiency. The lighting system inside the container is designed to mimic perfect growing conditions under natural sunlight. For its popular basil, the system recreates growing conditions in Italy in 1997, which was said to have produced the best basil crop in decades. Data is collected and analyzed at every step of the growing process. (Peggs is a former data scientist.) This allows adjustments to be made to optimize quality and yield—increasing profits for the farmer.

Square Roots is extending its work into other cities and plans to expand the diversity of food that can be grown in its high-tech system. This expansion plan fits well with the evolving vision of cities in the twenty-first century. The containers are portable and stackable and can be placed in nearly any suitable space. They're ideal for abandoned parking lots as cities transition to car-free mobility systems. An individual farm can be one container or twenty, depending on the space available and the goals of the farmer. The containers run on electricity, which means they can be part of the "electrify everything" movement's aim of substantially reducing the carbon footprint of city energy use. The container farms can also be designed to tap the electric grid at nonpeak periods of the day, saving energy. They can even provide food on demand, much like a transportation service, ready at a moment's notice. And since the food is local—you could even call it "ultralocal"—it's always fresh and tasty.

Rooftop farms, greenhouses, and container farms are just a few of the many innovative ways cities are trying to feed themselves. The ultimate goal is to get residents reengaged with their food where it produced. "People who grow their own food are likely to understand the processes of nature on agriculture," says Michelle Hong of Rooftop Republic. "We are aiming to change the concept that food is something that we only engage with at the supermarket. Only by addressing this disconnection—this broken relationship—will we be able to change people's mindsets and behavior and help them make more informed decisions about their food." ●

The Nature of Cities

By 2030, 70 percent of Chinese, 85 percent of Americans, and 86 percent of the United Kingdom will inhabit urban areas. These numbers vary throughout Africa but average about 50 percent. Before cities came into being six thousand years ago, life revolved around the confines and customs of a settlement or village. A relatively small number of residents and local customs dictated who you would meet, greet, and marry, since you knew everyone and everyone knew you. Strangers were rarely seen, familiarity was the norm, and xenophobia endemic. Although cities came about because of agricultural surplus, why cities were created is not clear. As they were established, cities reversed most of the attributes of village life. You could meet (and possibly marry) a stranger, and unfamiliarity became normal at the marketplace. There was no longer the intimate sense of place that hosted the village—the grassy knolls, riparian corridors, hilltops, or forested enclaves—nor was there interaction with mammals, predators, songbirds, insects, snakes, invertebrates, herbs, fungi, or plant medicine. Domesticated animals and look-alike dwellings were left behind in favor of souks, piazzas, temples, stadiums theaters, and locked doors. The diversity, relative leisure, and close human proximity of the urban environment created an explosion of art, music, religions, politics, law, invention, and scholarship.

The layout and design of cities tended to be haphazard or deliberate, depending on the proximity of a building to the town center. Urban planning was an exercise in place making, with people and officials collectively determining neighborhoods, the width of streets, the location of the central market, the breadth of wide-open plazas, the locations of drainages that kept the city from flooding, and where bridges needed to cross streams and rivers. Many of the ancient cities are underneath us today. Excavation and tunneling in Rome, London, Paris, Guangzhou, Tokyo, Mexico City, Cairo, Athens, and Istanbul has revealed layers of homes, plazas, pottery, graffiti, bones, tools, and tombs.

The spread of industrialism and today's urbanization have produced cities dominated by the car, concrete, and steel. The first skyscraper, built in Chicago in 1885, employed steel I-beams to support a nine-story-tall masonry structure. As construction techniques evolved, masonry facades were ditched, and steel-and-glass towers rose above the city center. Downtowns that were bustling during the day became dangerous ghost towns and wind tunnels at night. Although many older cities created urban

---

Vertical Forests was designed by Stefano Boeri as part of a broader Greener Cairo vision, which envisages six decarbonization strategies for the Egyptian metropolis aimed at achieving the ecological conversion of the city. In addition to planning new architectural forms, the vision includes a large-scale campaign for making the thousands of flat roofs of the city green. It also includes increasing urban vegetation through the creation of a system of green corridors that cross the old capital and join a larger orbital forest, making Cairo the first city in North Africa to deal with the challenge of climate change and ecological conversion.

parks, including New York, Barcelona, London, Paris, San Francisco, Munich, and Melbourne, newer cities were developed quickly after World War II without the same foresight seen in London's Hyde Park or the Tuileries in Paris. In the United States, historic downtowns were razed to make room for "progress," which took the form of automobiles, steel, concrete, and asphalt. In modern cities, what green there was could be found on verges, highway plantings, vacant lots, and small parks. Gardens were scarce, as were trees, birds, and flowering vines. For a child growing up in the marginal areas of many big cities, there is no landscape. It is a cityscape of sidewalks, traffic, and noise. Wildlife consists of mice, rats, and cockroaches. Nature does not enter into the child's life except in the form of weather.

Psychologist Peter Kahn uses the term "environmental generational amnesia," where successive city-raised generations become further removed from wildness, and children do not see stars at night or midsummer fireflies, hear crickets at dusk, watch foxes gamboling in the grass, or gape at courting hummingbirds dive-bombing at forty-five miles per hour to impress a potential mate. The cities are wonder-free zones. The absence of biodiversity easily leads to a lack of concern and indifference to nature. The presence of extensive plantings, parks, forests, and greenways in an urban environment is a needed response to what lepidopterist Robert Pyle describes as the "extinction of experience." He cautions that distance from nature "breeds

apathy toward environmental concerns and, inevitably, further degradation of the common habitat. . . . People who don't know don't care. What is the extinction of the condor to a child who has never known a wren?" In that environment, what is global warming to a child but the need for an air conditioner?

Today, the idea of a great city is changing. Urban planners sit alongside botanists, agronomists, geographers, foresters, urban farmers, ornithologists, architects, ethologists, doctors, research institutes, microbiologists, developers, civil society, and landscapers in order to move forests, orchards, vines, woody perennials, and human mental and physical well-being into the city in ways that were never before envisaged. Rather than a city on the hill, it is a city in a garden, forest, wetland, or salt marsh that is being designed and put forward. Vacant lots, parklands, and remnant woodlands are seen as precious because they are not developed. Cities are regretting their hardscapes. In 2012, Beijing suffered devastating flooding as storm water overwhelmed drains and coursed through city streets with no way to soak into the ground. It was a similar story across the nation and only getting worse under climate change. In response, China is creating "sponge cities" full of green, natural places where rain and flood water can be absorbed, including rooftop gardens, restored wetlands and lakes, new parks, and additional trees and other vegetation. Porous materials are being used in sidewalks and streets. The captured water is stored in underground tanks

where it is available for use. By 2030, China plans to have 80 percent of its cities absorb two-thirds of the rainwater that falls on them. Cities in India, Russia, and the United States are implementing "sponge" projects as well.

Singapore's National Park Board plants more than fifty thousand trees per year. Trees remove nitrogen, sulphur and carbon dioxide. Trees and greenery cool people's temperament by reducing stress. Data show that people who live around more trees experience less mental illness. They also have more robust immune systems. Sadiq Khan, the mayor of London, wants to make half of the capital city completely green by 2050, build four inner-city urban forests, and make London the world's first National Park City. The ability to cool cities and people in a heating world is growing in demand. Between 2012 and 2015, the city of Beijing planted an estimated 54 million trees under the government-led One-Million-Mu (666 ^{km2}) plan. The planting has reduced annual dust storms to the low single digits and has flipped the land from barren, rocky soil to communities with pine and willow trees. Vancouver's urban forestry plan includes strategic tree planting to increase canopy cover in low-income communities and the diversifying of tree species for resilience. Vancouver's plan views the city's urban forest as a living asset.

Italian architect Stefano Boeri designed the first vertical forest, completed in the Porto Nuova district of Milan in 2014, near the Garibaldi railway station. "Bosco Verticale" consists of two residential towers, of nineteen and twenty-seven floors, with eight hundred trees, five thousand shrubs, and fifteen thousand vines and perennials rooted within 8,900 square feet of terraces, giving residents the experience of living in a forest city. The selection of plants was carefully determined by overall sun exposure, including the height of the floors and shading from the adjoining tower, which resulted in a diverse multispecies forest extending 360 feet into the sky. Had the plantings been on flat land, it would be equivalent to five acres of forest. Each residence has a unique microclimate, as the trees and plants create humidity, sequester carbon dioxide, and emit oxygen. And each terrace has a unique microbiotic community. Exposure to diverse air and touchpoints in nature connect us to more microbes, and we know that greater exposure to forests' microbial communities improves our internal microbiome. The trees and plants buffer city noise, and in the quieter spaces of the terraces, twenty varieties of nesting birds can be seen and heard. A mature tree can release a hundred gallons of mist per day into the atmosphere, lowering surrounding air temperature for residents and the neighborhood. Instead of the classic urban heat island, wherein cities experience hotter temperatures than nearby rural areas, Bosco Verticale is an island of cool. It is projected to convert 44,000 pounds of carbon dioxide into oxygen annually.

Innovative city planning is now placing the city within a forest rather than the other way around. Stefano Boeri's Liuzhou Forest City, in the southern province of Guangxi, China, will be the world's first urban forest city. Mounded to imitate the surrounding mountain landscape, it will cover an estimated 437 acres and be home to thirty thousand residents, forty thousand trees, and a million plants. Residential terraces and sides of buildings will be adorned with greenery, offering habitat for the local animal species. It would absorb ten thousand tons of carbon dioxide and fifty-seven tons of microparticle air pollutants each year. The city would be connected to Liuzhou via a highly efficient railway and roads that are reserved exclusively for electric cars.

Boeri campaigns worldwide for urban forestry and sees cities, which are the greatest single source of greenhouse gas emissions, as a resource to address climate change. He wants to greatly multiply the presence of forests and trees in our cities. Boeri is an active part of imagining and designing the Great Geneva project in Switzerland. It is a system of eleven urban centers, including Geneva and Annecy and two lakes, constellated around the central Salève mountain. Their goal is to establish the planet's first biodiversity metropolis.

Perhaps the greatest gift of urban trees and plants applies to learning and appreciation. The surfaces and structures of the city are static, fixed, hard, largely unchanging. Trees and plants are the opposite. They can die for lack of water or bloom gloriously in the spring. They evolve daily. They are living beings, as is the viewer. Leaves, needles, and bracts change and fall to the ground. Trapeze artists—squirrels—fly from tree to tree. It is called nature, and as the world warms and more parts of the living world fail, flood, overheat, decline, or burn, the daily life of a tree and its inhabitants can connect city dwellers to a sense of what is being lost in the world. We know that 40 percent tree canopy cover in a city can reduce temperatures by up to 9°F. That dynamic applies to the entire planet, making global warming and its reversal less abstract and more understandable. An urban childhood can prepare the student for the coming world in a living, vibrant, diverse, green city. If our children are educated in the absence of nature, they hardly know what they see when driven to the country. The words *forest*, *hill*, *river*, and *field* can exhaust their vocabulary of terrain. The unnamed is the unseen. The regeneration of the world is fundamentally about engendering more life, one of the deepest and most profound joys innate to human beings, and that seed can be planted within every child in every garden, riverbank, and parkland. ●

Paradise Park in the morning, Jena, Thuringia, Germany.

Urban
Mobility

Cities were invented six thousand years ago—and no one knows why. After roaming the planet in small bands for millennia, many humans decided to settle down in villages, where they created agriculture, beer, spun clothing, copper tools, and hand-molded pottery. Those early cities faced the challenge of managing mobility—that of pedestrians, wheeled carts, animals carrying goods, livestock moving through the streets to market, and even boats plying canal systems in Venice and Bangkok.

In the early part of the twentieth century, cities began to embrace gas-powered vehicles. The first Model T rolled off the Ford assembly line in 1908, and by 1927, 15 million had been sold. The rise of the automobile changed city infrastructure design and patterns of mobility. America's first freeway, connecting downtown Los Angeles to Pasadena, opened in 1940. A master plan for freeways for all of Southern California was adopted seven years later. A national interstate highway system broke ground in 1956. Globally, the number of vehicles has risen from 340 million in 1976 to about one and a half billion cars and trucks today, and this is projected to reach two billion by 2030. China has overtaken the United States as home to the most cars.

Planners encouraged this newfound passion for freedom and convenience by transforming cities to meet the needs of cars and trucks at the expense of other forms of transport, especially getting around on foot. The transfor-mation comes at a cost. Excessive levels of tailpipe pollution are common in virtually all major cities, provoking a host of health problems, including lung disease. Noise pollution is ubiquitous. Traffic accidents claim nearly a million and a half lives annually. Traffic congestion paralyzes many metropolitan areas. The average American spends about an hour a day in their car, often alone. It is worse in other parts of the world. Commuting does not make people happy. New roads become congested within seven years. Public transportation systems do not always succeed. Cities failed pedestrians and cyclists with inadequate amounts of sidewalks, bike lanes, parks, shade trees, or safe ways to cross busy streets.

Cities and urban areas generate 70 percent of global greenhouse gas emissions annually, roughly one-third of which are from ground transportation. These numbers can be deceiving, however. Most vehicle emissions are linked to the length of commutes. The carbon footprints of spread-out cities are generally higher than densely urban areas. New York City has the smallest carbon footprint per person in the United States, followed by San Francisco. It matters how cities grow and how we get around. Although cities occupy less than 2 percent of the planet's land base, they are home to more than four billion people and will add another two and a half billion by 2050, mostly in Asia and Africa. Tokyo is the world's largest city, at 37 million people, followed by New Delhi (29 million) and Shanghai (26 million). The number of megacities with more than ten million people is projected to reach forty-three by 2030.

To counter these trends, cities are implementing new mobility strategies that get people out of their vehicles and back on their feet, bikes, or electric scooters or aboard a bus, train, or subway. Some cities are banning cars and trucks outright in certain areas, while others are limiting them to specific days. These include Paris, Bogota, Madrid, Dublin, London, Hamburg, Chengdu, and Hyderabad. Fourteen nations are banning the sale of gas- or diesel-powered cars altogether, with Norway setting 2025 as its deadline. For Israel, Germany, Denmark, Ireland, the Netherlands, Slovenia, and Sweden it is 2030. One by one, cities are changing their mobility focus and investing in bicycling infrastructure, pedestrian zones, reduced parking spaces, expanded open spaces, and reinvigorated public transportation systems. Particular attention is being paid to making cities walkable again. Walking is the cheapest, simplest, healthiest, and most renewable form of mobility—and often the most rewarding.

Milan announced it will convert more than twenty miles of streets to pedestrian and bike-only pathways. Oslo is banning parking spaces and incentivizing people to purchase electric bicycles to traverse its hilly terrain. The decision by Ghent, Belgium, in 2017 to close twenty of its busiest streets to cars has been popular. Guangzhou, in southern China, home to fourteen million people, boasts some of the highest levels of walking in the world, a result of the creation of a nature corridor along the Pearl River that links miles of pathways to sporting venues and tourist destinations. Barcelona has been regenerating public spaces since the 1980s, when it began replacing blocks of industrial buildings with parks and other amenities. The city is planning to repurpose many streets to people-centric "superblocks."

In Austria, Leonore Gewessler, the minister for climate, energy, and transportation, is instituting an ingenious plan to reduce emissions and accelerate a car-free society. Residents anywhere in the nation can pay three euros a day for unlimited access to all forms of public transport—buses, trains, and subways throughout the country. Given that the average cost of owning a car in the EU is more than six hundred euros per month, a citizen could save five hundred euros a month.

The car-free movement is merging with rapidly developing new trends in urban mobility, many of them centered on innovative technology. Vehicle electrification—especially as energy sources supplying the electrical grid become cleaner and more decentralized—significantly reduces greenhouse gas emissions and air pollution. Autonomous driving technology could decrease traffic accidents and fatalities as well as reduce the emotional stress of commuting, while providing expanded mobility options for underserved communities. Ride-hailing and sharing services, including fleets of vehicles in cities, could reduce the overall number of private vehicles in use. Public transportation is being reinvigorated with new technology, including the use of autonomously driven buses and data-linked, on-demand deployment systems that are useful for customers and transit authorities alike.

Integration is key to the future of urban mobility, particularly at transportation hubs where cars, trains, buses, and people come together. Smart systems enable different modes of transportation to communicate with one another digitally, creating opportunities for people to chart an efficient course to their destination. While apps and data collection make this integration possible, the goal is to work at scale, which requires a constant demand among citizens for mobility services. For planners, the key step is convincing people to give up their vehicles, particularly for the last leg of their commute into an urban core. Arguments in favor of this new mobility system include money savings, environmental concerns, the improved reliability of public transit, the health advantages of physical exertion, and the attraction of a more pleasant urban experience. For city residents, studies demonstrate a strong correlation between car-free mobility and feelings of neighborliness and a sense of belonging to a community. As a result, people feel more socially engaged and become more politically active. They are also more likely to use public transportation and support the adoption of electric vehicles and renewable energy.

In some ways, this new mobility is a return to the original idea of a city. Athens, Lisbon, Jerusalem, Milan, Hanoi, Beijing, Delhi—these early cities created shared spaces that live on to this day—shared streets, work areas, places of worship, markets, public areas, and sports venues. People mingled with one another, jostling in the alleyways, in the great ebb and flow of crowd-based living. This experience can be re-created today with multiple benefits. By shifting the focus from cars to people, cities can regenerate themselves as vibrant places to live and work. In doing so, mobility planners and city residents are asking questions that have echoed down the centuries: What are our streets for? What is the best way to use communal space? How should we get around? What decisions will make our lives more pleasant and our city more equitable? As our cities grow and become more densely populated in coming years, it will be imperative to create mobility systems that improve lives, are safe and enjoyable, and accomplish shared goals. ●

Electric bikes intended to expand transport options for citizens at affordable prices in Rome. Rome was chosen as the first city in Italy to launch one of the largest sharing-mobility networks.

The Fifteen-Minute City

Imagine a city where everything you needed could be found within a fifteen-minute walk or bike ride from your home, including fresh food, healthcare, schools, offices, shops, parks, gyms, banks, and diverse entertainment. The paths to get there are safe, tree-shaded, car-free neighborhoods where people can get to know each other. When interconnected, neighborhoods revitalize and strengthen community, lower greenhouse gas emissions, and improve livability with clean air and proficient public transit systems.

It's called the 15-minute city, and it's not imaginary. Paris mayor Anne Hidalgo has implemented an ambitious plan to restrict vehicles on streets while increasing options for walking, cycling, and people-first economic development in every part of the city. In 2016, cars were prohibited on a congested stretch of road along the Seine River, freeing it for pedestrians. A massive construction effort is underway, with the goal of completing one thousand kilometers of bike paths in the city, including along major thoroughfares such as the Champs-Élysées. Eventually there will be a bike lane in every street. To make room,

sixty thousand parking spots for private cars are targeted for elimination. During the COVID pandemic, these efforts accelerated. The city has also offered financial support for new businesses in targeted neighborhoods, increased green spaces, encouraged urban agricultural projects, and expanded the use of school buildings beyond standard business hours. It is part of the city's plan to become carbon neutral by 2050, but it is also much more.

The 15-minute city concept embraced by Mayor Hidalgo was developed by Carlos Moreno, a professor at the Sorbonne in Paris, who believes that the daily necessities of life should all be accessible by a short trip on foot, bike, or public transit. His research, which originally focused on reducing greenhouse gas emissions from the transportation sector, led him to a vision of modern cities as mosaics of walkable neighborhoods that meet the domestic, professional, and entertainment needs of residents. A key is mixing as many different activities as possible in an area. He also advocates using schools, libraries, and other spaces for multiple purposes, including during off-hours. When commute times are cut and the need for private cars is reduced, streets are freed up for pedestrians, pulling people out of their homes and into nearby retail and leisure activities. In 2019 alone, Paris reduced vehicle traffic by 8 percent. This has another benefit for residents: cleaner air. Auto pollution contributes to a host of human health maladies, including lung disease, heart conditions, and air pollution that is linked to a reduction in the cognitive function of children. Traffic noise is associated with elevated levels of depression and anxiety. In a 15-minute city, both are significantly reduced.

The movement is worldwide. In 2015, Portland, Oregon, adopted a Climate Action Plan that sets a goal of achieving easy access to basic needs by bicycle or foot for 80 percent of its residents, with an emphasis on low-income neighborhoods. In Spain, Madrid announced plans to transition to a 15-minute city model as part of its postpandemic revival, inspired by Barcelona's superblock system, in which vehicle access is curtailed in favor of public space for pedestrians. Chinese cities are including 15-Minute Community Life Circles in their growth plans, where communities are linked to a vehicle-free center. Melbourne, Australia, is testing a slightly expanded version: Plan Melbourne 2017–2050 aims to create "20-minute neighborhoods" in which people can meet most of their daily needs. Seattle announced in September 2020 that it would consider the 15-minute city concept as a guiding principle for the next version of the city's Comprehensive Plan. In doing so, Seattle joins a global effort called the C40 Mayors Agenda for a Green and Just Recovery, which emphasizes creating 15-minute cities to ensure livable communities.

A key to the 15-minute city concept is the most neglected form of transportation available to humans: walking. While we have relied on our two feet as the primary means of transport for most of the nearly four million years that we've have had the capability to walk upright, the recent predominance of motor vehicles and mass transit in cities has reduced walking to a largely recreational activity. Cities are dominated by streets built to accommodate cars and trucks, optimizing speed and convenience in urban planning and design. Individual vehicle ownership, while offering advantages in mobility, led to the prioritization of parking lots over parks, enabled urban sprawl, and locked in a polluting transportation system.

The 15-minute city, however, means something very different to someone who is blind, disabled, unable to walk, cannot afford to live in a dense urban core, or lives in an area that has limited access to transportation due to income, race, or age-related disparities. Walkability has paradoxically become a privilege, especially in less-affluent neighborhoods. Sidewalks are essential, not optional. So are regular, frequent, fixed-route paratransit services, and not just in dense, wealthy neighborhoods. To be inclusive,

the 15-minute city needs to invest heavily in pedestrian infrastructure that enables people to make safe and reliable transit connections. Parks, greenery, curb ramps, and pedestrian-friendly intersections should be required features of every neighborhood. And cities need to focus on affordability, especially housing. Too often, people have to choose between living close to transit services or living farther away in a home or apartment they can more easily afford. For the 15-minute city, access must be as important as speed and ease of mobility. As cities work to increase equitable access to resources, including markets, stores, and schools, cities are proving that they are more sustainable and can strengthen the bond between people and their communities.

There are common core elements to the 15-minute city:

- It gives residents of all neighborhoods access to essential goods and services, particularly fresh food and healthcare.
- It encourages every neighborhood to include housing of different sizes and levels of affordability (including in former office buildings) and accommodates many types of households, enabling people to live closer to where they work.
- It advances mixed-use retail and office space, co-working opportunities, teleworking, and digitization of certain services, all of which limit the need for travel.
- It spotlights underserved and lower-income neighborhoods for investment and encourages residents and businesses to become involved in measures for improvement.
- It can be tailored to a city's unique culture and conditions, to respond to specific local needs.
- It requires frequent and reliable public transport connections to other neighborhoods.
- It promotes "street-facing" uses at ground level, which help streets thrive.
- It encourages the flexible use of buildings and public space, including buildings designed to be easily converted to different uses.
- It boosts the need for attractive streetscapes, green spaces, and other amenities.

As the age of the car fades, cities are being redesigned to serve their residents, discovering how much healthier, vibrant, and resilient they can be. ●

La Rambla de la Llibertat, in the old town of Girona, Catalonia, Spain.

Carbon Architecture

Carbon architecture is a design movement that replaces the raw materials used in building construction with bio-based materials that sequester carbon. Instead of building with rocks (steel and cement), it makes buildings out of fiber. It employs plant materials that draw carbon dioxide from the atmosphere in order to transform the building industry from a major driver of climate change into a carbon sink. This is construction that cools instead of heats the planet. The population of the earth will increase by 25 percent in the coming three decades, which will require massive amounts of steel and concrete for residences, commerce, and workplaces if they are conventionally built. Carbon architecture can flip cities into carbon sinks instead of carbon sources.

The raw materials used in carbon architecture are primarily wood, dirt (clay), bamboo, straw, and hemp, engineered to compete with steel, cement, brick, and stone in durability, fire resistance, and structural strength. The initial green building movement focused on reducing operating emissions—those produced in heating, cooling, and powering a building. That makes sense, since approximately 29 percent of total U.S. emissions come from buildings. The carbon content required to manufacture steel, glass, coatings, cement, and brick is the embodied carbon, and it was not considered as important until more recently. Today there are thousands of net-zero buildings that consume no more, or even less, than the energy they

---

The twenty-four-story high HoHo Tower complex in Vienna, Austria, is currently the world's tallest timber building. It houses a hotel, apartments, a restaurant, a wellness center, and offices. Most of the building was prefabricated and assembled on-site. The construction system was kept deliberately simple, consisting of stacks of four prefabricated building elements: supports, joists, ceiling panels, and facade elements. About eight hundred wooden columns made of Austrian spruce carry the floors. It is designed to achieve "passive-house" energy efficiency.

produce on-site. Carbon architecture goes further; it creates buildings that sequester carbon before the power is turned on. The goal is buildings constructed of biologically derived material to transform cities, low- and mid-rise buildings that can capture and hold more carbon per acre than does a primary forest. Essentially, it is a transfer of carbon-sequestering materials into the built environment, panel by beam by floor by building—a complete transformation of what we now think of as a city.

Clay has been used in masonry buildings for millennia. Earthen clay bricks, called adobe in New Mexico, are standing after one thousand years in multistory residences in Yemen. Clays contain extremely fine particles that are electrostatically charged, which is why a gummy, sticky medium can be kiln-dried to make durable, watertight ceramics. Clay cannot replace cement in strength, but it can substitute for concrete in other ways, reinforced with mesh or bamboo for flooring, countertops, and bricks.

Think of straw as delicate, small hollow trees that support crops of rice, wheat, rye, oats, barley, and hemp. When these grains and seeds are harvested every year, what remains is carbon stored in tubular stalks. Billions of pounds of straw are produced annually in the world. Architects and material scientists want to transform 2 billion tons of fibrous cellulose into panels, blocks, and insulation. There are many ways to employ straw and hemp as a replacement material; however, building codes and the industry are risk-averse and conservative. Architects, engineers, and contractors are wary of postconstruction lawsuits resulting from component failure and pursue a safe strategy of what can be called "infectious repetitis." If construction is done in the customary way, there will be less risk. The advantage of straw is its abundance and cost. The story is better in Europe. France has been building with hemp since the early 1990s, and is the largest hemp producer in the European Union. In Spain, architect Monika Brümmer has created a significant market for her company, Cannabric, which manufactures bricks, blocks, insulating panels, felt, and boards made entirely of biofiber.

For centuries, the most abundant structural material for buildings has been wood. The 220-foot, 9-story Sakyamuni Pagoda, in Yingxian, China, was built nine hundred years ago and has survived wars, earthquakes, and dynasties. It contains no nails, bolts, ties, or metal—it is held together by hundreds of different joinery techniques.

Wood was used as logs, shaped logs, or sawn timber up until the twentieth century. Steel and concrete took over in an era of cheap fossil fuels and lack of awareness about its long- and short-term impacts. The advantage of steel and concrete is its strength, durability, and uniformity. Engineers could precisely specify materials needed for shear and load-bearing strength. The challenge with steel and concrete was weight. The higher the building, the greater the load and stress on lower floors, which meant greater use of steel to support it. As buildings got taller, material intensity increased exponentially. When steel and concrete were relatively inexpensive, intensity was not a consideration. Today the true cost of steel and concrete far exceed their nominal price: annual carbon dioxide emissions are around 3.7 billion tons and 2.6 billion tons respectively. The mining impact of iron and the theft and ransacking of sand from coastal shores for cement are but two of the direct impacts upon the environment.

In the past two decades, architects and designers, inspired by the possibilities of eliminating steel and concrete in low- and mid-rise buildings, created what is known as the "tall wood" movement. And the movement has taken off. The largest mass timber commercial building in the United States is the Carbon12 building, in Portland, Oregon. The number 12 refers to the atomic number of carbon, not its eight-story height. Completed tall timber and hybrid tall timber buildings can be found in France, Australia, Italy, Sweden, Canada, and the UK. The Mjøstårnet in Brumunddal, Norway, was until recently the tallest mass timber building in the world. It is a 280-foot, 18-story apartment and hotel development. (The two on-site 25-meter swimming pools were built entirely of wood.) Large glued laminated timber (glulam) was used for the internal columns, beams, and cross diagonals because of their pliability and fire resistance. Cross-laminated timber (CLT) was used for inner walls, elevator shafts, balconies, and stairs. The timber was connected with steel plates and dowels. It is imperative that the wood come from certified sustainable forestry practices.

The Brock Commons at the University of British Columbia, an eighteen-story student residence, is the third-tallest mass timber building, at 174 feet. Because mass timber buildings consist of prefabricated components, the Brock Tower structure was completed in less than seventy days. Perkins&Will architects have submitted plans for the eighty-story River Beech Tower in Chicago. Because mass timber buildings are innovative, involve newly designed structural components that are difficult to source in most areas, and are on a learning curve, they are more expensive. Several engineered and designed mass timber buildings have been delayed or canceled due to financing challenges.

The ecological advantage of wood is considerable, as steel and concrete combined are responsible for 12 percent of greenhouse gas emissions. Proponents calculate that the steel and concrete in a building that would emit two thousand tons of carbon dioxide in its manufacture would capture two thousand if built with mass timber. The challenges of withstanding fire is the most prevalent concern from those who have not studied mass timber technology. Fire can be addressed with drywall (gypsum);

however, a Yale study on mass timber building materials states unequivocally that glulam and cross-laminated timber can form a protective, charred layer in intense fire that prevents further combustion. Timber buildings are engineered to take into account how a theoretical fire might partially weaken structural timber. No material is inherently better at withstanding fire exposure. When subjected to high heat, steel becomes plastic and will bend, leading to structural collapse.

Engineered wood is constructed from numerous sources: small trees garnered from forest thinning, sawmill boards that are not commercially useful, plantation wood, dead trees from forest fires that have not begun to decay, and wood recovered from demolition. Smaller pieces of wood are glued together to create timber that has greater strength than a beam from a single large tree and, at this point in time, are far larger than those that could be obtained from temperate forests. Nevertheless, as mass timber buildings grow in popularity, deforestation could result. Thus far, the companies that choose to construct timber buildings want the source of the wood to match the intent of the building. Mass timber buildings have another advantage: they weigh 80 percent less than steel-and-concrete construction. Ninety percent of the weight of a high-rise building is the steel and concrete. Ninety percent of the greenhouse gas emissions from material are the steel and concrete. If and when mass timber buildings become less expensive than steel-and-concrete ones, and the demand for timber threatens to have a damaging impact on intact forest systems, there is a substitute for wood that is even stronger: bamboo.

Slats made from bamboo can be laminated, creating timber, beams, posts, plywood, flooring, and paneling superior to wood in overall strength and durability. And bamboo sequesters carbon far more quickly than fast-growing trees. Carbon offsets are easily identified and measured annually, giving bamboo an economic advantage over wood. And unlike trees, as many a gardener has discovered, bamboo does not die when cut. You can harvest bamboo from the culms indefinitely for decades.

The research and application of bio-based materials is slowed primarily by the inertia of the awareness, knowledge, and regulatory environment, including building codes. As is true for any industry that has long prospered making products in a certain way, there is resistance. But as with food and energy, a transition is underway as architects, engineers, and companies show the way to a world built biologically. ●

Caroline Palfy is the master builder, engineer, and project developer of the new Hoho Vienna. "We keep getting asked whether our timber resources are jeopardized by the current timber boom in the construction industry. In Austria, forests produce thirty million cubic meters of timber a year, of which 26 million cubic meters are logged. The remaining four million cubic meters remain in the forest, continually increasing timber stocks. In other words, one cubic meter of wood grows back every second, and thus the timber used for the entire HoHo Vienna project will have grown back in our country's forests in only one hour and seventeen minutes."

Food

Human beings have sought food for two million years. We migrated from our origins in Africa to Asia, Europe, and the Americas, developing tools, settlements, fire, and an intricate knowledge of plant and animal life.

Italian Cristoforo Colombo set forth across the Atlantic from Palos de la Frontera, Spain, to find a western gateway to India and China, a mission to bring back spices such as ginger, turmeric, nutmeg, pepper, cumin, and cinnamon, used in Europe to enhance flavors and aid digestion. When his three boats made landfall in Hispaniola, they encountered not Asians but the Taíno people, living peacefully on the island we know today as the Dominican Republic and Haiti. Columbus voyaged to the Americas on four occasions, having never found spices except a tree bark he claimed was cinnamon, believing to the end that he had found the western route to India. The word *Indian* is our ongoing testament to his ignorance. What he brought to the Taíno people was disease, pillage, slavery, rape, torture, and near extermination.

What Columbus and the Europeans who followed him discovered was a lush, edible landscape farmed by Indigenous cultures. The early explorers brought back new foods, particularly the potato, which reduced chronic hunger on the European continent. Another important food crop they "discovered" was corn, which is the largest grain crop by weight grown in the world today. Three root vegetables developed by Indigenous people in the Americas—potatoes (3,800 varieties in Peru alone), sweet potatoes (400 varieties), and cassava—are collectively the largest source of calories in the world. When you add cacao, tomatoes, avocados, peppers, cayenne, chiles, peanuts, cashews, sunflower, vanilla, pineapple, papaya, blueberries, strawberries, passion fruit, pecans, butternut squash, pumpkin, zucchini, maple syrup, cranberry, tapioca (from cassava), and several hundred varieties of beans, it is not difficult to concede that Amerindian farmers may have been the leading plant breeders in history.

Most of the world no longer needs to seek food. It comes to us in an extraordinarily complex and sophisticated system that has created unparalleled abundance. However, today's food system has become the single greatest cause of global warming, soil loss, chemical poisoning, chronic disease, rainforest destruction, and dying oceans. Because people love to eat and taste, the food system, for all its myriad ills and assaults, offers an extraordinary opportunity to regenerate soil, climate, community, cultures, and human health. It is the keystone solution, because it either upholds or degrades every aspect of human endeavor and impact addressed here: forests, farms, soil, oceans, cities, water, industry, and energy. The key to regenerating our food system is taste. It sounds absurd, perhaps, but due to commercially produced processed foods, we may have lost our taste and we may need to regain it.

The delicate fronds of taste buds waving on our tongues have been hijacked by the modern food industry. Just as language can be reduced to several hundred words and a few grunts, our nutritional literacy has been largely reduced to four intense flavors: salty, sweet, acidic, fatty—you get them all in french fries, Coke, and a hamburger. Purveyors include Kraft Heinz, PepsiCo, Mondelez, McDonald's, Mars, Nestlé, and other megacompanies known collectively as Big Food. These manufacturers know far more than we do about what happens in our mouths, about olfactory responses and mouthfeel, and how these tastes override the intellect and affect our brain and well-being. It is an outgrowth of a discipline called food chemistry. We became culinary hamsters in the gerbil wheel of supermarkets after World War II, feasting on ultraprocessed, fat-infused desserts, sugary ketchup, denatured white bread, and salty, heart-eroding snacks, exploited by our tastebuds, which evolved over millennia to heal and protect us, not to make us obese and diabetic. And the world followed suit. Junk food confers high status in much of the world. Eating at an American fast-food outlet is considered upscale and affluent. Obesity is epidemic in China, affecting approximately 16 percent of children under eighteen years. Childhood obesity is a near-perfect indicator of the onset of chronic disease and premature death later in life.

The taste buds in our mouths are not playthings to be seduced and manipulated; they are evolution itself, a teacher, a kindness, a guide. That quivering, moist, almost reptilian tongue in your mouth is a close friend and ally, a direct extension of billions of years of knowledge and evolution connected with and signaling to every cell in your body. It is how our body detects toxins; it's the first and most powerful expression of the immune system deciding what should become your body and what should not. It is how we developed into the only form of life that can say we are a form of life—and one that can consciously destroy its habitat and understand that there are biological limits to its desires and appetites. When we eat, we make a choice. When we choose a food, we either improve the world or harm it, honor our body or injure it, sustain the conditions that are conducive to life or degrade them.

This section describes how Big Ag and Big Food have degenerated our land, soil, food, environment, and health, and how regeneration can reverse all five conditions. The connection between soil, climate, and planetary well-being has become clearly evident to community groups, Indigenous nations, farmers (big and small), chefs, activists, nutritionists, restaurants, and NGOs worldwide, all of whom are working to reclaim the integrity and nutrition of authentic food, and create a new food system designed to support life on earth. ●

Wasting Nothing

Nearly one-third of the food produced for humans never makes it to our mouths. Some of it is left behind in the fields after harvest. Some is lost on the journey from farm to retailer because of spoilage in transport, lack of refrigeration, poor handling, or rejection by food companies for imperfections. Some is tossed away during processing. Unsold or uneaten food is discarded by stores, restaurants, and food service companies. At home, most excess food ends up in the garbage. Some food waste is composted, donated, or turned into animal feed, but more than 90 percent goes to the landfill in the United States. Some is incinerated. Meanwhile, 135 million people worldwide struggle daily with acute hunger and food insecurity and 800 million are undernourished. During the COVID pandemic, over 40 million Americans were estimated to have experienced food insecurity.

When we waste food by not eating it, we waste money—more than $200 billion annually in the United States and $1 trillion a year globally. When we waste food, we waste an opportunity to feed those who do not have enough. The nonprofit organization ReFed estimates that 50 million tons of food in the United States—representing nearly 130 billion meals—went unsold or uneaten in 2019. When we waste food, we also squander the resources that went into its production—the labor, transport, processing, packaging, and preparation. Food left on a farm after a harvest can be plowed under or recycled using biodigesters, but when food is deposited and buried in a landfill, it produces methane as it rots. Greenhouse gases generated by total food waste account for 9 percent of total global emissions, and 12 percent if you count landfill emissions. A 60 percent reduction in overall food waste would amount to a 7 percent drop in total greenhouse gas emissions.

Food loss occurs every step of the journey from farm to fork. In the United States, the foods most commonly discarded are grain products, followed by dairy products, then fresh fruits, vegetables, prepared food, and bakery goods. On farms, there is nearly always food left behind after the harvest due to labor shortages, high costs, imperfections, timing, or low prices. In the manufacturing process, food is peeled, shelled, pitted, stemmed, trimmed, and deboned. These by-products can be repurposed, but most are thrown away. At the retail level, customer demand for variety, freshness, and pleasing appearance means edible food is left on shelves. Restaurants and food service providers need fully stocked kitchens, but bulk purchasing and storage can also lead to waste. Portions left uneaten by customers must be discarded. On the home front, buying too much food, throwing it out too early, letting it spoil, and not freezing or composting leftovers are all causes of food loss.

In nations with low per capita income, food loss happens closer to the farm than the fork. In sub-Saharan Africa, more than 80 percent of food wastage happens during harvesting, transportation, storage, and processing, while only 5 percent is caused by consumers. In North America, by contrast, two-thirds of food loss and waste takes place at the consumer stage.

Achieving reduced food waste requires an effort at every step of the supply and consumption chain. At home, solutions include planning and preparing meals carefully; freezing and repurposing leftovers rather than tossing them out; and becoming less intimidated about expiration labels. (We often toss food that's still safe to eat.) At a grocery store, accept imperfect-looking produce for purchase. When eating out, if your eyes are bigger than your stomach, trust your stomach. Clean your plate. In 2020, President Xi Jinping of China announced measures to curb food waste among the nation's 1.4 billion people; under this Clean Plate Campaign, restaurants are urged to limit the number of dishes they serve and diners are encouraged to eat all the food they order. (Traditionally, leaving food on the plate was a sign of respect to the host.) China is cracking down on waste at banquets and official functions with fines. Its bounty of food is seriously impacting health. Obesity tripled in China between 2004 and 2014.

At the front end of the supply chain, solutions include increasing efficiencies and reducing spoilage during harvest, processing, and distribution. A company called Mori has developed natural edible coatings to prevent spoilage. It harnesses the unique attributes of silk to create a universal protective layer that can be applied to whole produce, cut fruits and vegetables, protein, and processed foods. The edible coating is safe, invisible, tasteless, and virtually undetectable, and can be easily absorbed back into the earth. Mori's coating technology increases shelf life significantly by slowing the key mechanisms making food spoil: dehydration, oxidation, and microbial growth. Not only does the protective coating reduce food waste, but it creates less demand for plastic packaging to maintain quality.

A company in Nigeria called ColdHubs produces solar-powered cold-storage systems with walk-in stations that can be installed off-grid almost anywhere, saving large amounts of food. Blockchain can improve transparency along the food supply chain in order to move products more efficiently to their destinations. At the retail level, better software can enhance inventory management, create dynamic pricing, align bulk orders with customer preferences, and alert outlets when a food donation is ready, as the EU organization FairShare and Irish organization Food Cloud are doing. Data analytics companies such as Leanpath can quantify and track the waste generated in the food service industry. Smart systems can help families order the right amount of food for their meals, aid entrepreneurs in developing new products, and get "imperfect" produce into the supply chain available to consumers. Premeasured meal kits reduce waste at home by providing precise amounts of ingredients.

Producers, stores, and consumers should consider donating surplus food to a food bank. The Food and Agriculture Organization of the United Nations (FAO) estimates that 821 million people go hungry every week. The polite term they use is *undernourished*, but regardless of what it is called, it seems to be increasing in many regions. Nearly 151 million children under five suffer from stunted growth. The Global Food Banking Network serves the hungry in nearly sixty nations, helping millions of people each year, including 46 million in the United States alone. By redistributing food, food banks annually prevent an estimated 10.5 million tons of greenhouse gases from being released into the atmosphere.

Another solution is "upcycling" surplus food into new products. Entrepreneurial efforts include creating soups from rescued produce; turning fruits into a powdered-sugar substitute; and using bread to replace malted barley in the beer-brewing process. In Colombia, a project is turning organic waste from the production of cacao beans into flavorings for drinks, sweets, and dietary supplements. Start-up beverage company Wtrmln Wtr creates a popular drink from watermelons rejected for sale for slight imperfections. The Barnana company produces healthy snack food from bruised bananas and plantains left behind on farms. UK-based Rubies in the Rubble upcycles rejected pears, tomatoes, and other produce into ketchups, relishes, and chutneys. Planetarians transforms the seed waste from producing sunflower oil into snack chips that are high in protein and fiber. It's not just food. Veles is a household cleaner made from 97 percent food waste and sold in a recyclable aluminum bottle.

Food waste can be recycled into renewable energy and soil amendments in agriculture. A start-up company in Massachusetts called Vanguard Renewables has placed million-gallon anaerobic digesters on farms in the region. Food waste is hauled to the site and fermented in the digesters, producing methane that can be used on the farm or sold to local energy providers. The fermented liquid, packed with organics, is used on the farm as a natural fertilizer, replacing chemicals derived from fossil fuels. Vanguard works with companies in the food supply chain, including Unilever, Starbucks, and Dairy Farmers of America, to redirect their waste stream to digesters and convert it to renewable energy, creating a pathway for greenhouse gas reduction.

Food is too precious to waste. Everyone around the world needs to value each bite for its flavor, its nutrition, its traditions, its story, and its impact on the natural world. ●

Eating Everything

How many edible plants do we actually eat? You might be surprised at the answer. Of the four hundred thousand species of plants on earth, two hundred have been widely domesticated. Three species—rice, wheat, and corn—provide 43 percent of the calories we consume as food or provide as feed to animals. Reconfiguring our diets to eat less of these staples and much more of everything else would have a big impact on our health, the natural world, and climate change.

How many plants *could* we eat? Bruce French knows the answer: thirty-one thousand and counting. He has spent fifty years creating a database of every edible plant in the world. It began during a teaching job in New Guinea, where French, an agricultural specialist from Tasmania, faced objections from students who wanted to know more about native edible plants rather than the Western ones he was promoting. Except French didn't know any. Investigating, he soon discovered that many native plants, both wild and cultivated, were more nutritious than introduced ones. They were abundant, too, though neglected. French quickly recognized the value of native plants for fighting malnutrition and decided to build his database to focus on five nutrients often lacking in global diets: protein, iron, zinc, and vitamins A and C. He also included information on plant origins, growing methods, and cooking tips.

His goal is to answer a critical question: What plants grow best where people live and help meet their nutritional needs? His answers form a long list of eating possibilities. There are more than 561 species of edible seaweed around the world, 387 types of ferns, 275 edible bamboo species, and 2,050 mushroom species. You can eat pepino dulce, a sweet cucumber-like fruit grown in South America; sprouting galangal, a root similar to ginger, found in Southeast Asia; rooibos, a medicinal plant from South Africa whose leaves can be made into tea; the roots of the yacón plant, a member of the daisy family that grows in the Andes; and lo han kuo (monk fruit), a fruit native to China that is much sweeter than sugar and used to treat diabetes. The list tickles the imagination: golden bootleg, creeping waxberry, lady's smock, bladderwrack, Persian catmint, Caesar's mushroom, joyweed, sand food, hungry rice, and cartwheels. Whether wild or cultivated, eating extends to all edible parts of a plant, including leaves, stems, flowers, fruit, inner bark, roots, oil, pollen, seeds, sap, and shoots.

Diversity is good for our health. As Dr. Mark Hyman points out, food is medicine. What we eat affects every aspect of what goes on in the vast ecosystem of interlinked functions inside our bodies. Consuming too much of the wrong types of food, such as refined sugar, starch, or highly processed substances, damages this ecosystem, causing diabetes, heart disease, and organ failure. Seeking cures, we turn to doctors and the pharmaceutical industry. Hyman believes modern medicine spends very little of its time correcting our poor food choices. Instead, we could change what we eat and repair our damaged ecosystems by putting proteins, fiber, vitamins, fats, and other nutrients to work restoring and maintaining our health. The microbes in our gut play a huge role in our well-being. Bad bugs thrive on sugar and starch. Good bugs love fiber, vegetables, whole grains, and fermented foods, such as sauerkraut. Food can provide essential minerals critical for enzyme function. The list goes on: omega-3s, polyphenols, phytonutrients, antioxidants, and much more. They improve our immune systems, supply energy to our cells, and help remove toxic substances from our bodies. The key is to replace highly processed substances with diverse, plant-based foods. They include laver seaweed (better known as nori), which is highly nutritious; black turtle beans, a "superfood" legume popular in Latin America; cowpeas, a nutritious, drought-tolerant crop native to Africa, used to make flour and stews; fonio, an ancient cereal from Africa, similar to couscous; and ube, a fast-growing purple yam from the Philippines. Other excellent foods include quinoa, spelt, lentils, wild rice, okra, spinach, spices, teas, and pumpkin, flax, and hemp seeds.

A remarkable study led by Eric Toensmeier of Yale analyzed the potential of a neglected and little researched class of foods known as perennial vegetables. These are crops that can be harvested year after year without reseeding, including herbs, bushes, vines, trees, cacti, palms, and other woody plants. There are more than six hundred types of cultivated perennial vegetables in the world, representing more than a third of all vegetable species and occupying 6 percent of global cropland. Some are well known, including table olives, asparagus, rhubarb, and globe artichokes. Many produce a crop during seasons when annuals cannot, and these include edible leaves. They can grow in conditions ill-suited for most vegetable production, such as desert, aquatic, and shade-dominated environments. More than a third of perennial vegetables are woody species that are well suited for agroforestry systems, especially on marginal land or depleted soils. They are a source of diverse nutrients. Expanded cultivation, particularly woody species, would sequester anywhere from 23 billion to a staggering 280 billion tons of greenhouse gases by 2050.

Eating everything protects the habitats of wildlife. There has been a 60 percent decline in wildlife populations since 1970, driven by habitat destruction for expanded agricultural activity, especially for soy, wheat, rice, and corn. Growing the same crops on the same land year after year drains nutrients from the soil, often leading to intensive use of fertilizers and pesticides that can hurt wildlife and damage the environment. Animal agriculture has a significant impact on wildlife habitat, particularly when land is cleared of native vegetation. Shifting global eating patterns to diets that are primarily plant-based will help alleviate this pressure.

Diversifying our food is a social justice issue. The damaging effects of our food system falls hardest on communities of color. Diabetes, heart disease, and cancer are amplified by poverty and a lack of access to affordable, nutritious food. In the United States, the rate of food insecurity for households headed by Blacks is double that of Whites, while one in five Latinx are food insecure, as compared with one in eight Americans overall. Native Americans are 17 percent more likely to be obese than Whites, and the rate of diabetes is higher among Blacks and Latinx. In South Dakota, the average life span of Native Americans is twenty-three years shorter than of Whites. Historically, Native Americans ate a highly diverse diet that included fish, wild game, herbs, fruits, beans, squash, maize, wild rice, tubers, and bread made from nutritious grasses. After their forced settlement cut off access to these foods, malnutrition became rampant. The U.S. government responded with an involuntary food program based on surplus commodity products high in fat and calories. Soon, the problem of malnutrition became the problem of obesity.

The legacy of poverty, discrimination, and cultural oppression has largely denied people of color access to healthy food and the wider variety of choices available to Whites. In response, a global food justice movement is working to correct systemic inequities and remove barriers to diverse food. There has been a revival of Indigenous foods and traditional cooking methods, led by Native farmers, educators, entrepreneurs, and social-media-savvy young chefs. It adds up to an emerging Native cuisine founded on nutritionally dense, regionally appropriate plants, animals, and fish that can seem exotic to many people. "I call this food 'ironically foreign,'" said Dana Thompson, co-owner of the Sioux Chef, an Indigenous culinary team, "because it's the food that grows right under our feet and is everywhere around us." ●

(Left to right, top to bottom) Waxberries, Pepino melons, Golden Bootleg mushrooms, and Yacon root.

Localization

What we eat and how it is produced has a profound impact on climate. When you drive a car, you know you are emitting greenhouse gases. In many cases, the bags of groceries in the backseat will have a greater climate impact than the car trip to and from the store. Recent studies show that 34 percent of total greenhouse gas emissions are caused by the food system. This includes production, transport, processing, packaging, storage, retail, consumption, and waste.

Localization of food is the step-by-step process of reestablishing regional growing and production of nutritious, authentic food for family, friends, and communities. People choose to localize food sources for multiple reasons: it addresses human health, childhood disease, agricultural pollution, right livelihood, social justice, soil erosion, malnutrition, urban food apartheid, and cultural and biological diversity. There may be no other single activity that encompasses a greater range of goodness for life, health, water, children, and the planet.

For most of human existence, people ate what they could hunt, gather, grow, or obtain by trade. Until the advent of railroads, agriculture remained largely local. However, as transport systems including long-haul trucking opened up distant markets, it made economic sense to grow grains—wheat, corn, barley, and rye—where they were most suited to soil and clime. With the development of refrigeration, fruits and vegetables followed suit. Food became a commodity, and lower costs trumped locale. Time-tested relationships between people and their food broke down, until they became largely vestigial. The economics of wheat, corn, soy, and vegetable oils such as canola favor large industrial farms. This was not only the beginning of Big Ag; it was also the birth of Big Food, an entirely new industrial food system that had never before existed.

Big Food is another word for mass-produced animal foods and ultraprocessed food-like concoctions made of soy, corn, fats, sugar, salt, chemicals, and starch. Also known as junk food, it makes up 60 percent of diets in the United States, and 54 percent in the UK. A more polite term is "food that does not nourish." Inadequate nutrition and disease are inseparable. Because of the ubiquity, addictiveness, and constant promotion of industrialized food, nearly 75 percent of Americans are obese or overweight, and one-third of Americans are either prediabetic or have type 2 diabetes. Obesity often leads to heart disease, cancer, diabetes, hypertension, dementia, arthritis, and more.

In the United States, a majority of people do not have access to good food or cannot afford it if they do. The American diet consists of big brands, big bread, big beer, big beef, big bacon, big cereal, big milk, big potato, big soda, and big corn, because it is heavily advertised and seemingly inexpensive. During the pandemic of 2020, the broader industrial food system broke down. American cattle ranchers reported losses of $13 billion. Dairy farmers poured hundreds of thousands of gallons of milk down the drain. The alternative to Big Food is ubiquitous food—localization of production, instead of centralization.

The reimagination of the industrial food system is manifesting itself in surprising and brilliant ways. There is

not just farm-to-table; there is also pier-to-plate. With their distributors and restaurant customers closed, fisherfolk cook their catch on the dock for consumption on the spot or to take home. Ranchers are delivering meat directly to consumers, as are fisherfolk. Farmers sell shares or subscriptions to their produce output and deliver it weekly. Instead of big farms growing one crop far from where you live, these are small farms growing many crops. In these CSAs—short for community-supported agriculture—the fruits and vegetables are often organic, ultrafresh instead of ultraprocessed, varying with the season, creating a relationship between a suburban or urban family and a specific farm family. Subscriptions provide a steady cash flow to the farmer at a higher premium than wholesaling to distributors or restaurants. With approximately ten thousand CSAs in the United States today, many are starting to add neighboring producers to their weekly deliveries, including eggs, bread, cheese, flowers, jams, and farm-fresh chickens.

Paradoxically, one community that is relocalizing its food is farmers in rural areas who grow corn and soy, primarily for beef and dairy. Most farmers who live in rural America are unable to get healthy, fresh food and produce. Seed grower and agronomist Keith Berns, who farms in Nebraska, created what he calls the Milpa Garden, a seed mix whereby farmers can create abundant supplies of fresh vegetables, beans, herbs, and fruit for their families and communities. Farmers pack their grain drill with a vegetarian grocery list of seeds: squash, beans, cabbage, broccoli, leafy greens, peas, sunflowers, cucumbers, herbs, tomatoes, radishes, okra, watermelon, cantaloupe, sweet

corn, and other edibles. The seeds are cross-drilled on a one-acre plot, creating dense plantings that crowd out the weeds. The flowering species attract insects that control pests, and plant density protects soil moisture. They have been called chaos gardens, but Berns, who is a regenerative, no-till farmer, prefers the term "milpa garden." Milpa means "cultivated field" in the classical Nahuatl language, still spoken in parts of Mexico; it's a term Berns learned from Charles Mann's book 1491. Berns based the milpa mix on the three-thousand-year-old Three Sisters method practiced by Mesoamerican farmers, who sowed (and still sow) corn, beans, and squash together, a polyculture methodology that spread north and was practiced by Native Americans before contact. The Milpa Garden mix is

(*Left*) Taylor and Jake Mendell in their greenhouse at Footprint Farm in Starksboro, Vermont. Founded in 2013, the Mendells raise organic vegetables on 1.5 acres of land. They serve 150 community-supported agriculture members, in this picture, with lots of carrots. Their crew includes their eight-year-old dog, Spud, and the newest member is one-year-old Baby Theo. They are avid members and supporters of the National Young Farmers Coalition, which helps young farmers across the United States learn from their peers, get land access, and even achieve student loan forgiveness.

(*Above*) Ron Finley is known as the gangsta gardener in Los Angeles, famed for planting gardens wherever he can find open land, including traffic medians, curbsides, and unused city property, in order to transform South Central LA's "food prison" into a fruit and veggie oasis. He is also famed for his summary of what fast food is doing to the Black community where he lives: "Drive-throughs are killing more people than the drive-bys."

closer to twenty sisters. After several months of harvesting—which is similar to a treasure hunt for the invited 4-H members, neighbors, food banks, and townspeople scrambling and discovering in the garden—the farmer can turn the field over to grazing animals that thrive on what remains. Some farmers purchase the milpa seed mix and sell the produce through farmers' markets, local grocers, and roadside stands. Tom Cannon, who farms in Oklahoma, is aiming to grow twenty to thirty acres per year, planted within a corn maze where people can forage, wander, and discover. The Cannons farm and live in the country, but Tom and his daughter Reagan were surprised that many of their customers no longer know how to prepare or preserve fresh food, so they are planning to provide recipes and cooking, pickling, and canning classes.

Hunger and food insecurity exist in rural America just as much as in urban America, and Keith Berns has dreams. He would like to see commodity farmers set aside 1 percent of their land for milpa gardens, which would be approximately two million acres spread across the country. It would increase U.S. vegetable production by 50 percent. Berns offers one acre of seed for free if the harvest goes to food banks, churches, homeless or women's shelters, or people in need. Physician Daphne Miller, who writes about the connection between health, culture, and agriculture, sees a change arising in the no-till, regenerative farming community. There is a newfound sense of purpose that transcends their commodity crops—the idea that regeneration means providing direct nourishment to families, neighbors, and community, something their soybean and corn harvests could never do. For farmer Tom Cannon, it is a paradigmatic change: "For years, I was trying to grow bigger. Now the challenge is to grow smaller and more local."

We are "really good at processing cheap food and selling uniform waxed apples at the same place we buy toilet paper," according to Julia Niiro, founder of the company MilkRun, but we are not so good at delivering tasty, sustainable food grown on local farms. She is bringing back the milkman in order to save family farms. Traditional milkmen delivered milk, butter, and eggs to people daily in horse-drawn carts starting in 1860 in Britain, a practice that spread to countries around the world. In America, milkmen delivered 30 percent of the nation's milk right up until the 1960s. Milk was bottled in glass that was constantly recycled, wooden cabinets were built into exterior walls for deliveries of milk and groceries, and money was left inside on an honor system. Cars, supermarkets, refrigeration, milk cartons, and suburbs killed the tradition, but Niiro thinks it is time to bring it back—only this time it will be milkwomen and men delivering nearby rural bounty. Cities and local farmers are only tenuously connected, except through farmers' markets. MilkRun is connecting over a hundred local farmers with thousands of Portlanders. It is able to provide six to seven times what they would make selling to grocery stores. Smaller farmers need money, but they can't raise their prices much. Creating ways to reconnect farms, farmers, cooks, schools, and people is what the new food system is about.

In California, a law passed that allows home chefs to prepare hot home-cooked meals, deliver them, have them picked up, or even offer them to the public in their dining rooms. There are now apps that connect people to providers in networks that offer more choices for eaters and more customers for cooks. People are able to stay at home and earn needed income, an economic opportunity that never existed for millions of truly gifted cooks, people who tend to emphasize recipes that are cultural treasures and family traditions. In California, famed chef Alice Waters pioneered farm-to-table in her restaurant in the 1970s, and the Edible Schoolyard Project in the 1990s. Students garden, grow, and prepare food for one another at school. Her latest project aims to localize farm-to-school programs—local organic farmers selling and delivering directly to local school cafeterias.

Localization is more than a health and climate issue; it can be delicious social justice. For Black and Brown communities, the term *food desert* is a White word for food

apartheid, a system where people of color have no agency in their food system—a kind of urban plantation served by liquor and convenience stores that sell bread and Twinkies, with no suitable food within walking or carrying distance. The struggle for food sovereignty goes back to the beginnings of America, when Indigenous and enslaved cultures (African Americans came from a continent that contained more than three thousand distinct cultures) were uprooted and deracinated from their lands, foodways, hunting grounds, and farms.

Virtually every person in America knows of the Mashpee Wampanoag Nation in Massachusetts, even if not by name. They were the people who shared their food and harvest with the hungry, disoriented, and confused Pilgrims in 1620. Yearly, we eat a meal that is derivative of their foodways, including yams, turkey, cranberry, beans, and pumpkin. Four hundred years after that mythologized Thanksgiving, in 2020 the Bureau of Indian Affairs told the Mashpee Wampanoag people that they were being "de-established" as a reservation, which meant they would have to pay punitive amounts of back taxes on the paltry 321 acres of their ancestral lands that remained to them. Their original lands had extended over hundreds of square miles in Rhode Island and Massachusetts that had been inhabited for twelve thousand years. Although the decision was overturned on a technicality, their status remains

uncertain. The lands they hold are crucial to food sovereignty—to their ability to feed themselves from their farms and traditional waters, where they harvest shellfish, crabs, and fish. For them and hundreds of cultures around the world, food sovereignty is cultural sovereignty, and cultural sovereignty is rooted in a deep understanding of the relationship between people and the land and its waters. It is the opposite of Big Ag. It is culturally sensitive, because cultures die or thrive depending on the sanctity, health, and integrity of ecosystems.

The hundreds of thousands of people in communities across the United States who are relocalizing access to healthy, clean food are not under the illusion that urban or community gardens are sufficient unto themselves. There is a grassroots movement building that wants to literally overturn the current food system, and this can be accomplished only through policy and law. It means ridding schools of fast-food chains and soda vending machines. It means a free lunch (at least) for every child in America. It means removing the inordinate influence of Big Food on school lunch programs, food stamps, and menus. If you live in a food system where most of the food sold makes most of the people sick and dependent on pharmaceuticals—a food system that benefits almost no one except distant shareholders of large companies—it is a system ripe for systemic change. If it is a food system that comes apart in a pandemic, then it is a brittle system that reveals the lack of food security, and that is what localization and food sovereignty address. Whatever you call it, and however localization continues to grow, morph, and permeate cities, towns, and communities, it has a profoundly beneficial impact on the greatest threat to food security—runaway global warming. Acts of localization regenerate the environment, water, children, oceans, soil, and culture. ●

(*Left*) Dave Chapman in his greenhouse at Long Wind Farm in Thetford, Vermont. Long Wind tomatoes are eaten with pleasure across New England early and late in the season. Dave is also the co-director of the Real Organic Project, a farmer-led movement that is trying to restore the original definition, purpose, and meaning of "organic," a term that has been greatly diluted and weakened by the influence of Big Food companies on the USDA. The Real Organic movement seeks to reconnect eaters and farmers around real food grown with love and passion. Soil grown, pasture raised.

(*Above*) Jamila Norman is an internationally recognized food activist and urban farmer. A University of Georgia graduate in environmental engineering, Jamila started the 1.2-acre Patchwork City Farms in 2010 in the Oakland City neighborhood of Atlanta. The certified organic farm features fresh vegetables, fruits, herbs, and flowers sold directly at her seasonal farm shop and through farmers' markets. In 2014 she served as a U.S. delegate to Slow Foods Terra Madre Salone de Gusto in Turin, Italy. She is a founding member of the South West Atlanta Growers Cooperative formed in 2010 to support local Black farmers and to create a culturally responsible food system for Atlanta.

Decommodification

When fourth-generation farmer Jonathan Cobb took over his family's twenty-five-hundred-acre farm north of Austin, Texas, he felt like the operation had become a cog in a large machine that produced crops like widgets for an industrialized food system that didn't care about the land or his family. He knew it hadn't always been this way. Back in the 1930s, the Cobbs grew diverse crops with minimal chemical inputs. After the war, however, the family adopted a new agricultural system of growing a single crop on the same parcel of land year after year, employing pesticides, insecticides, and synthetic fertilizer. The wheat the Cobbs grew disappeared into a local grain elevator, where it was blended with wheat from multiple nearby farms and stored until it could be sold and moved along the supply chain to a food company.

For Jonathan Cobb and his wife, growing a commodity meant a life lived on the margins. Crop yields had not kept pace with the rising costs of chemical inputs, and the price for their wheat was set by the commodity market, not by farmers. Supporting two families on one farm became nearly impossible. When a bad drought struck in 2011, the farm suddenly faced failure. Cobb had a hard

talk with his father. Rather than give up, they decided to break free from the industrial model and give regenerative agriculture a try instead. The Cobbs sold their plowing equipment and switched to no-till methods to grow their wheat. They planted cover crops and began raising cattle, hogs, sheep, and chickens. They converted worn-out cropland into perennial pastures, focused on reestablishing native grasses, and grazed their livestock with an adaptive, multipaddock management system. Within just a few years, the organic content of the farm's soils began to rise from previously depleted levels. The family direct-marketed their grass-finished beef, lamb, pork, and eggs to area customers. Economic and ecological health grew along with the bunchgrasses. For Jonathan Cobb, a sense of purpose returned to his life and farming became a source of enjoyment again.

Commodified food production is a positive feedback loop reinforcing practices that cause harm to people and the planet. The heavy use of machinery and reliance on a few plants grown as monocrops require increasing amounts of pesticides, insecticides, synthetic fertilizers, and genetically modified organisms to offset declining fertility, rising weed resistance, and eroding topsoil. The soil becomes a medium for holding chemical-fed plants upright, not a living reservoir of biological activity.

Farmers and customers have looked for ways to break out of the commodity system and support agriculture that expresses their values, such as pesticide-free products and regenerative land use. Take coffee, for example. It was a generic bulk commodity for decades, but today a great deal of coffee is marketed as fair-trade, shade-grown, rainforest-friendly, organic, and with named varietals—values that have become important to customers, thus encouraging a marketplace of growers, wholesalers, roasters, and retailers. Beer is another example. Craft beer was a cottage industry, largely localized, until a surge of interest during the 1990s expanded the marketplace for alternatives to mass-produced brews. The decommodification of beer created vigorous economic activity: a proliferation of craft breweries (five thousand) and brewpubs (three thousand) in the United States, accounting for one hundred fifty thousand jobs and over $80 billion in sales. Craft beer has an authentic story to go along with a more flavorful and diverse selection of product. By shortening the supply chain from the grower to the customer and cutting out intermediary steps and middlemen, breweries can deliver an original and meaningful story, reflected in thousands of original brand names, including Moose Drool, Hoppy Dreams, Naked Pig Pale Ale, For Richer or Porter, and He'Brew ("the Chosen Beer"). Stories, taste, and local involvement create community, conviviality, and connection.

To scale the decommodification of our food system, companies are establishing digital marketplaces that directly connect farmers and buyers. Their goal is to communicate to consumers the quality of the farm product, where it originated, and what practices were used to grow the food—the story behind the product. They work to match the qualities and values that customers seek from their food with farmers who desire to be paid for those same qualities, often at a premium price. However, the premium the buyer pays can be less than what she might have paid when buying through middlemen. Individuality and variability in crops, farms, and practices are rewarded in this decommodified system, which encourages everything on a farm to improve—soil, plants, animals, people.

One example is Indigo Ag, a company that provides extensive data collection and analysis on pricing, storage, transportation, carbon sequestration, and sale opportunities through its digital platform. Indigo began its business by studying microbes inside the tissues of crop plants in order to determine what enabled an individual plant to grow more successfully in a given climate or weather a bad drought and environmental stress. Indigo researchers discovered that each plant has a distinct set of microbes. Further investigation revealed a tremendous amount of variation from field to field, farm to farm, and practice to practice, all of which impacted the physical properties of each crop. Uniqueness, not uniformity, was the key to a crop's resilience and productivity. Diversity and specialization were good things—after all, plants in nature have been doing exactly that for millions of years. The commodity system, however, required that a food's individuality be removed so it could be piled like gravel on a train and sent away to be processed. But why would we want to eat something that was produced like gravel? That led Indigo Ag to ask the question: Why do we even have a commodity system anymore?

To create an alternative, the company started Indigo Marketplace, which connects grain farmers with buyers digitally. The unique values of the farm are matched to markets that desire them—sequestering carbon in soils as a solution to climate change, for instance. When a farm enrolls in Indigo's program, the company sends out experts to take soil samples in order to determine the baseline carbon content of the land. As the quantity of carbon rises over time in response to regenerative practices, the farmer will earn credits equal to the amount of carbon increased. Regenerative agriculture creates higher-value crops in varieties, taste, and mineral content. Companies no longer buy a commodity; they get a story, a narrative of people, place, and history that makes their brand unique. In 2019, Indigo launched a transportation division that digitally connects farmers, truckers, and buyers, eliminating the commodity pricing of Cargill and their distant grain silos.

Decommodification represents a sea change in our food system. It means growing food as an individualized product. Every farm and ranch has a unique story to tell. As Jonathan Cobb and his family discovered, every decision a farmer makes, the goals he or she sets out to accomplish, the seed they choose to use, the conditions of the land and soil under their care, and the traits of the plants they grow all become part of the story. Companies like Indigo Ag can connect these farms and their stories with markets that value them, benefiting both. A harmful commodity cycle started by nineteenth-century railroads is being broken and disrupted by transparency and traceability and will not be put back together again. Consumer preferences will remake modern agriculture and redirect it toward commonly shared goals, such as addressing climate, soil health, and human well-being. ●

Two bulk cargo ships wait to be off-loaded at the import and processing facility of East Ocean Oils & Grains Industries near Zhangjiagang, just above the mouth of the Yangtze River. The ships came from Brazil and Argentina loaded with soybeans, which is 85 percent of the feedstuffs off-loaded here. The soybeans will be stored here until they are converted into soybean oil and roughage for animal fodder.

Insect
Extinction

It is hard to love ants. They invade our kitchens, infest our homes, swarm gardens, and ruin picnics. They can bite and sting, causing a painful reaction. They aren't pretty. Their unlovability, however, masks an important reality: humans could not survive without ants and other insects. Literally.

Insects appeared on earth more than 400 million years ago, roughly the same time as the first land plants.

This wasn't a coincidence. Insects coevolved with many life forms and are essential to many important ecosystem functions. Take ants, for instance. Comprising more than fourteen thousand species spread across nearly every part of the planet and outnumbering humans by 1.3 million times, ants provide a variety of critical services that we rarely see or appreciate. They are highly efficient predators and can keep pest populations low. By tunneling, ants can move as much dirt as earthworms. Loosening and aerating the soil, they improve its water-holding capacity, which can be especially useful in degraded and dry land. Ants redistribute plants seeds and nutrients, often bringing them back to their nests, which can increase soil fertility as well as help new plants get established. They clean up the dead and aid in the decomposition of organic matter.

All of these activities have a positive impact on food webs and can lead to an increase in the density and diversity of other animal groups. Ants are also useful indicators of ecosystem health. Their presence in a farm field, for instance, that has been recently retired from intensive chemical agriculture is an early sign that the land is recovering. By altering soil structure, food sources, and seed dispersal, ants can help rehabilitate damaged land by creating the conditions for the arrival of other species in all types of ecosystems. Like humans, ants are highly social and well organized, which may explain their success. They live in complex societies, care for their young, and specialize their tasks. They are selfless, efficient, loyal, obedient, industrious, territorial, tribal, competitive, and omnivorous. They take risks, occasionally take slaves, serve royalty—and love a picnic. They are even mentioned in the Bible: *"Go to the ant, thou sluggard; consider her ways, and be wise"* (Proverbs 6:6).

It is advice that falls on deaf ears today. In general, insects garner more hostility than respect from humans, even though, in the words of biologist E. O. Wilson, they are among "the little things that run the world." Of course, some species do garner our affection. Butterflies are easy on the eye, dragonflies have magical qualities, and ladybugs are adorable. In recent years, we have begun to appreciate the importance of honeybees and other pollinators to our food systems. It is estimated that one out of every three bites of food we take comes from pollinated plants, constituting as much as 50 percent of a typical produce section in a grocery store. They include carrots, kale, lemons, mangoes, apples, broccoli, celery, cherries, avocados, cantaloupes, squash, strawberries, sunflowers, almonds, pears, and many more. In fact, more than 85 percent of all flowering plants on earth—more than three hundred thousand species—require pollinators. They include clover and alfalfa, crops vital to dairy cattle and thus to the production of milk, cheese, and yogurt. Then there's chocolate, which is derived from the seeds of the cacao tree, whose flowers are almost exclusively pollinated only by a family of tiny flies called biting midges. And don't forget honey, made by bees. Insects are even a food group themselves: think beetles, crickets, caterpillars, ants, cicadas, grasshoppers, and wasps, estimated to form part of the traditional diet of at least two billion people.

There are approximately 5.5 million species of insects on earth, totaling 80 percent of all animal life, of which only one million species have been formally described by scientists. Like ants, many insects influence the ecology of their home ground by recycling organic matter, moving and mixing soils, feeding on pests, and eliminating dead creatures. They are also a vital food source for birds and other animals. Some insects act as vectors for infectious diseases, such as mosquitoes, or damage crops. All insects, however, are a critical part of the tree of life, forming extensive and often mutualistic networks of relationships throughout the plant and animal kingdoms. Since the start of the Agricultural Revolution, the destinies of insects and humans have been closely bound together through our shared dependence on plants. The expansion of agriculture and the rise of civilizations required insects. No food, no progress. Today, insects are an essential part of agroecological solutions to the climate crisis, as influencers of carbon and nutrient cycles above ground and carbon sequestration in the soil.

For all the good things insects do, however, they are in grave peril. According to a report from the Intergovernmental Science-Policy Platform on Biodiversity and Ecosystem Services, of the one million species around the planet facing extinction within fifty years, half of them are insects. Although insect populations have been on the decline since the start of the Industrial Revolution, losses have accelerated alarmingly in recent years. There has been a 33 percent reduction of butterfly abundance over a twenty-year period in Ohio, a 46 percent loss of moths in Scotland during four decades of survey, and a 77 percent reduction of flying insects in only twenty-seven years in German study plots. A major scientific review in 2019 concluded that almost half of the world's insect species are declining, and one-third are threatened with extinction. Among the hardest hit are butterflies, bees, beetles, and tropical ants. These numbers are almost certainly underestimates, since the vast majority of insect species are understudied, undescribed, and often overlooked by researchers.

The causes of this dire situation lie squarely at our feet. Unchecked economic activity is driving rapid habitat loss, land degradation, and ecosystem fragmentation. Insects are harmed by air and water pollution, biocides, the direct and indirect effects of toxic substances, the spread of invasive species, global warming, overexploitation, and the coextinction of animal and plant species. Deforestation, agricultural expansion, and urbanization are among the principal drivers of natural habitat loss and fragmentation, directly impacting insect populations— activities that are expected to accelerate in coming decades. The loss of connectivity between suitable habitats can

Sweat bee (*Augochlora pura*) covered with pollen.

isolate many insects. Under climate change, the expected ecological shifts in land and water habitats will stress niche-dependent insect species. Those unable to adapt or migrate will likely die.

A major culprit in the accelerating insect crisis is pesticides, particularly as they are used intensively in industrial agriculture. Pesticide use has risen steadily year after year, reaching nine billion pounds in use annually worldwide, and nearly one billion pounds per year in the United States. Scientific documentation on the deleterious effects of chemical biocides is extensive and generally well publicized, a pattern set by Rachel Carson's famous clarion call *Silent Spring*, published in 1962, and amplified a decade later by the successful campaign to ban the insecticide DDT. Pesticides adversely impact insects directly, by killing them, and indirectly, by altering their habitats, whether by eliminating food sources or damaging the networks of ecological relationships that insects need. Toxic accumulation, called "low-dose effects," over time can also pose a significant threat to insect populations by disrupting their ability to reproduce, damaging their immune systems, and disrupting growth. Other types of pollution that cause physical harm or interfere with the natural behavior of insects involve synthetic fertilizers, industrial chemicals released from factories and mines, and light, noise, and electromagnetic disturbances, whose disorienting effects are poorly understood by scientists.

Other threats that can be laid at our feet are the introduction of invasive species, insect and otherwise, at the cost of ecologies and economies. Often the impact on insects is direct predation or competition for food resources. Ants and wasps can aggressively displace their rivals among natives and disrupt local environments with their behavior. Red fire ants, for instance, which are native to South America, were accidently introduced to Alabama in the 1930s and quickly spread throughout the South, damaging agricultural crops and stinging millions of people every year. While some invasive species are considered beneficial—the treasured honeybee originated in northern Africa or the Middle East—most are viewed as deleterious and are known to play an important role in driving local species to extinction. They also disrupt economies. There are a host of insects that kill timber-producing trees, including bark beetles, tent caterpillars, and borers of various stripes, with as many as two new species arriving in U.S. forests from foreign shores every year, literally wasting billions of dollars of valuable resources. The most vulnerable insects are those that have a highly specialized relationship with other species (especially if they are parasites), increasing their already high risk of extinction.

Compounding the invasion crisis are changes to ecosystems caused by climate change. It's not simply a matter of record-setting temperatures, prolonged droughts, or extreme flooding events. Shifts in rainfall patterns over the years, gradually rising average temperatures, lower soil moisture levels, and the cumulative diminishment of snowpacks all contribute to long-term changes in habitats that can directly impact the behavior of insects as well as their ability to adapt to new conditions. To their advantage, the short life cycles of insects mean they can respond to changes in their environment more quickly than other animal groups, but this is often counterbalanced by the reduction or loss of coevolved plant species as the planet warms. Insects that depend on freshwater habitats are especially challenged by climate change as water resources are increasingly stressed. Here again, insects are useful indicators of change. Dragonflies are highly sensitive to climate variations, for example, and sometimes referred to as "climate canaries."

What can we do? Work to protect natural areas and habitats, particularly those that have high concentrations of insect life, which often correlate with global plant and animal hot spots. A significant challenge for conservation involves expanding our mindset from its long-running focus on mammals, including charismatic game species and predators, to include invertebrates, friends in the insect world beyond butterflies and bees. Legislative and policy pressure is needed to help decision-makers, researchers, and funders support projects that address the impending insect extinction crisis. We need to reduce and eliminate the direct causes of the crisis, especially the use of biocides in agriculture, including weed killers. A transition to regenerative farm and forestry practices greatly improves the conditions for insects, especially as degraded land is restored. Working across a spectrum from "land sharing" to "land sparing," as well as a focus on habitat mosaics—combinations of trees, grasslands, and shrubs, including hedgerows on farms—are required to protect insect diversity. We must confront indirect threats as well, by reducing the impacts of invasive species, among other things.

A good place to begin might be to channel the enthusiasm of nineteen-year-old Charles Darwin, who opened a letter to his cousin in 1828 by writing, "I am dying by inches, from not having anybody to talk to about insects." ●

A four-spotted chaser (*Libellula quadrimaculata*) resting on a stem, covered in early morning dew.

Eating Trees

One of the most nutrient-dense foods that are useful to humans may be one you've never tasted. It's a leaf of the moringa tree. Native to the foothills of the Himalayas, the fast-growing moringa is a drought-tolerant species that can thrive on degraded land. Its small leaves, spread across a branch like a waving hand, are 30 percent protein and packed with nutrients, including vitamins A and C, calcium, and potassium—gram for gram, more than carrots, oranges, milk, and bananas, respectively. Moringa leaves contain all nine essential amino acids. They can be chewed fresh, cooked, or dried and added as a powder to staples such as bread flour. The tree's bark, flowers, seeds, and roots are also edible, providing additional protein, vitamins, antioxidants, and minerals, including iron, zinc, and copper. The therapeutic effects of moringa include prevention of disease, cell regeneration, reduced blood sugar levels, and anti-inflammatory properties.

Moringa trees are an example of a perennial tree crop—a plant that can be harvested year after year—in contrast to annual vegetables, such as broccoli, lettuce, and melons, which must be grown from seed after each harvest. There are more than seventy tree species that produce edible food. They play an important role in sequestering atmospheric carbon in their leaves, stems, trunks, roots, and associated soils for extended periods. Like a forest, they don't need tillage to grow, which means the carbon that has accumulated among the microbes, fungi, and mineral aggregates in the soil can remain intact. Carbon cycling is strengthened by the diversity of perennial species that can be combined into food production systems, including different types of trees, shrubs, herbs, palms, vines, and grasses, creating a rich network of pathways for carbon to make its way underground. The longer growing seasons of perennials, the decomposition of their leaves on the soil surface, and the varied depths of their roots, combined with their ability to grow in a wide range of landscapes under varying conditions, means they can hold carbon in the soil longer than annual plants. They also grow bigger year after year, adding carbon-infused mass to their trunks and stems and growing more green leaves to capture sunlight and photosynthesize.

Trees produce well-known nuts, fruits, beans, and syrups. On average, U.S. consumers in 2018 ate seventeen pounds of fresh apples per person, twenty-four pounds of fresh citrus, twenty-eight pounds of bananas, and nearly two and a half pounds of almonds, while drinking a total of 3.5 billion pounds of coffee (roughly two cups per person per day). Perennial tree crops are highly diverse, coming in all shapes and sizes, including shrubs and ground-level plants. Among the many fruit and nut species are dates, figs, blackberries, grapes, plums, pears, persimmons, limes, kiwis, pecans, peanuts, walnuts, and pistachios. Trees with leaves you can eat include the goji, which originated in China and whose nutritious berries have become very popular; the linden, a large tree whose flowers produce a delicious honey and can be brewed for a medicinal tea; and the beech, whose lemony-tasting leaves are often added to salads.

Chestnuts, a cousin to the beech tree, have been an important source of starch in human diets for millennia. There are four types of edible chestnuts—European, American, Japanese, and Chinese—each with a pleasant flavor. Considered a "perfect" tree for its multiple uses, including timber, the chestnut was a staple in American life until a parasitic fungus introduced in the 1900s killed four billion trees in the first half of the twentieth century—nearly every chestnut tree on the continent. Reestablishment efforts are underway, including the cultivation of resprouted trees, crossbreeding programs, and biotech research, in hopes that the American chestnut can be restored to its role as an iconic and prolific source of nutritious food. Historically, American chestnuts produced many hundreds, perhaps thousands, of pounds of nuts per tree every year. Modern varieties provide a multitude of essential nutrients. Every one hundred grams of roasted chestnuts contains 43 percent of our daily requirement of vitamin C, 25 percent of vitamin B6, and 16 percent of thiamine, plus a number of essential minerals.

One of the first researchers to recognize the value of tree crops was geographer J. Russell Smith. In 1929, after witnessing soil erosion around the world, particularly on hillsides, Smith wrote a book advocating for a "permanent agriculture" centered on trees and other perennial plants that did not require tilling or expose bare ground for part of the year. Inspired by chestnut farms in Europe, he wrote that "the crop-yielding tree offers the best medium for extending agriculture to hills, to steep places, to rocky places, and to the lands where rainfall is deficient." He also argued for two-story agriculture, with trees above and annuals crops or pastures below, though only where it naturally fit the land. He considered perennials to be underutilized and unexploited genetically, which meant they could be developed to be locally adapted and high yielding. At the time, his arguments fell on deaf ears, but today there is renewed interest in his ideas as perennial agriculture rises in the eyes of many as a way to meet climate and food security challenges.

An ancient idea gaining new traction is a food forest, which is a plot of land covered with edible perennial plants designed to mimic the edge of a forest. The plot can be as small as one-tenth of an acre or as large as hundreds of acres. On it, the farmer or gardener creates layers of plants based on their verticality, with tall trees sheltering smaller ones, interspersed with shrubs, bushes, herbs, flowering plants, mushrooms, and even vines. The goal is solar management: which plants catch the sun's rays when, for how long, and during what time of year. Shade-tolerant plants can grow under the protection of tree canopies, while sun-loving plants fill the open spaces. The design is deliberately busy—just like a natural forest edge—with each species filling a niche. The perennial crops can be a mixture of nuts, fruits, or leaves, depending on local conditions and the needs of the farmers.

Food forests are an ideal form of regenerative agriculture, because they are diverse, high yielding, multiseasonal, and relatively low maintenance. Like a forest, they don't need to be weeded, fertilized, or tilled and can produce food year after year for a long period. They naturally enrich the soil with carbon, feeding microbes. They can be designed to attract beneficial insects, including pollinators, with a mixture of flowering plants and trees. They create habitat for a variety of mammal and reptile species. The quantity and diversity of the edible yield of a food forest is determined by the farmer or gardener, though attention should be paid to biological needs and nutritional relationships among perennial plants.

Food forests are not a new idea. For centuries, Indigenous people have been growing food in perennial systems modeled after multilayered forests. Large parts of the Amazon forest, once considered mostly free of human alteration, show extensive signs of having been highly managed for tree crops and diversified agriculture thousands of years ago. Charcoal and plant pollen evidence indicates a cycle of forest burning followed by the planting of maize, squash, and root crops. Nonnative tree species were also raised for their produce, including cashews, cacao, palms, and Brazil nuts. Today, food forests—also called home gardens—are a prominent and expanding part of traditional and small-scale rural farming systems around the world. There is rising interest in food forests for urban environments as developers, community activists, and city leaders explore their potential to provide access to food while enhancing aesthetics and creating habitat for wildlife.

Trees, shrubs, and other woody plants are among the most effective natural agents of carbon sequestration on the planet. Perennial agriculture, whether used primarily as a system for growing food or forage, is a carbon-beneficial way to produce nutritious crops, restore degraded land, and provide economic returns for millions of people. Integrating tree crops into perennial food systems brings the best of both worlds together, particularly when applied to depleted soils and vulnerable farms that have easily eroded slopes. It's more than just food. Perennials provide medicines, building and craft materials, natural dyes, fuel, fiber for clothing, and items for household use. Raised together in a polyculture, as nature intended, we can have our trees and eat them, too. ●

A tractor spraying a slurry of kaolin clay on groves of Tuscan olive trees on the McEvoy Ranch near Petaluma, California, to prevent infestation by the olive fruit flies a month before harvest. In keeping with organic philosophy, they have interplantings of grapes (including Pinot Noir, Syrah, Montepulciano), as well as chickens for the employees. For weed control they use goats, as well as tractors and hand labor. Their primary product is a blend of oils from seven Italian olive varieties, and the pits and pulp are returned to the soil as mulch, along with a supplement of organic fertilizers, such as poultry feathers and manure. The McEvoy Ranch is 550 acres, and has 57 acres planted in olives and 7 acres in grapes; the rest is left as wildland and irrigation ponds.

We Are the Weather
Jonathan Safran Foer

Jonathan Safran Foer is the author of We Are the Weather: Saving the Planet Begins at Breakfast, *from which this essay is excerpted, and* Eating Animals, *a New York Times best-seller, published in 2009. His focus in both books is the impact of animal food on climate change. There is no agreement in the literature about the amount of greenhouse gas emissions generated by the meat and dairy industry. The UN's Food and Agriculture Organization, in a non-peer-reviewed study entitled* Livestock's Long Shadow, *calculated it to be 18 percent of total emissions. The impartiality of the FAO report was challenged because of its partnership with the International Dairy Federation and the International Meat Secretariat. In a March 2021 study published in Nature Food, emissions were broken down on every stage of the food chain up until 2016. Food system emissions totaled 34 percent of global greenhouse gas emissions, the largest single sector by far. The portion attributable to meat and dairy was not broken out, but previous studies showed that four foods have the greatest carbon footprint: beef, lamb, cheese, and dairy. Reducing meat and dairy consumption remains one of the top actions an individual, family, institution, cafeteria, or country can undertake with respect to food. In this excerpt from* We Are the Weather, *Foer writes passionately about the simplicity of the choice and the complexity of the resistance to the path to take if we are to transition to some semblance of climate stability.* —P.H.

The chief threat to human life—the overlapping emergencies of ever-stronger superstorms and rising seas, more severe droughts and declining water supplies, increasingly large ocean dead zones, massive noxious-insect outbreaks, and the daily disappearance of forests and species—is, for most people, not a good story. It not only fails to convert us, it fails to interest us. To captivate and to transform are the most fundamental ambitions of activism and art, which is why climate change, as subject matter, fares so poorly in both realms. Revealingly, the fate of our planet occupies an even smaller place in literature than it does in the broader cultural conversation, despite most writers considering themselves especially sensitive to the underrepresented truths of the world. Perhaps that's because writers are also especially sensitive to what kinds of stories "work." The narratives that persist in our culture—folktales, religious texts, myths, certain passages of history—have unified plots, sensational action between clear villains and heroes, and moral conclusions. Hence the instinct to present climate change—when it is presented at all—as a dramatic, apocalyptic event in the future (rather than a variable, incremental process occurring over time), and to paint the fossil fuel industry as the embodiment of destruction (rather than one of the several forces that require our attention). The planetary crisis—abstract and eclectic as it is, slow as it is, and lacking in iconic figures and moments—seems impossible to describe in a way that is both truthful and enthralling.

As the marine biologist and filmmaker Randy Olson put it, "Climate is quite possibly the most boring subject the science world has ever had to present to the public." Most attempts to narrativize the crisis are either science fiction or dismissed as science fiction. There are very few versions of the climate change story that kindergartners could re-create, and there is no version that would move their parents to tears. It seems fundamentally impossible to pull the catastrophe from over there in our contemplations to right here in our hearts. As Amitav Ghosh wrote in *The Great Derangement*, "The climate crisis is also a crisis of culture, thus of the imagination." I would call it a crisis of belief.

Although many of climate change's accompanying calamities—extreme weather events, floods and wildfires, displacement and resource scarcity chief among them—are vivid, personal, and suggestive of a worsening situation, they don't feel that way in aggregate. They feel abstract, distant, and isolated rather than beams of an ever-strengthening narrative. As the journalist Oliver Burkeman put it in *The Guardian*, "If a cabal of evil psychologists had gathered in a secret undersea base to concoct a crisis humanity would be hopelessly ill-equipped to address, they couldn't have done better than climate change." Our alarm systems are not built for conceptual threats. . . . The truth is as crude as it is obvious: we don't care. So now what?

Social change, much like climate change, is caused by multiple chain reactions that occur simultaneously. Both cause and are caused by feedback loops. No single factor can be credited for a hurricane, drought, or wildfire, just as no single factor can be credited for a decline in cigarette smoking—and yet in all cases, every factor is significant. When a radical change is needed, many argue that it is impossible for individual actions to incite it, so it's futile for anyone to try. This is exactly the opposite of the truth: the impotence of individual actions is a reason for everyone to try.

Polio couldn't have been cured without someone inventing a vaccine—that required an architecture of support (funding from the March of Dimes) and knowledge (Jonas Salk's medical breakthrough). But the vaccine couldn't have been approved without a wave of polio pioneers volunteering for a trial—their feelings were irrelevant; it was their participation in the collective action that allowed the cure to be brought to the public. And that approved vaccine would have been worthless if it had not become a social contagion, and therefore a norm—its success was the result of both top-down publicity campaigns and grassroots advocacy.

Who cured polio? No one did. Everyone did.

This book is an argument for a collective act to eat differently—specifically, no animal products before dinner. That is a difficult argument to make, both because the

topic is so fraught and because of the sacrifice involved. Most people like the smell and taste of meat, dairy, and eggs. Most people value the roles animal products play in their lives and aren't prepared to adopt new eating identities. Most people have eaten animal products at almost every meal since they were children, and it's hard to change lifelong habits, even when they aren't freighted with pleasure and identity. Those are meaningful challenges, not only worth acknowledging but necessary to acknowledge. Changing the way we eat is simple compared with converting the world's power grid, or overcoming the influence of powerful lobbyists to pass carbon-tax legislation, or ratifying a significant international treaty on greenhouse gas emissions—but it isn't simple.

We do not simply feed our bellies, and we do not simply modify our appetites in response to principles. We

tipping point of "runaway climate change," when we will be unable to save ourselves no matter our efforts.

We do not have the luxury of living in our time. We cannot go about our lives as if they were only ours. In a way that was not true for our ancestors, the lives we live will create a future that cannot be undone. With respect to climate change, we have been relying on dangerously incorrect information. Our attention has been fixed on fossil fuels, which has given us an incomplete picture of the planetary crisis and led us to feel we are hurling rocks at a Goliath far out of reach. Even if they are not persuasive enough to change our behavior, facts can change our minds, and that's where we need to begin. We know we have to do something, but *we have to do something* is usually an expression of incapacitation or at least uncertainty. Without identifying the thing that we have to do, we cannot decide to do it. Here are a handful of facts that connect animal agriculture and climate change.

- The current climate change is the first caused by an animal and not by a natural event.
- Since the advent of agriculture, approximately twelve thousand years ago, humans have destroyed 83 percent of all wild animals and half of all plants.
- Globally, humans use 59 percent of all the land capable of growing crops to grow food for livestock.
- There are approximately thirty farmed animals for every human on the planet.
- In 2018, 99 percent of the animals eaten were raised on factory farms.
- On average, Americans consume twice the recommended intake of protein.
- People who eat diets high in animal protein are four times as likely to die of cancer as those who eat diets low in animal protein are.
- About 80 percent of deforestation occurs to clear land for crops for livestock and grazing.
- Animal agriculture is responsible for 91 percent of Amazonian deforestation.
- Forests contain more carbon than do all exploitable fossil fuel reserves.

Climate change is the greatest crisis humankind has ever faced, and it is a crisis that will always be simultaneously addressed together and faced alone. We cannot keep the kinds of meals we have known and also keep the planet we have known. We must either let some eating habits go or let the planet go. It is that straightforward, that fraught. Where were you when you made your decision? ●

eat to satisfy primitive cravings, to forge and express ourselves, to realize community. We eat with our mouths and stomachs, but also with our minds and hearts. When I first chose to become vegetarian, as a nine-year-old, my motivation was simple: do not hurt animals. Over the years, my motivations changed—because the available information changed, but more importantly, because my life changed. There is a place at which one's personal business and the business of being one of seven billion earthlings intersect. And for perhaps the first moment in history, the expression "one's time" makes little sense. Climate change is not a jigsaw puzzle on the coffee table, which can be returned to when the schedule allows and the feeling inspires. It is a house on fire. The longer we fail to take care of it, the harder it becomes to take care of, and because of positive feedback loops, we will soon reach a

Energy

This picture of the sun was taken by Rainee Colarcurcio in 2019 during a solar minimum. Normally the sun looks like a fiery storm at sea, bedecked with sunspots and solar flares that extend a quarter million miles into space. Every eleven years or so, during a solar minimum, sunspot activity calms down. The black object seen traversing the sun is the International Space Station crewed by nine people from Russia, Italy, and the United States.

Global warming will not be stopped and reversed without ending the use of fossil fuels. The combustion of coal, gas, and oil creates 82 percent of carbon dioxide emissions. To reduce fossil fuel emissions by 45–50 percent by 2030 from 2010 levels, as recommended by the UN Intergovernmental Panel on Climate Change, is monumental in scope. To substantially reduce carbon emissions, we need to divide the subject of energy into three categories: source, usage, and application. Where does energy come from? What is it used for? And how do we make use of it?

We are privileged. Never in history has civilization been blessed with so much abundant energy. Until fossil fuels, energy was provided by fire, animals, and enslaved people. China destroyed most of its ancient forests thousands of years ago. Three-quarters of the world was enslaved in some manner in the seventeenth century, a perverse and barbaric form of "energy." The way we measure the value of energy is by how much work it can do—in British Thermal Units, calories, or joules. If you calculate the amount of work a barrel of oil can do compared with a human being, each of us, even the poor, has one or more fossil fuel "servants" at hand. The average Indian household has five, and the average American four hundred. A warm house, the instantaneous action of a clothes dryer, the quick hop to the store in your car—we rarely calculate the time and personal energy required to do these tasks absent our carbon servants.

Although we are fortunate to have abundant supplies of energy, carbon-based energy creates massive amounts of pollution. And not merely pollution, but poison that destroys life. Fossil fuels are toxic to air, lakes, oceans, soil, plants, people, and animals. They create smog, distribute particulate matter into the air, and cause lung and respiratory disease. Coal and oil contain benzenes, mercury, cadmium, lead, methane, sulfides, pentanes, butanes, and dozens more toxins. In addition, most of the energy we use, whether it be coal, gas, or oil, is wasted, meaning the energy does no useful work. Energy, in its thermal or electrical form, powers systems that are badly designed and poorly engineered, including our buildings, cars, and factories. According to the National Academy of Engineering, the United States is approximately 2 percent efficient, which means that for every one hundred units of energy employed, we accomplish two units of work. Energy efficiency is slightly better in some Asian and European countries, but worse in other parts of the world. Because energy is used wastefully all over the world, we can radically reduce energy use and still create the same product or accomplish the same task.

With the discovery of abundant deposits of coal, gas, and oil in the eighteenth and nineteenth centuries, society turned away from animals, trees, and charcoal. Fossil fuels are convenient, concentrated, and inexpensive. They consist of dead plants and organisms produced by ancient sunshine in coal swamps, marine deposits, forest bogs, tar sands, and shales, some deposits dating back 650 million years. Coal, gas, and oil constitute 84 percent of our primary global energy use today. The climate crisis has been caused by hundreds of millions of years of carbon captured by plants, trees, and marine phytoplankton being released back into the atmosphere in a geologic fraction of a second. Carbon dioxide emissions from fossil fuels total 35 billion tons per year. The goal is to reduce that number to 11 billion tons by 2030. Currently, renewable energy in all its forms—solar, wind, hydropower, and geothermal—provides 5 percent of primary energy. At this point, solar and wind farms harvest only a fraction of the incoming heat and photons that reach the earth daily. The revolution in renewable energy is not about changing the ultimate source of our energy. Except in the case of nuclear power, the primary source of energy has always been the sun. The renewable energy transition harnesses the abundant energy of the sun, enough to not only power our civilization, but with excess energy generation available to help pull carbon dioxide back out of the atmosphere. This is a momentous turning point in civilization. ●

Wind

Thomas Edison built the world's first coal-burning electric plant on Pearl Street in New York City in 1882. To serve its customers, it created the first electrical grid—wires strung on poles—and was illuminating ten thousand lamps for 506 customers within two years. His electrical power plant was highly profitable, and more were soon built across America by both Edison and Westinghouse. As the power plants got larger, so did the grid. On-demand electricity revolutionized business, industry, homes, and cities, regardless of whether the electrical generating plants were powered by coal, gas, hydro, oil, or, starting in the 1950s, nuclear. Electric and gas utilities were closed systems, publicly sanctioned monopolies. Access to the grid was tightly controlled, which prohibited decentralized renewable power generation from wind and solar.

In 1976, an obscure and forgotten act of defiance changed the utility world. It took place on a rooftop two miles north of the original Pearl Street plant. A group of renewable energy activists and architecture students, including solar pioneer Travis Price, were restoring an abandoned, fire-damaged apartment block on the blighted and dangerous East Eleventh Street, an area known as Strippers Row, where thieves dismantled parked cars at night and sold the parts during the day. A wind turbine student from Hampshire College named Ted Finch joined what would become known as the Eleventh Street Energy Task Force. He had studied wind power under Professor Bill Heronemus, a former U.S. Navy captain and visionary civil engineer, who coined the term *wind farm*, the idea that an array of wind turbines situated close together could act as a single power plant for the grid. Having predicted and witnessed the energy crisis of 1973, Finch proposed the first offshore wind farm, a 13,695-turbine installation off the coast of Massachusetts. It was dismissed as wildly unrealistic and absurdly expensive (which it was, at that time).

In New York, Finch was impressed by the gusty winds that came off the Hudson River. It was 1974, one year after the worldwide oil crisis that saw energy prices quadruple. Like many people, the Energy Task Force was looking for ways to greatly reduce residential energy use, particularly because the local utility, Con Edison, had the highest electrical rates in the country. Finch proposed putting a wind turbine on the roof to directly generate power for their building. They approached Con Edison for permission, but the utility was appalled by the idea, saying it

was illegal and dangerous and would likely destroy its equipment. Finch could have ignored Con Edison and let wind power the building independently, but his goal was to force utilities to open up the grid to local renewable energy generation. Con Edison told him he must first file the proper paperwork. However, there was no one at the utility who could tell him what the paperwork was, if the paperwork existed, or who could provide it if it did exist. The Eleventh Street Energy Task Force went ahead without permission.

Finch built a thirty-seven-foot steel tubing tower, but he had no crane or hoist to lift it upright. Thirty-five friends and neighbors assembled and slowly pulled the tower up from four directions using ropes attached to the top. Once it was standing, the structure then had to be placed directly onto the anchor bolts. It was an act of faith, especially on a windy day. If the tower had toppled, the potential damage to people and property would have been untold. A used,

twelve-foot-diameter Jacobs wind turbine that Finch resurrected from a Midwest farm was finally installed. After the group converted the direct current to alternating current, the turbine was turned on, the rotors spun, and two thousand watts of electricity flowed into the grid. For the Energy Task Force, who'd had their power turned off by Con Edison for late payments, watching the event was joyous. No one had ever seen a utility meter run backward.

As Finch well knew, Con Edison's fears and warnings were meaningless; nothing happened to the grid. Though its towering power plant was two blocks away, the utility did not know about East Eleventh Street for a week, until they read a story in the *New York Daily News*. A front-page photograph showed the operating wind turbine, with the Con Edison plant in the background. They sued. More new stories appeared, and it became a cause célèbre. Ramsey Clark, the former attorney general of United States, showed up at Eleventh Street and said it was as important

a civil rights case as voting, and he offered to defend the group for free. Public sentiment turned against the utility and Con Edison eventually conceded that people had the right to produce their own power and sell it back to the grid. Utilities, technically, no longer had a monopoly. This

---

Block Island Wind Farm, with five turbines in 90 feet of water two miles off the shore of Rhode Island, is the first offshore wind farm in the United States. The wind farm has a total generating capacity of 30 megawatts and will produce over 125,000 MWH each year, enough to power 17,000 homes. Approximately 10 percent of the capacity is for use on Block Island, and the rest will be sent to the mainland via underwater cable. The cable will also allow Block Island to get power from the mainland at times when there is not enough wind power. The turbines are 360 feet above sea level, with blades 240 feet long. The wind farm is permitted for a twenty-year life span, and when decommissioned, it is required that the supporting foundations be cut off at sea bottom. So far, local fishermen say that the towers have increased habitat for fish and other forms of marine life.

When the Eleventh Street wind turbine went up, energy from the most productive wind turbines cost twenty times more than energy generated by coal, gas, or nuclear. Today, wind power (along with solar) is the least expensive form of newly installed electrical generation in the world. Coupled with solar energy generation, storage, and transmission, wind can completely replace fossil fuels before 2050. The reason it will happen has less to do with mandates and climate concern than it does with cost. New nuclear power plants are four to seven times as expensive, and that's not counting decommissioning costs, insurance, ongoing maintenance, or the cost of federal loan guarantees. Coal plants are two and a half times more expensive, not counting the transportation costs of securing the coal. Gas plants are one and a half times more expensive. In 2019, wind provided 7 percent of U.S. electricity and 5 percent of worldwide electricity.

Atmospheric scientist Ken Caldeira points out that 2 percent of the planet's wind would power all of civilization. This is supported by the International Energy Agency, which calculates that offshore wind could generate 420,000 terawatt-hours per year, more than fifteen times global electricity demand in 2019. Wind became dominant in cost and production through meticulous and remarkable engineering. Turbines have a theoretical maximum limit when it comes to how much energy they can capture. Called the Betz limit, it shows that no more than 59.3 percent of the kinetic energy of the wind can be converted by a rotating turbine. It makes no difference what the wind speed is, the number of blades employed, how long they are, or how fast the turbine rotates.

The two biggest factors in wind energy's decline in cost have been the size and scale of wind turbines and the increase in their capacity factor. Capacity factor is how much energy a turbine captures within the Betz limit. If a wind turbine has a theoretical output of five megawatts and produces two megawatts, its capacity factor is 40 percent. Output is less than rated capacity for many reasons. The wind turbine might be situated in a less-than-optimum wind regime, because wind is intermittent and inconsistent over time, or there might be different "cut-in" speeds—the wind speed, in miles per hour, required before the turbine actually rotates and generates electricity. Average capacity factor of new installations has leaped from 25 percent in the early 2000s to 50 percent today, owing primarily to better wind turbine and wind farm design, tower heights that can access stronger winds, the size and area of the rotors, lower cut-in speeds, greater reliability, and improvements in gearboxes, computer simulation for given wind locations, and blade materials. Estimates are that capacity factors can continue to increase, toward 60 percent onshore, and even greater offshore.

was followed by federal legislation that encouraged renewable energy producers and made it mandatory for utilities to purchase and distribute their power. Today, the Empire State Building and its related buildings are powered entirely by wind.

There was a deeper takeaway from this change. When utility companies were first created around the country, they started out and then remained vertical monopolies; this included their transmission lines. Electricity generated in one area was captive and could not be shared. If an adjoining region was experiencing brownouts or power outages, and the neighboring utility had a surplus of energy, that was tough luck. Utilities had to be coaxed and even mandated to open up their electric transmission lines to other power suppliers. Once transmission lines were connected across the country, the best sources of wind energy, which are predominantly in the Midwest, from North Dakota to Texas, could be sold to the biggest markets, which are primarily on the East Coast.

Fossil fuels had such massive success because of their abundance, cost, and convenience. Those advantages are now in doubt. What is not in doubt is the rate at which onshore wind energy is dropping in cost. What might further shake up energy predictions is the likelihood that offshore wind could follow onshore wind to become the third-least-costly form of electrical energy.

Henrik Stiesdal, who is called the "godfather of wind," wants to make offshore wind as practical and affordable as onshore. "I had some bad moments thinking about climate," he recounts. "The politicians will not solve it. We need to solve it ourselves." Today, three-quarters of offshore wind farms are placed in the shallow waters of the North Sea, clustered mainly around Germany and the UK. Similar waters exist off China and the Eastern Seaboard of the United States, but in order for offshore wind to fully develop, it will need to go into deeper waters, farther off the coasts of California, Portugal, the UK, and Japan. There, large turbines could harvest more powerful winds and would be unseen from shore. The potential is significant. The waters off California contain eleven times more harvestable energy than what is required by the state. The problem is how to anchor skyscraper-size wind platforms weighing a hundred tons in thirty-foot waves when the seafloor is a thousand feet below.

Stiesdal's legacy is impressive. He began experimenting with model turbines in high school, during the 1970s. He was making turbines in his family's rural home, with each successive prototype becoming larger and more effective. The feeling of torque from the small spinning turbines was what hooked him. He could feel the power of the wind in his hands and arms. When the turbines reached the size of usefulness for a residence, an executive from a local agricultural equipment manufacturer called Vestas dropped by and was impressed. A licensing agreement was signed for his prototypes, which used the same three-blade configuration found on every major wind turbine to this day. Today, Vestas is the largest wind turbine manufacturer in the world, with $13 billion in sales and factories in twelve countries. More than seventy thousand of its wind turbines have been installed in eighty-one countries.

Offshore wind has always been possible, but considerably more expensive due to siting, maintenance, underwater transmission cables, and the corrosive saltwater environment. In Europe, this consideration was offset by local resistance to towering, noisy wind turbines sited in rural areas. The relatively shallow North Sea was the logical answer, and thousands of turbines have been installed there. In fact, the largest wind turbines in the world are being installed in the North Sea. The largest of all, the twelve-megawatt Haliade-X, was constructed offshore of the Netherlands by General Electric in 2019. It is 853 feet tall, with three-hundred-foot blades producing thirteen megawatts, which is sufficient energy for twelve thousand homes. The Haliade-X has a 63 percent capacity factor and develops 45 percent more energy than its closest rival. But more rivals are coming soon: in 2024, Vestas will be installing fifteen-megawatt offshore turbines that can power twenty thousand homes.

Stiesdal is designing wind turbines for the open seas. Floating platforms that imitate offshore oil-drilling platforms are complex and costly to make. His early wind turbine designs succeeded because they could be produced quickly and inexpensively. Stiesdal has designed a floating foundation that can be industrially manufactured and repeat the economies of scale achieved by Vestas for onshore wind. The materials are the same, and most of the labor is done by robots. Skeptics doubt that costs will drop low enough or subsidies will be high enough to support deepwater wind turbines. But for Stiesdal, the question is not *How can we afford it?* The question is *How can we afford not to?*

In the year 2019, 27 trillion watts-hours (terawatt-hours) of electricity was produced worldwide. Wind turbines produced 1.4 terawatts, just over 5 percent of global electricity production. To achieve 100 percent renewable electricity production by 2050, assuming 50 percent wind and 50 percent solar, wind turbine power production would need to quadruple by 2030, double again by 2040, and increase by another 80 percent by 2050. The projected usage of electricity as determined by the International Energy Agency and the World Bank assumes that our homes will use the same amount of electricity as they do now, that economic demand for goods and services will triple, that there will be three billion cars on the road, and that the world will not change its materialistic and consumptive behavior in thirty years. On the other hand, usage numbers do not anticipate a completely electrified world of transport and buildings. Either way, it is difficult to imagine that in the coming decades more people like Ted Finch, Bill Heronemus, and Henrik Stiesdal will not show up. They are here already. We may not know who they are yet. ●

Ted Finch and team after erecting his hand-built 30-foot steel tower on the roof of an East Village tenement in New York City in 1974. It was the first installation of renewable energy regeneration in the city of New York, opposed by Con Edison, whose oil-burning generating plant can be seen just peeking above in the background.

Solar

The economic potential of solar energy was called into question and second-guessed for decades. The only remaining debate is how soon will solar (and wind) erase fossil fuels from the face of the earth. The cost of solar today is lower than what the International Energy Agency once predicted it would be in the middle of the twenty-first century. Since fossil fuels account for 85 percent of carbon dioxide emissions, this is arguably the most important breakthrough in the world of climate and energy. How did photovoltaic electricity get so cheap?

It started in 1839, when nineteen-year-old Edmond Becquerel was experimenting in his father's laboratory. Anodes and cathodes made of bismuth, platinum, lead, tin, gold, silver, and copper were scattered about the workshop, and for some unknown reason, Edmond dipped two platinum electrodes into an acid bath and shone a light upon one—and electricity was generated. He had discovered the photovoltaic effect—how materials exposed to light under certain conditions could create an electric current. Looking back, it was an historic event. At the time it was largely ignored, a novel science experiment, irrelevant to an exploding industrial age, a world in the throes of a long and heated romance with coal and later gas and oil.

In the 1870s, Edmond's photovoltaic experiment was repeated by several scientists using selenium. In 1883, American entrepreneur Charles Fritts created a working photoelectric module for a rooftop in New York City. It produced six watts of power in the noonday sun and cost a pretty penny—the solar cells were made of gold-plated selenium. But Fritts had a vision. He was optimistic that his photovoltaics could compete with the just-completed coal-powered electrical generation plant a few blocks away on Pearl Street. Although Fritts's modules attracted great enthusiasm in many quarters, including from Werner von Siemens in Germany, it was puzzling for many scientists. How could a photovoltaic module create more energy than the latent heat energy of sunlight?

The answer came from a Swiss patent office in 1905, when a twenty-six-year-old theoretical physicist named Albert Einstein published a "daring" paper stating that light had an attribute that had not been identified, that the energy of light was not only continuous but also discontinuous. It contained what Max Planck called quanta, known as photons today, a stream of subatomic particles that varied with the wavelength of light. The shorter the wavelength (ultraviolet), the greater the energy. For the next fifty years, trial-and-error experiments were undertaken by labs, inventors, and entrepreneurs to increase the efficiency of solar cells. Patents were filed, experiments published, materials modified, silicon crystals grown. Finally, in 1954, a breakthrough solar cell was made by Bell Labs for a NASA satellite. It was constructed of silicon instead of selenium, and it made headlines worldwide. *The New York Times* wrote that "limitless energy of the sun" might be a possibility. The enthusiasm was both prescient and

premature. To power a house with the Bell solar cells would have cost $1.43 million. Today, solar-generating capacity in the United States powers one in eleven homes. The manufactured price per watt in 1955 was $1,785; today it is ten cents, a 99.99 percent reduction in cost.

Charles Fritts's guess that solar energy would compete with coal was correct. But it took the oil shock of 1973, the 1997 Kyoto Protocol, Germany's generous solar energy subsidies in the 2000s, and rising global awareness of climate change to make solar truly take off. The watershed moment on cost came with the vast Chinese solar factories. Including solar farms, they are one of the energy sources visible to astronauts from space, another being the tar sands of Alberta, Canada. The growth of solar energy has been consistently underestimated. Institutional predictions about the cost and growth of solar energy from the International Energy Agency in Paris, to McKinsey & Company in New York were much too conservative. Richard Swanson, the founder of solar panel manufacturer SunPower, explained why. According to "Swanson's law," the cost of solar photovoltaic modules drops 20 percent for every doubling of cumulative volume. Swanson was conservative. The drop is over 30 percent. As of 2019, the unsubsidized cost of electrical energy for a utility-scale solar power plant, including construction, installation, and operation, was 44 to 76 percent less than that of a coal-powered plant, 69 to 81 percent less than existing or proposed nuclear power plants, and 16 to 46 percent less

than a natural gas plant. This was never predicted. And the cost is still falling.

In May 2020, renewably generated electricity in the United States topped coal for the first time. Three-quarters of the coal-powered plants in the United States could be shut down, mothballed, and replaced with solar, and it would save money for the owners of those plants while lowering electricity rates for consumers. Coal has become a fossil fuel in the true sense of the word. It is dead.

Starting out as a power source for satellites, solar energy now powers boats, indoor lighting in India, schools in Tanzania, streetlights in Detroit, remote charging stations, wearables (sewn into clothing and backpacks), highway signage, cell towers, concerts, electric cars and bikes, solar windows, car surfaces, vaccine refrigerators, and airplanes. There is solar paint that will capture energy from the sun and transform it into electricity, solar trash bins that compress waste when fully charged, and sports stadiums run entirely on solar energy. Solar farms and

---

(*Left*) The Nasu-Minami Photovoltaic Solar Power Plant located on a former golf course in the mountains of Tochigi Prefecture, Japan.

(*Above*) A worker pulling a cable along a floating solar power farm at sea, off Singapore's northern coast, just across the Johore Strait from the Malaysian state of Johor. Thousands of panels stretch into the sea off Singapore, part of the land-scarce city-state's push to build floating solar farms to cut greenhouse gas emissions.

arrays are found everywhere, from the vast concentrated solar plant in Chile's Atacama Desert to panels nestled between rice paddies in Japan. In the Netherlands, 73,000 photovoltaic modules, 13 floating transformers, and 192 inverter boats were assembled and placed upon a lake. The solar array probably set the world record for speed of installation, at up to one megawatt per day, all built and constructed by electrically charged (solar) boats and vehicles. The waterborne solar array can withstand waves, high wind, and snow.

The largest floating solar farm in the world covers a flooded former coal-mining region in Anhui Province in China. Sinn Power of Germany has developed a maritime platform that will generate energy from waves, wind, and solar all at the same time. It is modular, can be connected in series, could provide power for islands, and could amplify power gathered at offshore wind farms, lowering overall capital costs. There is a solar-power plant operating a stone's throw from the 1986 nuclear reactor explosion in Chernobyl, Ukraine. Australia, the world's third-largest exporter of fossil fuels in the form of coal, may finally be turning to the sun. Singapore, which relies on liquid natural gas for 95 percent of its electricity, is working with Sun Cable to lay down a twenty-eight-hundred-mile-high voltage direct current cable under the Java and Timor seas that would replace 20 percent of its electricity with power from a thirty-thousand-acre solar array in the Northern Territory.

Germany, with its relatively modest amounts of sunshine, is a world leader in shifting from fossils to high shares of solar power, deriving 46 percent of its electricity from renewable sources. It has made the transition without any disruption to consumer and industrial power. The government has woven renewable energy into the fabric of civic life. The agricultural village of Wildpoldsried, in southern Bavaria, resolved to renew its community in 1997, which meant building a new sports hall, theater, pub, and retirement home. The condition demanded by its citizens: the projects would not incur any municipal debt. By 2011 it had accomplished that and more: a new school, four biogas digesters, seven windmills, a district heating plant, several small hydro installations, and a natural wastewater system, all with no indebtedness. It was financed by producing five times more renewable energy than the village uses and selling it at a considerable profit to other communities. Wildpoldsried plugged the leaks and closed the loop. The leaks in a community are the funds that are always leaving, to purchase gas, electricity, fuel, food, and the like—funds that never return. The loop is the circle of production and economic activity that every community requires. Plugging the leak means making more efficient use of energy, generating heat and electricity locally, producing and eating local food (and not wasting

it), and charging electric vehicles with renewable energy. Closing the loop means saving money. When money stays within a community, it creates employment and people prosper. When too much income leaves, communities suffer. In essence, locally produced renewable energy can regenerate a community.

Wildpoldsried would not have been possible had there not been a German energy transition plan (Energiewende), a centralized and agreed-upon path to achieving nationwide carbon neutrality by 2050. There was no precedent for Germany's commitment. The issues faced included a car industry based on diesel and gas, higher energy costs for industries competing internationally, power grids not fully suited to intermittent supply (sunshine) or power surges, the decommissioning of its nuclear power plants, disruption to the largest utilities, and the legacy of coal mining and its labor force. It has been a difficult, political, and sometimes heated process every step of the way. Although much can be learned from the path Germany forged, transitioning to total renewable energy is a process where there are losers: coal miners, gas and oil producers, and utilities that stall and impede the transition to solar and wind energy in many regions and countries.

Ridding the world of fossil fuels is the most formidable barrier to reversing global warming. That goal requires two actions. First, substitutes for coal and liquid fuels must be quickly expanded: solar, wind, and energy storage for electric vehicles and building electrification. The second crucial activity is stopping the institutional inertia that continues to finance coal-powered plants, oil and gas exploration, pipelines, fracking, and liquid natural gas terminals in the world. If we do not, they will operate and emit for decades. JPMorgan Chase is the banker that is doing the most to finance fossil fuels, including Arctic oil and gas, offshore oil and gas, fracking, coal mining, and the tar sands. Right behind them are Wells Fargo, Citi, and Bank of America. Thirty-five banks poured $3.8 trillion into the fossil-fuel industry within five years of the Paris Agreement. Once investors have sunk capital into a fossil fuel project, it will continue to operate, it will influence and/or corrupt legislators, and it will claim to be a job creator. As a job creator, solar energy tops fossil fuel employment by five to one. And critically, banks prevent renewable energy projects from getting financed, because markets are locked into coal and gas for years to come.

The world is still going backward with respect to carbon emissions, not forward. In the past thirty years more fossil-fuel-generated greenhouse gases were emitted than in the prior 230 years. "Optimistic" projections by international agencies show solar energy growing from 3 percent of world electricity generation to 22 percent by 2050, with overall renewable energy (solar and wind) at 50 percent of world electrical generation. These are not optimistic projections, given our current situation. Renewable energy production can and should be 95 to 100 percent of world energy production. The EU has committed to a carbon-neutral energy system by 2050 and thinks that can be accelerated. More than sixty German companies, from Puma to ThyssenKrupp, want economic stimulus to be directly tied to a green transition, saying in essence that no more money should be spent on the "old economy." Salzgitter, one of Germany's biggest steel manufacturers, wants the government to support transitioning steel away from being coal-powered to being powered by renewably created hydrogen fuel.

The impact and requirements of the extraordinary reversal of how we produce energy are immense. It is doable. Action to stem and reverse emissions is a government, corporate, and social mandate. Every level of human agency needs to engage in order to make this happen, from a small home to a giant steel plant making hydrogen fuel. In order to prevent global warming from exceeding 1.5°C, exponential growth will be needed in solar manufacturing and installation. From 1993 to 2002, installed solar capacity tripled. Eight years later, in 2010, it had increased thirty-threefold. By 2019 it had increased another fifteenfold. For solar power to replace 50 percent of fossil-fuel-generated electricity, solar production and installation will need to be eight times greater by 2030

than it was in 2019. By 2040 it will need to double, and then double again by 2050. This equals three solar panels for each and every human being on the planet. Wind will need to do the same in terms of energy output. If the world achieves significant advances in energy efficiency and constructive changes in reducing overconsumption of what is unnecessary, energy requirements could be reduced.

The cessation of fossil-fuel combustion is the threshold that we must cross in order to remain on the planet and develop as people, cultures, and a civilization. The first coal-powered electric plant in the United States was engineered by Thomas Edison and built in New York City in 1882. The last coal-powered plant in New York State, the Kintigh Generating Station, in Somerset, closed in March 2020. It is not a question of where we are going as a civilization. It is a question of how rapidly we can achieve the complete conversion to a culture energized by current solar income. ●

The East Golmud solar development started in 2009 and consists of several different solar companies. The existing and planned solar parks will make Golmud China's largest producer of solar electricity, covering 120 square kilometers by 2030 and generating one billion watts (a gigawatt) of electricity. It is located north of the Tibetan Plateau at 9,200 feet and is raked by strong winds that create crescent-shaped, concave sand dunes.

Electric Vehicles

Nearly two centuries after their invention, electric vehicles have reached a tipping point, signaling the end of the internal combustion engine. In the United States, the first useful electric vehicle debuted in 1889 on the streets of Des Moines, Iowa. Chemist William Morrison created a battery-powered vehicle resembling a horseless carriage that reached a top speed of fourteen miles per hour. By 1900, electric vehicles (EVs) accounted for roughly a third of all vehicles in the United States. Electric car sales stayed strong for another decade, until Henry Ford released his petroleum-powered, mass-produced Model T, flooding the market. Within a few years, the internal combustion engine had triumphed. EVs were forced to the sidelines, where they remained until the energy crisis of the 1970s rekindled interest. Automakers rolled out a variety of electric car models, but consumer concern about limited top speed (forty-five miles per hour) and range (less than fifty miles on a charge) curbed sales. With a booming global economy and oil prices dropping to historic lows in the 1980s, the future looked bleak for the electric car. Times have changed.

In 2010, there were seventeen thousand electric vehicles in the world; today there are more than ten million, with China accounting for the largest share. In 2020, more than three million EVs were sold worldwide, a 43 percent increase over the previous year. Norway became the first nation in the world where more electric vehicles were purchased than petroleum-powered ones. Volkswagen, the largest automaker in the world, has announced that it will stop designing cars with internal combustion engines in 2026 and fully transition to electric vehicles. General Motors said it will debut thirty new EV models by 2025 and cease making anything but electric vehicles by 2035. Volvo announced that it will sell only EVs by 2030. Ford will electrify its bestselling F-150 truck and iconic Mustang, promising that both will outperform their petroleum-powered predecessors. Tesla Motors, which produced its first electric car in 2006, sold almost five hundred thousand EVs in 2020, and its Model 3 has become the bestselling electric car on the planet.

According to the Intergovernmental Panel on Climate Change (IPCC), greenhouse gas emissions from the transportation sector have more than doubled since 1970, rising at a faster rate than any other sector, with approximately 80 percent of this increase attributable to road vehicles. Petroleum-powered cars, trucks, buses, and other vehicles generate roughly 16 percent of total greenhouse gas emissions worldwide. In 2019, gasoline consumption in the United States averaged 390 million gallons a day, accounting for nearly half of all petroleum use in the nation. Diesel fuel accounted for another 20 percent. These numbers are a major reason why governments are hastening the transition to EVs. In the fall of 2020, the worst fire season in California's history motivated Governor Gavin Newsom to issue an executive order requiring sales of all new passenger vehicles in the state to be zero-emissions by 2035. A few weeks later, New Jersey decided to follow California's lead. China requires a quarter of all vehicles sold in the nation to be electric by 2025.

A key will be transforming the source of electricity produced by power plants. The grid is becoming increasingly renewable as electricity flows into it from wind, solar, geothermal, hydro, and biomass. A Tesla Model 3 EV charged by solar energy generates 65 percent less greenhouse gas than an equivalent petroleum-powered vehicle over the course of its lifetime. The company's battery-making Gigafactory in Nevada is soon to be powered entirely by renewable energy. Electric vehicles in Norway run on electricity generated by hydropower. To deliver the electricity needed to run a huge fleet of EVs globally, as well as to meet the rising energy demands of electrifying other aspects of daily lives, a vast overhaul and upgrade of aging power grids is required. New transmission lines and distribution centers will need to be constructed in order to reach decentralized sources of energy, including remote wind and solar farms.

Tailpipe exhaust contains particulate matter, nitrogen oxides, and volatile organic compounds. Electric vehicles do not cause air pollution. A full transition to electric cars by 2040 will result in significant reductions in asthma, heart disease, and lung cancer. Paris officials have prohibited all polluting vehicles from the city center and intend to make it car-free altogether. Another clear advantage of electric vehicles is upkeep. With fewer than twenty moving parts, EVs require 50 percent less maintenance than cars with internal combustion engines.

To support a growing electric vehicle fleet, expansion of charging infrastructure is underway. Companies such as ChargePoint and Electrify America are working to connect and expand the charging network in the United States. Tesla has created its own Supercharger network consisting of nearly two thousand stations worldwide, with over twenty thousand outlets. Volkswagen is investing $2 billion in charging infrastructure. This raises a frequently cited concern among potential customers: *Will charging my EV cost more than refilling the gas tank of my car?* Although numbers vary depending on driving conditions and types of vehicles, the quick answer is no. By one estimate, if you drive a thousand miles a month in your EV and pay ten to twenty cents per kilowatt-hour to charge it at home, your bill will run between thirty and sixty dollars a month. By comparison, an internal combustion car averaging thirty miles per gallon, over the same amount of miles at current gas prices, would cost more than a hundred dollars a month.

As the web of electrical chargers grows, it will reduce another major concern: range anxiety. Today the typical range is two hundred miles on a charge, and it is forecast to reach four hundred miles by 2028, roughly the same distance a large passenger vehicle gets on a tank of gas.

Electric cars are powered by lithium-ion batteries, which have relatively low energy density, which slows recharging and limits driving range. New developments are changing these numbers. An Israeli company recently announced it had developed a battery that could be fully recharged in five minutes—approximately the same amount of time it takes to fill a gas tank.

Batteries were once responsible for a third of the price of an EV, but their price fell nearly 90 percent from 2010 to 2020 and is approaching a point where carmakers can offer an EV at a price comparable to or lower than an internal combustion vehicle as early as 2023. Many batteries require rare minerals, including cobalt, nickel, manganese, and lithium, whose sources are often concentrated in just a few countries. As demand for EVs rises, so will the need for more minerals. The amount of lithium used in battery manufacture in the United States, for example, is projected to nearly triple by 2030. The effects of rising demand will need to be addressed. Mining operations adversely affect wildlife populations, groundwater supplies, and ecosystem health. They can stir opposition from local populations. Although old batteries can be repurposed for other energy storage, including for use in homes and businesses, many end up in landfills. The material in lithium-ion batteries can be recycled. To create a sustainable supply chain, automakers have joined the Global Battery Alliance. When batteries are ready to be retired, the alliance ensures that the materials will be recovered and reused.

Exchanging internal combustion engines for electric ones will not achieve needed climate goals on its own. Changes in travel behavior are also required. This means more people using e-bikes, ride-share, or taking public transportation on electrified buses and trains. Low-occupancy EVs emit far less greenhouse gas over their lifetime than comparable petroleum-powered cars, but they may emit more per passenger than high-occupancy electric public transit vehicles, depending on occupancy.

A convergence of advances in battery technology, a steady reduction in sticker price, improving performance of EVs, and concern about climate change are quickly creating the conditions for market disruption. It's happened before. In 1903, Scott Montagu, a member of the British Parliament, predicted that the emergence of motorized vehicles would have little impact on the use of horse-drawn carriages and wagons. A decade later, horse-drawn carriages were surrounded by a sea of automobiles. We are on the verge of another disruption equally as profound. ●

The Lightyear One is a prototype EV from the Netherlands. It features five square meters of high efficiency solar cells under the safety-glass roof, enabling the car to travel up to 43 miles a day without charging, and 450 miles when fully charged. On a summer day, the rooftop panels will increase its range by 30 to 40 miles.

Geothermal

In the 1940s, American inventor Robert C. Webber accidently burned his hand on the deep freezer in his cellar and came up with an idea that would help change renewable energy.

Webber touched the outlet pipes of the freezer expecting them to be cold. They were scalding hot instead. He realized that the pipes were dispersing heat collected from inside the freezer, keeping it cold, so he connected the pipes to his boiler—and created more hot water than his family could use. Being resourceful, he ran the excess hot water through another pipe and used a fan to blow the heat into the house. When this experiment proved effective, Webber decided to tap another source of heat: the ground below his cellar. He knew the soil stayed warm all year, even in winter. He buried loops of copper pipes and added Freon gas, which would absorb some of the ground heat as it passed through the pipes. He then released the collected heat mechanically in the cellar and used it to warm the house. The experiment was so successful, Webber sold his coal-fired furnace the following year.

Heat pumps move heat from one place to another thermodynamically. Through the use of a closed-loop system, low-temperature heat is raised by an electrical compressor to a temperature sufficient to heat a building. A warmer source creates higher efficiencies. The heat source can be inside a building or outside—air, for instance. During winter, an air-source pump moves heat from the outdoors into your house (cold air still has thermal warmth). Water, such as a pond, can also be a source of energy.

Webber's innovation was to use the *ground* as a heat source.

The earth's crust is a massive solar battery. Half of the radiation energy from the sun that falls on the surface of the planet is stored there, warming it. Daily and seasonal high and low temperatures do not penetrate more than a few feet below ground—what is commonly called the frost line. Below this line, the temperature of the ground is nearly stable year-round—fluctuating between 45° and 70°F worldwide. At thirty feet below the surface, the temperature of the ground is constant, maintained by the sun's warmth. This is geothermal earth heat generated by the sun. It is ideal for heat-pump technology. In the winter, a pump transfers heat from the ground to a building, and in

the summer it can transfer heat from the house back to the ground. It is baseload energy, too—available all day, every day, and at a constant rate.

Want a hot bath in the middle of winter? Geothermal can provide it.

Heat pumps use much less energy than conventional furnaces and can reduce electricity consumption for heating and cooling by over 60 percent. They require a small amount of electricity to operate, which can easily be provided by renewable energy. Although they are expensive to install, ground-sourced heat pumps are safe and quiet, don't pollute, have low operating costs, and can last twenty years or more. There are no chimneys, gas meters, propane tanks, noisy air-conditioner units, or combustible parts.

Heat pumps primarily displace natural gas and coal, the most abundant sources of heating and cooling in most of the industrialized world. About half of the homes in the United States use natural gas to warm the air, heat water, cook food, and dry clothes. In 2019, the residential sector accounted for about 16 percent of total U.S. natural gas consumption. Heating and cooling buildings generates one-tenth of all greenhouse gas emissions in the United

States annually. According to an analysis by the Rocky Mountain Institute, replacing a gas furnace with a heat pump will reduce carbon emissions in forty-six states, comprising about 99 percent of all households. Even if you can't power your heat pump with solar energy, electricity generation from the grid continues to decarbonize each year as natural gas and renewables replace coal. This

---

(*Left*) Aerial view of Grand Prismatic Thermal Spring in Yellowstone National Park, Wyoming. The bright colors are natural, coming from thermophile bacteria in the water. People can be seen walking on the raised trail at the top of this frame, giving scale to this large natural feature. The spring is 320 feet across and 160 feet deep. The blue color is a result of light refracting through the clean mineral-rich volcanic waters. The yellow and orange hues are chemosynthetic mats (not requiring sunlight) of thermophilic algae and bacteria, which assort according to their temperature preferences. The water is heated to 71°C by a magma chamber of the Yellowstone Supervolcano.

(*Right*) A view of pipework and the geothermal power plant located in Reykjanesvirkjun, on the southwestern tip of Iceland. The power plant generates 100 megawatts of electricity from two 50-megawatt turbines, using steam and brine from a reservoir at 290°C to 320°C, which is extracted from twelve wells that are 2,700 meters deep.

is a reason why cities recently began passing legislation requiring all new buildings to be 100 percent electric.

One of the electricity-producing renewable sources of energy is a traditional form of geothermal: very hot water.

This geothermal energy is left over from the formation of the planet, 4.5 billion years ago, and the heat generated from the decay of radioactive minerals. Temperatures at the earth's inner core can exceed 5,000°C—roughly the same as the surface of the sun. The next layer is molten rock, which is encased by a two-thousand-mile layer of silicate material called the mantle. This is surrounded by the earth's crust, a layer of solid rock between three and fifty miles thick. Cracks in the crust, often found where tectonic plates meet, allow magma to rise close to the surface, heating reservoirs of water. These reservoirs are accessed by drilling bore holes. The water and steam are then piped up to a power plant, where they are converted into electricity. It can also be used as a heat source to warm buildings. When the water has cooled, it is sent back to the reservoir for reheating, which makes it a renewable resource.

In the 1970s, a technique called enhanced geothermal was developed, which injects highly pressurized water into superheated rock via deep wells, fracturing the rock along fissures. Water injected into these fractures is heated by the rocks and then pumped back to the surface via a second well, where it is used to generate electricity at a power plant. The water is then recirculated through the wells to be heated again in a closed-loop system. Advanced techniques, including directional drilling, have allowed companies to expand the size of the heated area as well as tap hot rocks at greater depths, enlarging the amount of geothermal energy available.

Iceland produces 30 percent of its electricity from geothermal sources linked to the island's extensive volcanic activity. Japan and New Zealand tap hot pools, geysers, and steam vents. The United States produces the most geothermal energy in the world. Its 18 billion kilowatt-hours of electricity generation each year is equivalent to over ten million barrels of oil. Other countries that employ geothermal include Kenya, Costa Rica, Indonesia, Turkey, and the Philippines.

Current geothermal technology provides less than 1 percent of generated energy worldwide, because it depends on access to extremely hot water. Easy-to-reach underground pools hot enough to power plants are limited to countries with volcanic zones. The cost of drilling for hard-to-reach sources is expensive. It takes time to develop a resource—as many as ten years from planning stage to power generation. A large solar array can come online in less than a year. A wind farm might take four years. Deep drilling for geothermal confronts complex technical challenges, including intense pressure, dangerous temperatures, and encounters with corrosive fluids. Enhanced geothermal water injection has been linked to earthquakes, including a 5.5 magnitude temblor in Pohang, South Korea, that shook buildings and required emergency housing for seventeen hundred residents. The high pressure created by injection of water into fractures can activate unknown faults that can trigger an earthquake.

In recent years, geothermal technology has evolved. Climeon, a Swedish company, has developed a small 280-cubic-foot modular unit that generates electricity from low-pressure, low-temperature (as low as 80°C) geothermal sources, which are highly abundant around the planet. It employs heat exchangers, which transfer ambient underground heat that turns a customized turbine generator.

Each unit can produce about one hundred fifty kilowatts of electricity, enough to power one hundred European homes year-round. The units provide electricity at rates that make them competitive with wind and solar power in Europe. Since the units are modular, a customer can install as many as needed to meet their energy requirements. The company's standardized units can work almost anywhere and at any scale. Climeon's first geothermal power plant opened in Iceland, followed by two commissions in Japan, including a pilot project at one of the nation's traditional hot spring resorts. Climeon is exploring prospects in Taiwan, New Zealand, and Hungary, which has large geothermal potential for the company's low-temperature technology.

Climeon's units can also utilize waste heat from factories and other industrial sources. In many industrial practices, around half of the energy used is transformed into waste heat. One of Climeon's clients is a steel factory where water is used to cool the hot metal. Normally, the 90°C water is discarded into the environment, but in this case the heated water powers a cluster of the company's units, generating electricity.

Geothermal energy is abundant, inexhaustible, reliable, affordable, efficient, climate-friendly, and multifunctional. When the wind dies down or the day turns to night, geothermal keeps going, providing baseload power for wind and solar. It has joined a fast-growing movement focused on regeneration. Humanity is "rapidly moving toward using less and making better use of things," says Berglind Rán Ólafsdóttir, managing director of a geothermal company in Iceland that is working on emission-free production. "We can utilize the gifts of nature in a way that minimizes impact. Most important is reducing the carbon footprint. We intend to abolish it completely by 2030." ●

Electrify Everything

In his book *Electrify*, Saul Griffith makes a telling analogy between the COVID-19 pandemic and global warming. We knew years ago that a pandemic would arise at some time in the future and that we needed to prepare for it. The United States and most countries had not prepared. We were cautioned decades ago about global warming and we have not prepared for that either. Like global warming, a COVID-19 infection starts with a fever. In the parlance of epidemiology, we were told we needed to flatten the curve, the rate at which the virus spreads, in order to reach equilibrium and a downward rate of infection. Flattening the greenhouse gas emissions curve entails carbon neutrality (equilibrium) as the threshold to achieve drawdown, after which we can begin to address the reversal of global warming. Just as viral pandemics need vaccines, so too does global warming. The difference is that we had 70+ percent of the climate cure for many years—a complete change-over in our energy infrastructure, the total electrification of the energy grid, and elimination of every form of fossil-fuel combustion. Famed climate journalist Bill McKibben writes, "The first principle of fighting the climate crisis was simple: stop lighting coal, oil, gas, and trees on fire, as soon as possible. Today, I offer a second ground rule, corollary to the first: definitely *don't build anything new* that connects to a flame."

Achieving this is not a pipe dream or children's crusade. It is pure physics and straightforward economics. Electrifying every flow of energy will reduce the cost of energy for virtually everyone, rich or poor. It will require equitable financial tools so that the transition is affordable for all parties.

The Haw River House is a 2,600-square-foot, net-zero home located in North Carolina. Its rooftop solar array provides all of its electricity. Insulation, passive house design, energy-recovery ventilators, and solar reflective shades improve energy efficiency and help maintain a constant temperature. A geothermal heat pump handles the rest of the heating and cooling needs. It is also water independent; a small well supports a rainwater collection and purification system that, when full, can provide water for 230 days.

No one has studied the precise carbon footprint of the United States more assiduously than physicist Saul Griffith. However, his work in the United States is applicable to every country in the world. When he refers to electrifying everything, it means replacing a fossil-fuel economy with wind, solar, hydro, electric vehicles, heat pumps, and a masterfully designed electric grid that allows energy to flow from source to sink and back, easily and effectively. Batteries large and small for electricity storage will be ubiquitous. In a connected grid, cars and houses can be energy receivers at night and energy providers during the day. According to Griffith, if we electrify the whole of the world economy, we will need less than half of the primary energy we currently use.

At first blush, a gas-fired generator appears far more effective than a wimpy solar panel. Looks deceive. Renewables are more efficient. Coal- and gas-fired generation plants convert heat to electricity using boilers, steam, and turbines, resulting in overall energy loss 68 percent for coal plants and 42 to 50 percent for many natural gas turbines. Solar and wind convert energy from the sun more directly. The energy of a photon is transferred to electrons in the semiconductor. There is no combustion. Wind turns a turbine. The wind is free, no heat required. Because of these efficiencies, conversion from combustion to renewables will reduce overall energy use in the United States by *23 percent* for starters. And renewable energy is already the least expensive form of newly generated energy and is continuing to get cheaper. That cannot be said of any other form of energy generation.

A greater amount of energy can be saved by electrifying cars, trucks, and trains. Eighty percent of a car's energy heats the air after it heats the engine block, muffler, and tailpipe, any of which can cause third-degree burns. Thus, 20 percent of the energy goes to the wheels. In electric cars, 90 percent of the energy goes to the wheels. If electric cars were powered by renewables, an additional 15 percent of primary energy needs would be eliminated.

The amount of energy required to produce fossil-fuel energy is sizable. Approximately one million tons of coal are transported to the east coast of China daily. The Daqin Railway is the busiest freight line in the world, with coal trains exceeding four miles in length. The elimination of fossil fuels eliminates the Daqin Railway. Electrification makes fossil fuel exploration, mining, drilling, extraction, pumping, refining, and transporting unnecessary, saving another 11 percent of primary energy. Not included is the significant energy used to create the steel for the mining and drilling equipment, oil tankers, liquid natural gas terminals, railway cars, refineries, and gas stations. Nor does this include the energy required for remediating and treating the damage, pollution, and health impacts caused by fossil fuels.

At home, in offices, and in industry, heat pumps can replace gas, electric, or oil burners for warmth and cooling and for heating water. Heat pumps use electricity to extract heat from either the air or the ground, and produce three times as much heat per unit of energy as gas, oil, or resistance electrical heating. This will save another 5 to 7 percent of primary energy if implemented. LED lighting, which is five to ten times more efficient than conventional lighting technologies, with bulbs that last five to ten times longer, reduces overall primary energy use by 1 to 2 percent.

For industrial processes that require high temperatures and thus large amounts of energy, including iron smelters, blast furnaces, and cement plants, as well as certain types of transportation, such as shipping, trucking, and air travel, using electricity to produce hydrogen fuel may be the best option. Hydrogen is the most abundant element in the universe. Pound for pound, it holds almost three times the amount of energy as a fossil fuel. It can exist as a gas or liquid. But to be useful as energy it must first be separated from its source material. One source are hydrocarbons, including methane, but the waste product is carbon dioxide. Another source is water, whose byproduct is oxygen. Splitting hydrogen from oxygen requires a fuel cell called an electrolyzer and a large amount of electricity. If the power source is renewable—solar, wind, hydro, geothermal—then the resulting product is called green hydrogen. This clean energy source can be made wherever water and electricity are available. Although more expensive than fossil-based hydrogen, costs are plunging as renewables become cheaper. Many governments see green hydrogen as an important part of the world's future energy mix. The European Union is investing in clean hydrogen. In 2020, Saudi Arabia announced a $5 billion hydrogen plant powered by wind and solar. The head of the International Energy Agency believes green hydrogen is where wind power was ten years ago.

The Asian Renewable Energy Hub, generating twenty-six thousand megawatts of wind and solar capacity for Australia and Asian markets, is being proposed for sun-baked Western Australia. Covering twenty-five hundred square miles, it will generate the power used to produce ammonia for clean hydrogen for domestic and Asian steel production, mineral processing, and manufacturing.

When added up, electrifying everything results in a 60 percent reduction in overall energy use in the United States while providing the same desired or needed products and services. The impacts in the rest of the world are similar. However, we can reduce energy use further. That 60 percent does not include building retrofits that reduce energy use by 40 to 80 percent, smart thermostats, and more efficient appliances. If cars were made of carbon fiber instead of steel, and used regenerative braking systems that capture the energy lost when slowing down, automobile energy use could drop by another 50 percent or more. It does not include a connected world where people work locally instead of undergoing lengthy commutes. It does not include circular material flows creating significant reductions in energy requirements. In other words, the 60 percent reduction in overall energy use does not take into consideration that we need to cease consuming the world faster than it can be regenerated.

While electrifying everything ultimately reduces overall energy use, it requires doubling the amount of electricity needed, from 2.3 terawatts to 4.8 terawatts by 2050, a difficult task. If we live in 2050 as we do today, a great deal of energy will still be wasted. Half a million people dancing and gambling on the world's three hundred fifty cruise ships at any given moment is an extraordinary use of valuable energy. Unless we are mindful of our impacts, the potential of renewable energy sources will be overwhelmed by future energy use. Electric cars are not a panacea. Driving a five-thousand-pound EV to get Chinese takeout for dinner is a waste of energy, even if renewable. Lithium-ion batteries used to power EVs require rare minerals and extractive mining.

Electrifying everything is transformative. We don't need breakthrough energy technology to achieve signifi-cant reductions in carbon emissions. We have the tools we need right now. We don't need carbon capture-and-storage schemes at power plants to keep their emissions from rising into the air. We can stop burning fossil fuels instead. We don't need to make big personal or economic sacrifices to achieve climate goals. We can still have our cars. But they must be electric. Energy from the power grid must be all renewable. But we will need only half as much. Electrifying everything will be a mammoth job and it must be done quickly. It is an opportunity. Twenty million jobs will be created in the first decade of transformation and millions of people will have permanent employment in the new energy economy. Costs will come down. Benefits will accrue. Skies will clear. Cities will be quieter. Homes and offices will be smarter. Life will go on. Better than before. There is enough future global capacity for solar and wind to provide a decent standard of living for all of humanity if good examples can be established for sustainable energy consumption. ●

Heat pumps pull heat out of the air or ground. Working like an air conditioner in reverse, heat pumps can supply a household or entire building with all the required heat and hot water and reduce energy use by 50 percent. If the electricity used to power the heat pumps is renewably sourced, they produce a 95+ percent reduction in greenhouse gas emissions.

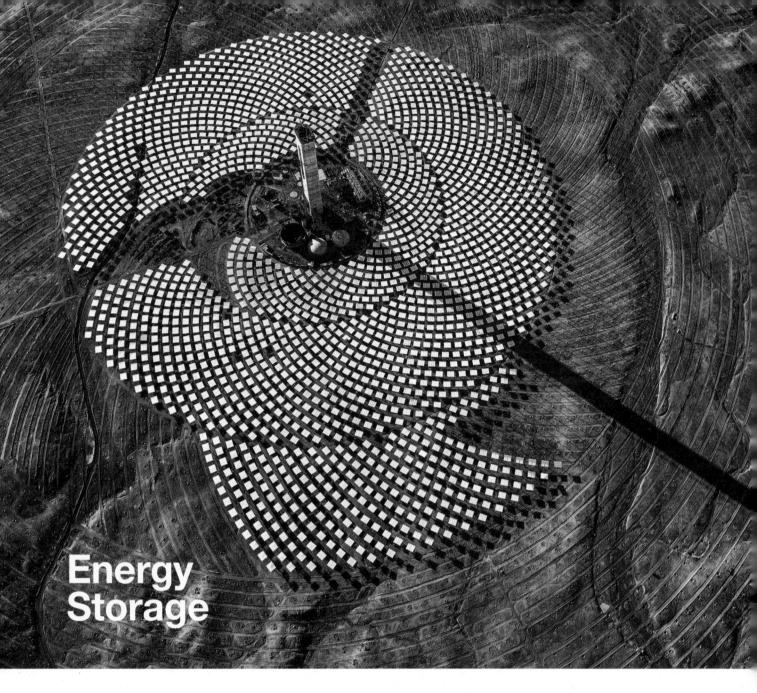

Energy
Storage

Wind and solar, the two primary renewable energy sources, are intermittent. The wind doesn't always blow, the sun doesn't always shine, and they certainly don't get stronger to meet additional demand. To ensure a reliable electrical grid, utilities are required to have flexible power—electrical generation capacity that is always available regardless of time, season, or weather. A totally renewable electrical grid would require energy storage capacity of about 4.4 million gigawatt hours per year by 2050, 275,000 times more than current storage capacity.

Currently, most renewable energy storage capacity is in the form of pumped storage hydropower. When electrical energy is abundant or not needed, it is used to pump water up to a reservoir. When energy is needed, water is released, spinning a turbine to generate electricity. How-ever, hydropower storage has limited capacity because it needs nearby storage sites at a higher elevation and it usu-ally requires the creation of artificial lakes or dams, which can harm the local environment. Some are evolving this technology: by replacing water with a denser fluid, RheEn-ergise can build smaller pumped storage systems on lower inclines that generate the same power as traditional dams. When built underground, these systems free up the land above for renewable energy sources like solar or wind or other development. While it is an essential energy storage solution, pumped storage hydropower is limited by its geographic constraints.

The storage technology that powers most modern devices, from smartphones to electric vehicles, is the lithium-ion battery. Until recently, battery storage was too costly to implement at a large scale, but that is no longer the case. The cost of battery storage dropped by 90 percent between 2009 and 2019, and is expected to drop another 75 percent by 2050. As a result, electric vehicles are expected to achieve cost parity, if not become cheaper, by 2024. Lithium-ion battery storage can be implemented almost anywhere and has a rapid response time. Pumped hydroelectricity takes a few seconds to respond to demand. Batteries respond in milliseconds, enabling them to cover rapid spikes in demand. This makes lithium-ion batteries an ideal replacement for what are known as peaking plants—gas-fired power plants that are turned on to address unpredicted spikes in demand. Battery storage of the future won't be limited to massive storage facilities. Lithium-ion technology is used in both large-scale energy storage and electric vehicles. When electric vehicles dominate the market, the world will have billions of lithium-ion batteries connected to the grid, ready to share energy when it is needed.

However, lithium-ion batteries will not meet all of our storage needs. Although there have been dramatic breakthroughs in these batteries, including EV batteries capable of fully charging in five minutes, as well as new recycling technologies, the core materials have significant environmental costs, the batteries degrade with use, and some lithium mines are sites of human rights abuse. Even without these issues, lithium-ion batteries are effective only at providing energy for hours at a time, while in some geographies, seasonal storage will be required to make up for the lack of sunlight in the winter months.

Engineers and scientists have created inventive solutions for both of these issues. First, some are creating batteries using other materials. Ambri is developing a liquid metal battery for grid storage using liquid calcium and solid antimony that could be a third of the price of competitive lithium-ion batteries, with minimal degradation. Researchers at the University of Southern California have created a new type of flow battery that uses iron sulphate, a waste product from the mining industry that can release energy over a longer period of time than traditional lithium-ion batteries. Form Energy is installing an "aqueous air battery system" in Minnesota that "leverages some of the most abundant materials on the planet" instead of using lithium and other metals. Others, like Noor Midelt in Morocco and Norwegian company Energy-Nest are implementing thermal batteries made from molten salt or crushed volcanic rock. Such systems work by using excess energy to heat insulated reservoirs, which can be used later to run steam turbines. The molten salt batteries are designed so that heat is lost slowly, enabling inexpensive storage—estimated to be thirty-three times less expensive, per kilowatt-hour, than batteries. MGA Thermal is exploring another option: to replace the coal in coal power plants with a block of blended metals, roughly half the size of a toaster, that is stackable like Legos, designed to store extraordinary amounts of heat. Instead of burning coal to boil the water in a steam turbine, the alloys can be heated by renewable energy, and added to or removed from the boiler to scale energy generation up or down to meet demand, replacing coal entirely while utilizing the same infrastructure.

Instead of water being pumped uphill, concrete can be raised upward. The Swiss company Energy Vault has created a massive six-armed crane connected to wind turbines or solar farms that uses their excess renewable energy to lift thirty-five-ton composite concrete blocks into a giant tower. The gravitational force of the descending blocks spins the turbines, generating electricity. The technology can be used in almost any topography, unlike storage technologies that rely on physical height differences in the landscape. Researchers at the University of Malta are taking the concept of hydroelectric storage one step further. Instead of pumping water uphill, they pump water into a chamber, causing the air inside the chamber to be compressed. When energy is needed, the air is allowed to expand, pushing the water back out and through a turbine to generate electricity.

Finally, some are looking to create a form of energy similar to fossil fuels in its portability and energy density. The leading candidate is hydrogen—the same element that powers our sun. While hydrogen can currently be derived affordably only by using fossil fuels, green hydrogen, created by splitting hydrogen from oxygen atoms in water, is becoming cost-effective as the cost of renewable energy continues to plummet. Germany is going all in on green hydrogen, investing $1.54 billion in hydrogen and fuel cell technologies by 2026 to become a world leader in hydrogen technology. Green hydrogen will not only be a powerful new form of energy storage but will be crucial for transforming fossil-fuel-intensive industries like steel and cement to carbon neutrality.

All the energy storage technologies have one thing in common: they are analogous to solutions found in nature. Plants store energy from the sun by turning it into sugar; a geyser erupts when enough water has pressurized its chamber. These technologies aim to preserve the energy found in everyday phenomena. ●

This is the first stage of the now completed Cerro Dominador concentrated solar power plant situated in the Maria Elena Commune in the Atacama Desert, Chile. The molten salt technology employed in the plant can store up to eighteen hours of electrical generation capacity, which allows for a continuous flow of solar energy twenty-four hours a day. The completed plant covers 1,750 acres and contains 10,600 heliostats that automatically track the sun.

Microgrids

When Pacific Gas & Electric cut power to customers in northern California during the fall of 2019 to reduce the risk of sparking a wildfire, it plunged two million people into darkness. The lights stayed on, however, for the members of the Blue Lake Rancheria tribe, located near Eureka. The tribe's casino hotel was able to provide rooms for critically ill patients from facilities that had lost electricity. The gas station and store were among the few businesses to stay open. Ultimately, the tribe helped more than ten thousand people during the crisis, roughly 8 percent of Humboldt County's population. Why did the lights stay on? The tribe had built its own power grid.

The Blue Lake story starts with a massive earthquake near Japan in March 2011. The resulting tsunami crossed the ocean and flooded the California coast near Eureka, forcing many residents to take refuge at the tribe's resort. Realizing afterward how vulnerable they were to a power blackout, tribal leaders decided to build a state-of-the-art microgrid on the reservation with financial support from the state. A microgrid is an assembly of storage batteries, distribution lines, and power sources, including wind, hydro, geothermal, and solar. Although usually connected to the regional grid of electricity transmission, microgrids are run as independent utilities. If power is cut on the big grid, they become "islands" that can supply electricity on

their own. To increase self-sufficiency, the Blue Lake tribe partnered with a German firm to install smart software that integrated weather forecasting with projected electricity demand, creating a sense of certainty in uncertain times.

The advantage of microgrids was reenforced in the fall of 2012 when Hurricane Sandy struck the northeast, knocking out power to more than eight million people. Lights stayed on where microgrids operated, including the Food and Drug Administration's White Oak research station and parts of New York University's campus. Princeton University's cogeneration microgrid provided electricity to four thousand apartments, three shopping centers, and six schools for two days after the storm struck.

Globally, nearly 800 million people lack access to electricity, more than 60 percent of them in rural areas. Around 14 percent of households on Native American reservations lack electricity, largely as a result of their remote locations. That's one reason why microgrids are now being planned on tribal lands in Oklahoma, Alaska, Wisconsin, and California. In Nigeria, 77 million people—roughly 40 percent of the total population—do not have reliable electricity. The situation is particularly acute in Nigerian agriculture. Electricity is essential to farm activity, including milling grains, refrigeration, and pumping water for irrigation. Traditionally, power has been provided by diesel machinery. However, fuel costs can exceed annual incomes. Microgrids and solar home systems have the potential to change these dynamics significantly by reducing costs to farmers and boosting productivity while improving human well-being.

Microgrids are not a new idea. The first operational one was created by Thomas Edison in 1882 at his Pearl Street power plant in Manhattan. Before centralized grids were established, small microgrids served cities, providing energy to hospitals, universities, schools, and prisons. These microgrids mostly relied on fossil-fuel-generated heat and power systems, including steam. Today, microgrids are getting another look as risks to conventional grids rise under stresses caused by weather events amplified by climate change. According to the World Bank, 55 percent of U.S. power outages between 2000 and 2017 and over a third of Europe's outages were caused by extreme weather events. In addition to their reliability, microgrids are more efficient than centralized grids. U.S. grids lose 6 percent of generated energy as it travels over high-voltage transmission lines. Grids in India lose up to 19 percent.

As states, cities, and companies set carbon emissions reduction and elimination goals, microgrids are increasingly seen as a way to deliver renewable energy to customers. Microgrids now often source their energy from solar panels or wind turbines, which have fallen dramatically in price and are being used for a wide variety of climate-friendly electrical purposes, including vehicle charging stations in cities. In 2018, Illinois regulators approved Commonwealth Edison's microgrid cluster plan for Chicago, one of the first in the nation designed to integrate microgrids with renewable energy resources.

The U.S. Department of Defense is the single largest consumer of petroleum in the world. To counter this oil dependency, the department has begun shifting from diesel generators to electricity produced by on-base microgrid utilities powered by renewable energy, including at large institutions such as the Navy base in San Diego. Switching to a microgrid system at the Marine Corps depot on Parris Island, South Carolina, is expected to save $6.9 million in utility costs each year and reduce energy demand by three-fourths.

As the range of potential uses for microgrids expand, new types of systems are being developed, including ones based on technologies that allow devices to communicate with each other over the internet, increasing efficiency. While microgrids are usually customized to meet specific needs and conditions, the industry is also developing modular microgrids that can be manufactured in standard units and installed quickly. New technology has also lowered the cost of battery storage while increasing capacity. On the horizon, microgrids could use hydrogen-fuel-cell technology to lower its carbon footprint even more.

New technology has also inspired novel concepts. In Bangladesh, four million households in rural communities have bypassed the conventional options for electrification by installing solar home systems instead—among the highest total in the world. However, these systems have limited capacity and remain prohibitively expensive for large sections of the population. Enter a technology called swarm electrification. In Shakimali Matborkandi, a village south of the capital, Dhaka, a microgrid corporation called SOLshare has installed a peer-to-peer sharing grid that employs the company's smart electrical meter, which allows owners of solar systems to buy and sell electricity directly from other community members. This technology is called "swarm" because it scales up easily and quickly. Individual households link together first, and as the total grows, they are collectively able to take on more electrical tasks. Households that cannot afford a solar system can still participate by installing a SOLshare electrical meter and then buying electricity from their neighbors.

This type of microgrid has an extra climate benefit. By connecting solar home systems together, SOLshare unlocks up to one-third more solar energy. Usually, any electricity generated by a solar home system that is not immediately used is lost. When a community is linked together, with some producing excess energy and others consuming it, the community can more fully take advantage of its solar panels. SOLshare estimates that its systems across Bangladesh collectively reduce eleven thousand pounds of carbon dioxide per year. It also increases resiliency. If one household's solar home system stops functioning, they can continue to purchase electricity from their peers.

Although microgrids face a variety of challenges, including relatively high costs for construction, regulatory obstacles, financial incentives that advantage fossil fuels, and opposition from traditional utilities, they have the potential to make a major contribution to ending climate change. By tapping nearly unlimited quantities of renewable energy from nearby sources and redistributing it locally, microgrids can empower communities to be self-sufficient and resilient to climate extremes while reducing greenhouse gas emissions at the same time. ●

Kahauiki Village in Hawaii is a 144-unit community that provides long-term, affordable residences for homeless families and a suite of on-site services and facilities. Funded through a public-private partnership, it was built using low-cost, maintainable, and sustainable construction solutions, including repurposing emergency homes originally built for the 2011 Tohoku tsunami victims. The community is mostly energy independent, powered by a 500-kilowatt solar-powered microgrid with 2.1 megawatt hours of battery energy storage. The system is supported by some gas appliances, a generator, and a trickle of backup power from the grid to charge the batteries in extended overcast conditions.

Industry

Every industry is a system, and every industrial system is extractive, whether it be energy, food, agriculture, pharma, transport, clothing, or healthcare, among others. Extraction takes resources from the living world, which causes harm. The result is less life. Extraction is thus degenerative. Every industrial system is a direct cause of global warming, not only because of greenhouse gas emissions but because of the damage to soil, water, oceans, forests, air, biodiversity, people, children, workers, and cultures. Harm is not the intention of companies, but in order to become regenerative, a company must first recognize that it is innately degenerative. This is not an accusation; it is a biological fact, and represents a huge opportunity.

The focus by industry on climate is directed to greenhouse gas emissions caused by production, transportation, and operations. That makes sense, because industry accounts for 30 percent of global energy consumption. In China it is about 50 percent. Greenhouse gas emissions are created by an extraordinary variety of activities, from machining to smelting, rail systems to refining, air freight to office towers. Energy-intensive processes involve chemical, physical, electrical, and mechanical procedures. The external impacts of industry include air and water pollution, toxic emissions, poverty wages, loss of biodiversity, deforestation, destruction of Indigenous cultures, and advertising that encourages overconsumption of cars, electronics, travel, alcohol, tobacco, fast fashion, and junk food.

The worldwide business community, although initially slow to respond to the need for significant action on climate, biodiversity, and social justice, has been moving more resolutely in recent years, with its primary focus on boosting efficiency, reducing energy use, utilizing more renewable energy, eliminating toxins, recycling, employing circularity to decrease waste, and purchasing carbon offsets.

In the past, improving carbon footprint metrics focused on specific processes, functions, and outcomes within a business. In this sector of the book, we focus on the entirety of an industry. If a specific product is harmful or unnecessary, how it is made, how it involves the circular economy, or how much energy is supplied by renewables is irrelevant. The top line of every profit-and-loss statement will determine the future of humanity, not the bottom line. Does it emit or sequester carbon? Does the top line cause loss of life, habitat, and natural resources, or increase life, habitat, and the regeneration of nature? Does it foster or degrade social equity? The expertise of the modern industrial world is not in question. The goals and assumptions are in question. We have one job to do on this planet: to protect and enliven it for the future. Either a corporation is doing this or it is not.

One company can serve as an example of why focusing on its parts and pieces can obscure its overall impact and the deeper issue of whether its products or services are needed. PepsiCo operates the largest trucking fleet in the world, including 11,245 tractors, 3,605 trucks, 18,648 trailers, and 17,000 pickups. The top-selling products transported in those trucks are Pepsi, Mountain Dew, Lay's potato chips, Gatorade, Diet Pepsi, 7UP, and Doritos, all known, even in polite company, as junk food. Junk food is defined as food that has low nutritional value, is sold in convenient packaging, and needs little or no preparation. It contains high amounts of fat, salt, sugar, and starch, which cause chronic disease including obesity, type 2 diabetes, heart disease, stroke, hypertension, and more. Despite overwhelming evidence that sugary drinks are harmful to young children and teens, PepsiCo continues to increase its promotion of soft drinks on social media, websites, apps, television, and at sporting events. Black and Latino children see twice as many soft drink ads as White children. Ads use Black and Hispanic celebrities such as MichaelJordan, Penelope Cruz, Jennifer Lopez, Nicki Minaj, LeBron James, Cardi B, and Serena Williams. Pepsi works with other soft-drink companies to prevent bans on extra-large sodas or soft-drink taxes. Pepsi is committed to using 100 percent renewable energy in its U.S. direct operations. The question for Pepsi and many other companies is: Renewable energy to what end?

To credibly address the climatic crisis, there needs to be a corporate shift beyond initiatives, commitments, offsets, and endorsements of social justice. A solar-powered soft-drink factory does not address the root cause of the climate crisis. Pepsi's commitment to climate ignores the well-being of children. The immensity and inertia of giant food companies creates a sense that they are locked into making products that are inherently detrimental. Maybe, maybe not.

Some of the biggest companies in the world announced in 2020 that they're committed to becoming regenerative companies. They will need to determine what that means for every aspect of the business. Cynicism is understandable, given the practice of greenwashing. But underneath these commitments are people, just like the person reading this page. People who have children, families, and communities, people who see the looming crisis writ large. Many large and thoughtful corporations are adopting better standards for measuring their performance. This sector examines the challenges and what can be done. There is no point in tiptoeing around the causes of global warming. We are either in a crisis or we are not. At the same time, there is no gain in blaming and shaming either. We know what needs to be done. The question is how do we come together, get it done, and make it right? ●

Textile factory in Huai'an, Jiangsu Province, China.

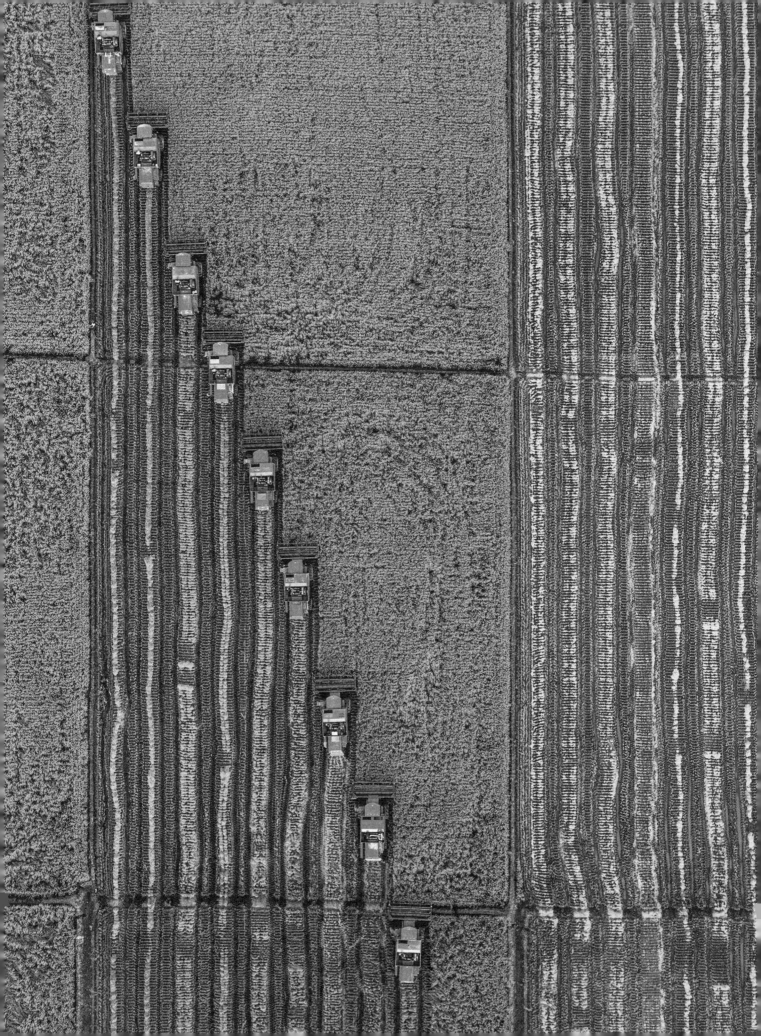

Big Food

The $15 trillion food industry is the largest industry in the world and is a monumental contributor to climate change. Transforming the food industry holds extraordinary opportunity for humankind and is fundamental to regeneration. Industrial food is grown using disruptive, unsustainable chemical agricultural methods that damage soil, people, and nature, and pollute the water. Farmworkers are poorly paid, have few if any rights, experience high rates of injury, are exposed to pesticide poisoning, are rarely covered by health insurance, and are generally exempt from labor laws. Highly processed food products are causing a global epidemic of human metabolic disease, including obesity, diabetes, hypertension, stroke, and heart disease. The food we grow, make, and eat is damaging our bodies, farming communities, and the planet.

The food system comprises a highly integrated system that feeds everyone. Components include growing, packing, processing, distributing, selling, storing, marketing, consuming, and disposing of food. Within the system, large multinational corporations dominate. Four chemical companies—Bayer, Corteva (formed by the merger of Dow and DuPont), ChemChina, and BASF—control 70 percent of the global market for seeds, fertilizer, and pesticides. Four corporations—Archer Daniels Midland, Bunge, Cargill, and Louis Dreyfus control more than 70 percent of the global grain trade, including livestock feed. In the United States, half of the national grocery market is controlled by four companies, with Walmart capturing nearly one-third. The ten biggest global food companies largely determine which staple foods are grown and what most people eat.

These multinational firms and their market power are collectively termed Big Food. Because they sell across broad geographical areas and markets, Big Food requires its products to be identical in composition, flavor, and texture. Consistent taste necessitates uniformity in seeds, plants, and animals—the raw ingredients. To satisfy their corporate customers, farmers grow crops in monocultures with little genetic diversity on hundreds of millions of acres of farms ranging from one square mile to forty square miles in size. Monocultures stress the soil and require increasing amounts of fertilizers, herbicides, and pesticides to maintain profitable yields as the soil becomes less fertile. This increases stress on farmers, who try to increase production the only way they know how: by applying more expensive inputs. Burdened by record debt, trade wars, climate change, and low commodity pricing, farming is one of the professions in the world with the highest rates of suicide.

While farmers large and small struggle to break even on a year-to-year basis, the top ten food companies never have a bad year. In 2019, their revenues exceeded $500 billion. The bulk of Big Food's sales come from ultraprocessed foods, or, in Michael Pollan's words, foodlike substances: Mini Oreos, Kraft Macaroni & Cheese, Honey Buns, Gatorade, M&M's, Doritos, Kraft Singles, Cini Minis Cereal, Gobstoppers, Tombstone Pizza, Spam, Cap'n Crunch, Count Chocula, bologna, and so on. Nearly 60 percent of the calories consumed in the United States come from ultraprocessed foods. Harvard Medical School defines ultraprocessed foods as foods that "are made mostly from substances extracted from foods, such as fats, starches, added sugars, and hydrogenated fats. They may also contain additives like artificial colors and flavors or stabilizers." It might be fair to say that most consumers cannot explain what is listed on the ingredient panels. That means people do not know what they are eating, because it is not food. It is a chemistry experiment.

Ultraprocessed foods are addictive because our taste buds were hacked long ago by food chemists. Sugar is an addictive substance. Two twelve-ounce cans of Mountain Dew contain nearly a half cup of it. So too is monosodium glutamate (MSG), which is added to processed food under fifty-five different names and forms. Snacks, chips, and processed meats are laden with addictive salt. Soft drinks and energy drinks are laced with addictive caffeine. Ultraprocessed food consists mostly of fat, carbohydrates, protein, salt, and "natural flavors" comprised of any of more than a hundred chemical additives. Natural flavor is the fourth-most-listed food ingredient in the approximately eighty thousand food products available, but they are anything but natural. These flavors are about smell more than taste, since that is how the body primarily senses flavor. Some of these same food ingredients, like propylene glycol, BHT, BHA, tertiary butylhydroquinone, and polysorbate 80, can be found in your shampoo and hair conditioner, where they are known as parfum.

Foods made of starch, sugar, salt, and fat release dopamine and serotonin, rewarding the same pleasure centers as do cocaine and heroin. While these foods provide instant gratification, they do not nourish. They malnourish, which is why they are called junk food. The body senses the deficiencies and craves more food to compensate for food depleted of nutrients, grown on depleted soil.

The first day of rice harvest in Heilongjiang Province is a showcase for growing China's primary staple. Heilongjiang is China's most productive province for rice, and also grows the highest quality. Here on Erdaohe Farm was Longjing 46 or "Dragon Rice 46," which was planted in early May and yields 866 pounds per acre.

This nutritional hunger drives obesity. Overweight people are virtually always nutrient deficient. People with modest incomes will buy more of the very foods that malnourish them, because that is all they can afford. We picture starvation as small kids with skinny arms and sunken cheeks. Obesity is starvation, too.

The true cost of junk food is multiplied several times over by the resultant healthcare costs. The United States spends two times more on healthcare than on food, yet globally 50 percent of disease and mortality is attributed to what we eat. In 1990, no state in America had an obesity rate higher than 20 percent. In 2020, no state had an obesity rate lower than 20 percent, and many states surpassed or were closing in on 40 percent. More than 80 percent of food advertising dollars promote fast food, sugary drinks, candy, and unhealthy snacks. Ten of the twenty bestselling breakfast cereals contain between 40 and 50 percent sugar. Makers of soft drinks spend twice as much advertising to minority children than to White children.

Big Food knows about the deteriorating health of their customers, but for legal, commercial, and reputational reasons won't admit their role. Coke paid millions of dollars for "studies" that contradicted peer-reviewed science, promoting the fallacy that obesity is caused by lack of exercise and that sugar is part of a balanced diet. Meanwhile, food companies lobby the government to prohibit regulation. In 2018, when the $900 billion Farm Bill was being debated in Congress, experts in health, food security, and poverty urged the USDA to remove soft drinks from the authorized purchases allowed by the Supplemental Nutrition Assistance Program (SNAP), the government aid program that supports 42 million low-income Americans, also known as the food stamp program. The soft-drink industry was unilaterally opposed. In preparation, the industry wrote a playbook on how members of Congress should object to the proposed restrictions. Members of the House read into the record prescripted phrases like "food police," "unpatriotic," "nanny state," "there will be confusion at the checkout stand," "denial of freedom," and "a violation of the recipients' constitutional right to pursue happiness."

A study tracked nearly half a million SNAP participants over a decade, showing that recipients had twice the rate of cardiovascular disease and were three times more likely to die from diabetes than nonrecipients. The study showed that if SNAP was modified to incentivize the purchase of healthy foods—vegetables, fruits, whole grains, nuts, fish, and plant-based oils—and disincentivize sugary beverages, junk food, and processed meats, it would prevent 940,000 cardiovascular events and 147,000 cases of diabetes annually. The resulting health cost savings per year would be $429 billion, six times greater than the $70 billion cost of the SNAP program itself. The proposed incentives and disincentives were simple: healthy food would cost 30 percent less, and soft drinks and junk food would cost 30 percent more. If every family shifted their purchases accordingly, it would increase their spending power by $21 billion per year. The cost would be more than paid for by decreased Medicaid and Medicare

expenses. Who could oppose a simple, life-enhancing, money-saving pivot in the SNAP program? Big Food could. Their strategy is visible during the first ten days of every month, when food stamps are issued. During these periods, Pepsi and Coke increase their advertising for soft drinks and junk food in poor and disadvantaged neighborhoods. The combined compensation for the two CEOs of Pepsi and Coke in 2017 was $42 million.

We have come to believe that big is better, safer, cheaper, and more reliable when it comes to food and farming, that localizing the growing and manufacturing of food is a bygone pipe dream. The biggest myth is that if we do not deploy industrial agricultural techniques, the world will starve. Industrial agriculture dismantles the life, structure, and health of soil. Weakened soil needs more chemicals and produces more runoff, erosion, and nutritionally defective food. Breakfast-to-dinner saturation of ultraprocessed foods is a worldwide health disaster. More than two billion people are overweight, and 600 million are obese. India's diabetes rate has doubled in the past thirty years. When Kentucky Fried Chicken opened its first outlet in Beijing in 1987, the diabetes rate was 1 in 25. Today it is 1 in 10, and there are 4,200 KFC outlets, 3,300 McDonald's, and 2,200 Pizza Huts. Half of all Buddhist monks in Thailand are obese, because they depend on donated food. There is only one country in the world that has not seen a rise in obesity, heart disease, and diabetes: Cuba. They have no fast-food restaurants or ultraprocessed foods per se. They spend 11 percent of their gross domestic product on healthcare. The United States spends 20 percent.

Only one country has taken on the junk-food and soda companies: Chile. It became the second-most-obese country in the world, behind the U.S. Half of Chile's six-year-olds and 75 percent of its adults were overweight or obese. Chile decided to take action. Having Dr. Michelle Bachelet, pediatrician and former health minister, elected president was key. Today, black stop signs are printed on packages of foods high in sugar, saturated fat, calories, or sodium. Cartoon characters are banned from junk-food marketing. Tony the Tiger disappeared from Kellogg's Frosted Flakes. Trinkets are disallowed as a means of selling candy or sugary foods. Junk-food commercials are banned from 6:00 a.m. to 10:00 p.m. on TV and radio. Junk food and sugary foods cannot be sold or served in schools. An 18 percent tax on soft drinks was mandated. Food companies had to reinforce messages of healthy eating and activity to promote their brands. When these initiatives were first proposed, there was a tsunami of opposition from domestic and foreign food-company lobbyists. Researchers report that kids are now warning their parents about what to avoid. It has been transformative.

The task at hand is to feed the world, reduce greenhouse gas emissions, and create a just, regenerative food system that honors its essential workers so that they work in healthy environments and receive a living wage. This becomes ever more crucial as global warming unleashes more migration due to crop loss, drought, and floods. In both Africa and South and Mesoamerica, people move north to escape poverty and hunger. A crucial response to both global warming and social justice is the transformation of climate-impacted lands so that they are more resilient and resistant to drought and flood, more productive for families and farmers. Due to droughts, overgrazing, deforestation, and industrial agriculture, 25 percent of the earth's land area is degraded. An estimated 700 million homeless and hungry migrants will be with us by 2050 unless we act. We can start the regeneration of farmlands, forestlands, wetlands, and grasslands and increase their resilience.

The number one solution to human health and regenerative agriculture is to stop purchasing ultraprocessed food. Individuals, companies, co-ops, cafeteria, and hospitals need to embrace real and regional food where possible—boosting community-supported agriculture, local farmers, and organically farmed food; educating your friends and colleagues; and engaging in food-related activity that brings people together. This is not about banishing candy or ice cream cones. It is about banishing commercially sponsored ignorance. In order to fully realize the potential of a transformed food system, we must start at square one and recognize that hungry people are everywhere and we need to feed one another. A sustainable food system that provides all people with nutritious, delicious, healthy food is the ultimate act of regeneration. To paraphrase chef José Andrés, we need to treat every hungry and homeless person as a deserving guest. This will help us design the food system of the future—a food system that addresses global warming while providing dignity and purpose to all of its essential workers. When we provide clean, healthy food, we are engaged in the act of healing land, body, and climate. Food is the one sector that touches every aspect of culture, climate, health, and ecology, and it is a sector where every person has influence. ●

CP Group's chicken facility processes 120 million chickens per year (today 200,000, but during preholiday periods up to 400,000) with over two thousand employees working a single eight-hour shift. Ninety percent of their chicken is for domestic consumption, and 10 percent for other parts of Asia. All parts of the chickens are used; even the chicken fat is used in paint, and the feathers are processed into powder for animal food. The internal organs, feet, and heads are sold for human consumption. In 2013 there was a poultry health scare/scandal in China that depressed demand, but by 2015 chicken consumption had increased 20 percent. This facility supplies most of the major fast-food brands in China, including McDonald's, KFC, Burger King, Pizza Hut, Papa John's, Tyson, Walmart, Metro, and Carrefour, among others.

Healthcare Industry

The mission of the healthcare industry is to maintain a person's well-being and restore their health if they become injured or sick. Climate change amplifies mental and physical health issues and is being called the greatest health threat of the century. There has been a 50 percent increase in the past two decades in heat-related deaths among older people. Rising levels of airborne pollution are increasing respiratory and cardiovascular diseases. Warmer temperatures lead to higher rates of vector-borne diseases, which heightens risks associated with pregnancy, including reduced birth weights. Climate-related natural disasters have been rising dramatically and have been experienced by 1.7 billion people in the past decade. The negative effects of climate disruption on public health affect poor people and communities of color disproportionately, and could push 100 million more people into extreme poverty by 2030.

The healthcare industry itself is a significant contributor of greenhouse gas emissions, led by the carbon footprint of hospitals and pharmaceutical companies. Health care is directly responsible for nearly 5 percent of global carbon dioxide emissions—over two billion tons each year, more than half of it caused by the United States, China, and the European Union. In the United States, the healthcare sector is responsible for 10 percent of carbon dioxide emissions and significant amounts of waste.

There are two different healthcare industries, and they overlap. One is made up of public and global health professionals. Whether in hospitals, clinics, war zones, or refugee camps, doctors, orderlies, and nurses are there, working difficult circumstances and under great duress, ministering to those in need regardless of income, demographic, race, gender, or the cause of outbreaks, disease, or injury. In the case of epidemics, they are literally the front line ensuring that regional outbreaks of Ebola, dengue fever, cholera, HIV, and Zika do not become widespread or global. Frontline workers have been and continue to be the tireless heroes and sheroes in virtually all countries, espousing and teaching about nutrition, preventive care, prenatal care, and vaccines. The tradition is rooted in Christian traditions, and in America includes Rebecca Lee Crumpler, the first Black woman to earn a medical degree in the United States, who ministered to newly freed slaves in the 1860s, and Sara Josephine Baker, who took her New York University M.D. to Hell's Kitchen in New York City and educated people on the basics of nutrition, infant care, and sanitation.

However, there is another healthcare industry, and it is failing. It is the allopathic medical system that, abetted by Big Pharma, focuses on symptoms instead of causes. It prefers that patients take drugs for months, years, or a lifetime. The numbers tell the story. The worldwide adult obesity rate has multiplied by six since 1975. In the United States, 73.6 percent of adults over the age of twenty are overweight or obese and it leads the world in rate of obesity, at 42.5 percent. The number of adults with elevated blood pressure—the leading risk factor for cardiovascular disease, which kills more people each year than any other cause—doubled between 1975 and 2015. Deaths from noncommunicable diseases in low-income countries increased from 23 percent in 2000 to 37 percent in 2015. In 1980, people with diabetes numbered 108 million globally; in 2019, 463 million people were living with diabetes. Life expectancy in the United States is declining, partly as a result of nearly five hundred thousand opioid-related deaths from 1999 to 2019. Suicidal thoughts and behaviors among Americans aged eighteen or younger rose by nearly 300 percent between 2009 and 2018, and 200 percent among eighteen-to-thirty-four-year-olds. There is a health crisis. And a climate crisis. We need to address them both. We need people to be healthy for the earth to be healthy, and vice versa.

In the late nineteenth century, medicine made exceptional breakthroughs, because it focused on the environmental and social conditions that prevent sickness, including hygiene, sanitation, improved sewage systems, clean water, proper ventilation of factories, and the prohibition of child labor. In the twentieth century, medical and pharmaceutical technology transformed the healthcare industry into a system emphasizing the treatment of the symptoms of disease, not its prevention. Research into the complexity of disease grew explosively, financed largely by pharmaceutical companies looking for effective drugs. Statins were such a drug. Statins lower the illness and mortality rate for those at risk of cardiovascular disease. How to lower the levels of problematic LDL cholesterol was the question drug companies solved. But was that answering the right question? Cholesterol helps repair damaged blood vessels, and that can cause blockages. The relevant question is: Why do more than half of adult Americans have damaged blood vessels?

Stanford-educated physician Dr. Molly Maloof calls modern medicine timeworn medicine. We know the reasons blood vessels get damaged: elevated blood sugar, hypertension, stress, and air pollution. The siloed approach to symptoms can reduce risk and discomfort for a patient, but it does not create health. The United States consumes 48.5 percent of all pharmaceutical drugs in the world and spends 20 percent of its gross domestic product (GDP) on healthcare. When it comes to emergency medical treatment for accidents, burns, and traumatic injury, the system may be the best in the world. In terms of health outcomes, the United States is ranked between eleventh and twenty-ninth in the world, depending on which metrics are used.

The interlinked effects of rising temperatures, environmental degradation, poverty, displacement, disease, and extreme weather require a comprehensive medical response rather than a symptomatic one. Human health is an outcome of healthy food and diet, which is dependent on healthy soil, which is an outcome of healthy agriculture. Highly processed foods are low in nutrients and high in unhealthy fats, salt, and sugar. Diet constitutes the direct cause of obesity, diabetes, and virtually all metabolic and cardiovascular disease. It is said that the United States cannot afford a public healthcare system, but it seems the United States can afford a public sick system, also known as Big Food.

According to the medical journal *The Lancet*, only half of all nations have drawn up national health plans for the climate crisis, including vulnerability and adaptation assessments for local and regional healthcare systems. Over half of the cities surveyed for the report expect climate change to significantly affect public health infrastructure, such as hospitals. Preparing national health plans for the climate crisis is essential but insufficient. An outstanding and rarely mentioned solution to climate change is universal healthcare. It should be instituted as a fundamental human right, not because of climate. Health is complete physical, mental, and social well-being and is inseparably linked to poverty alleviation. To address the climate crisis, populations need to be motivated, engaged, and active. That will not happen if people are sick, if they cannot feed their children, if they lack housing, or if they cannot find living-wage jobs. Regenerating climate and the planet means stopping the degenerative and debilitating predicaments facing the poor. Rwanda, ranked the 167th "poorest" nation in the world in terms of wealth per capita by the World Bank, has universal healthcare, an outcome of its convulsive experience with genocide. Its new regime devoted itself to healing the country in as many ways as possible. The poorest people in Rwanda get free care from a community-based healthcare system. The wealthiest pay eight dollars a year. It is a decentralized system comprised of seventeen hundred local health posts, five hundred health centers, forty-two district hospitals, and five national referral hospitals. The health minister, Dr. Agnes Bingawaho, engaged Paul Farmer and Partners in Health to design and create their healthcare system. The United States, with the largest GDP in the world, is ranked 8th in terms of wealth per capita. There is no free healthcare for the poor, the unemployed, children, the elderly, or migrants. Over 30 million citizens are uninsured, and those who are insured can find themselves bankrupted by medical bills that exceed policy limits. In some states, citizens and in particular poor adults are blocked from receiving government insurance. In Rwanda, Dr. Bingawaho

Khadiza, a twenty-two-year-old Rohingya refugee, holds her twelve-month-old son, Mohammad Haris, who is being treated for malnutrition at the Médecins Sans Frontières (Doctors Without Borders) hospital in the Kutupalong refugee camp in Cox's Bazar, Bangladesh. The Kutupalong camp is home to about four hundred thousand Rohingya refugees who were forced out and fled Myanmar.

mandated that every citizen in Rwanda be medically screened so as to detect disease or preconditions early. Preventative care in the United States is rarely if ever accessed by the majority of the population.

Providing care to patients and working to solve climate change can go hand in hand. A recent study showed that in the decade following the opening of an affordable health clinic in rural Indonesia, illegal logging in an adjacent national park decreased 70 percent. In addition to the health benefits of the clinic itself, including a decline in infectious and noncommunicable diseases, the clinic gives discounts to patients based on community-wide reductions in logging, simultaneously promoting public health and preserving forests. It also accepts barter as payment, which means patients are able to access healthcare by trading tree seedlings, handicrafts, and labor. This system was designed in collaboration with local communities. Researchers found that villages directly adjacent to the clinic, with the highest rates of clinic use, also had the greatest decrease in logging. "The data support two important conclusions," said Monica Nirmala, the director of the clinic from 2014 to 2018. "Human health is integral to the conservation of nature and vice versa, and we need to listen to the guidance of rainforest communities who know how to live in balance with their forests."

An emerging field is regenerative medicine. It employs techniques, phytonutrients, diet, supplementation, microbiota, and exercise that address the body as a system, not as a set of disparate organs and functions. The primary area of application is the human microbiome. There are thousands of species of bacteria that live within us (the human body is a community)—more bacterial cells than human cells—and according to the Human Project, when it comes to genes, bacterial genes outnumber human ones. Their function in the body has opened entirely new vistas in medical science and treatment. Our digestive system is a teeming ecosystem of microbes that break down proteins, fats, and starches into assimilable nutrients and that modulate blood sugar levels, regulate our mood, boost our immune system, stop pathogens, and expand our cognitive abilities. Serotonin—a neurotransmitter that affects our mood, learning, and memory—was thought to be produced solely in the brain stem. We now know that 90 percent is produced in the gut. Strains of microorganisms are being cultured and introduced into the ecosystem of the microbiome that heal, that can be used to regulate, reduce, and eliminate symptoms currently treated with drugs. Rather than interfere with bodily processes to eradicate something, as do drugs (if they did not interfere with the body, there would be no side effects), probiotic regenerative medicine makes the microbiome more diverse and responsive. This is exactly what regenerative agriculture does to soil in its true form. How to think and imagine holistically about every facet of human well-being is the offer, invitation, and gift of climate change. We don't have the answers . . . yet. They come about when we ask the question. ●

A Médecins Sans Frontières physician and workers at the Batanfungo Hospital in the Central African Republic, a hospital that serves thousands of displaced people and casualties caused by the violent clashes between the fourteen armed factions and militias operating in the country. In three months, Médecins Sans Frontières provided 370,000 outpatient consultations and treated more than 270,000 malaria patients in their twelve facilities.

Banking Industry

People place money in banks to save for the future. Banks loan the savings to companies that imperil the future. For example, in the United States, oil and gas pipelines are being laid across ecosystems, the lands of protesting farmers, and the sacred territories of Indigenous people. In Australia, new mines have destroyed sacred Aboriginal sites and pristine watersheds. Coal-fired power plants are being financed throughout China, India, and numerous countries in Africa. Banks make this possible. Between 2016 and 2019, four major banks in Australia financed fossil-fuel projects that would cancel out the nation's emissions reduction target twenty-one times over. In the five years after the 2015 Paris Agreement was adopted, thirty-five banks from Canada, China, Europe, Japan, and the United States lent and invested $3.8 trillion on fossil-fuel projects—nearly half of which went to extracting more fossil fuel. And the amount lent and invested was greater in 2020 than in 2016. Science urgently calls for a radical reduction in carbon emissions; banks finance the growth of carbon emissions.

Two leading development banks, the International Finance Corporation (IFC) and the European Bank for Reconstruction and Development (EBRD), have invested $2.6 billion in deforestation, industrial agriculture, and concentrated animal feeding operations (CAFOs) for pork, poultry, and beef production, disregarding the studies and data showing how harmful CAFOs are to climate, land, air, water, people, and animals. Both banks have stated commitments to reducing climate impacts, yet animal agriculture is one of the highest-emitting industries. Asset manager BlackRock committed to ceasing its support of coal but did not mention deforestation. The extinction of species, forests, and a livable world is tantamount to the extinction of the future. And these activities are still easily funded.

Banks fund fossil-fuel companies not necessarily because they can get a higher return, but because the bankers are more familiar and comfortable with those investments. Returns on coal, gas, and oil investments have been very profitable over the decades, and the inertia of familiarity can take precedence over conscience and purpose. In a recent analysis, the backgrounds of six hundred prominent bankers were analyzed. It showed that only a small handful had any experience with renewable energy investments. More than seventy had worked for major corporate emitters, including fossil-fuel companies. None had worked for renewable energy companies. In a study of 39 international banks, research showed that 565 bank directors had prior employment or affiliation with fossil-fuel and polluting industries. A similar analysis found that three out of four directors on seven major U.S.

Tar sands deposits being mined at the Syncrude mine north of Fort McMurray, Alberta, Canada. The tar sands are the largest industrial project on the planet, and the world's most environmentally destructive. The synthetic oil produced from them is three times more carbon intensive than conventional oil supplies. The tar sands are responsible for the second-fastest rate of deforestation on the planet, second only to the Amazon rainforest. They produce millions of liters of highly polluted water every day, which leaches out into the Athabasca River and has serious health impacts on First Nation peoples living downstream.

banks have ties to the fossil fuel industry. Banking is largely a culture that has a short-term bias. The major difference between fossil fuel and renewable energy investments was not profitability. It was time. Historically, fossil-fuel investments have made money quickly, whereas renewable energy projects have provided steadier returns over a longer term.

However, fossil-fuel investments are no longer the surefire moneymaker they once were. To get energy, you need energy, whether to drill an oil well, mine a coal vein, or construct a solar panel. The ratio between how much energy is required to obtain a barrel of crude oil and how much energy a barrel of oil would produce is called EROEI—energy return on energy invested; it is the energy payback, and it is an important number. When solar panels were first introduced commercially, the EROEI was low, 3:1. This meant it would take eight years to pay back the amount of energy expended to create the solar panel. Oil and gas returns were between 40:1 and 60:1.

A recent University of Leeds study counted all of the energy required to transform oil, gas, and coal into fuel, heat, and electricity. When the energy utilized in transportation, shipping, refining, storage, pipelines, generators, and terminals is included, the EROEI for fossil fuels is closer to 6:1 for oil and 3:1 for coal- or gas-generated electricity. Banks do not calculate energy in, energy out. They count money in—and money out. As we go deeper into the earth to find oil, or use bitumen deposits such as the Alberta tar sands (a mixture of sand, clay, and extremely heavy crude oil), we use more and more energy to achieve declining results. The EROEI for oil was about 1000:1 at the beginning of the twentieth century and is now 6:1. The Alberta tar sands oils are 3:1, and possibly 1:1 if you include the tar sands' full life cycle. This does not count the long-term damage from the toxic slurry left behind.

We face a net energy cliff, a point in time when it takes as much or more energy to produce the energy needed. Energy production becomes parasitical rather than constructive, and in the case of the tar sands, that may already be the case. The flow of capital hides overall impacts and deficits.

Contrast this with renewable energy. Today, solar and wind are 7:1 and 18:1, respectively. The EROEI on renewable energy is increasing over time, due to demand, economies of scale, and technological breakthroughs in manufacturing and efficiency. Wind is now the least expensive form of electrical generation in the world, rather than coal or combined-cycle gas. In 2020, Imperial College London and the International Energy Agency analyzed stock market data to determine the rate of return on energy investments over a ten-year period. In France and Germany, renewables yielded ten-year returns of 171.1 percent, as compared to minus 25.1 percent for fossil fuels; and in the United States, renewables returned 192.3 percent over ten years, versus 97.2 percent for fossil fuels.

Demands that banks stop financing fossil fuels are having an effect. In 2017, ING Group forbade transactions linked to any aspect of the tar sands. BNP Paribas announced that it would not finance tar sands or shale oil. These banks were joined by Société Générale, HSBC, Royal Bank of Scotland, UBS, Norges Bank (with its $1 trillion wealth fund), and others. The World Bank stopped lending to oil and gas extraction in 2019, six years after it stopped financing coal-fired power stations. Crédit Agricole pledged to stop financing companies that operate coal power plants and mines, and Deutsche Bank stopped issuing new credit lines to coal companies. Storebrand, a Norwegian fund (worth more than $112 billion), has divested from coal stocks, several oil companies including Exxon-Mobil and Chevron, and miner Rio Tinto. Some financial

institutions have begun divesting from industries that pollute. Storebrand is the first to divest from companies that support anticlimate lobbying. They are divesting from BASF, a German chemicals company, and Southern Company, a U.S. electricity supplier, because they lobby against climate action.

In 2019, renewables made up more than two-thirds of the additions to the electricity sector, and renewable energy generation has increased at least 8 percent each year for the past ten years. Despite great financial returns, the amount of money invested in renewables is not where it needs to be to meet 2030 goals. If banks do not support investments that lead to net-zero carbon emissions, including regenerative agriculture, carbon positive buildings, afforestation, and proforestation, it won't happen. If traditional banking doesn't, an emerging banking system might.

Financial technology—or fintech—has changed the way people pay, lend, and invest money. It employs innovative digital technology to compete with big banking. In the right hands, it can promote green finance and social justice. There are now more than seven thousand fintech companies in the world, employing smartphones for digital banking—personal finance that is easier, cheaper, responsive, and inclusive. Fintech is reaching more people, more quickly, in more parts of the world than was thought possible, owing to the near ubiquity of smartphones. There are fintech financial hubs around the world in smaller cities and regions, shifting the balance of power away from the money center banks located in Zurich, New York, Hong Kong, and Frankfurt. Seven of the top twenty fintech countries are Lithuania, the Netherlands, Sweden, Estonia, Finland, Spain, and Ireland—not places we would associate with international banking. The smallest of companies can set up payment systems without a bank at all. Square and Stripe allow users to turn their phone, tablet, or computer into an online payment terminal. A source of the funds big banks employ is the "float" of monies received from a credit card holder that are not yet paid out to the vendor. That is largely erased in fintech.

Branch is a fintech company operating in Kenya, Tanzania, Nigeria, Mexico, and India that applies machine learning to establish creditworthiness for its four million customers. It has made more than 21 million loans, a typical loan being fifty dollars. Paga is a Nigerian company that, like PayPal, enables peer-to-peer transfer and receipt of payments. These and many other fintech companies are fulfilling needs that larger banks do not and cannot. And they are digitizing institutional processes, lowering costs, and improving transparency and accessibility. Tala grants instant ten-dollar to five-hundred-dollar microloans to underserved customers in developing countries. The idea for Tala started in India, where founder Shivani Siroya started personally loaning small amounts of money to people she had observed being productive and diligent in their business and personal life. It was a friend who pointed out that her underwriting methodology was based on her observations of daily life, not credit scores, since her market is the 1.7 billion people who have no credit rating. Her breakthrough was realizing that daily life, in many respects, is embedded in our mobile phones, which show usage, receipts, payments, and communication patterns and habits.

In America there are a number of independent and green banks. One of the oldest is Amalgamated Bank, a publicly traded, union-owned bank that serves more than a thousand unions. Its roots go back to 1923, when labor organizer Sidney Hillman decided to create an institution that offered workers the same services and opportunities that the wealthy were receiving from their banks. It is a member of the Global Alliance for Banking on Values, a people-focused international network of banks devoted to sustainable environmental, economic, and social development. It includes the Opportunity Bank Serbia, Ekobanken in Sweden, Vancity in Vancouver, and BancoSol in Bolivia. Another member is Beneficial State Bank, in Oakland, California, cofounded by Kat Taylor in 2007. With more than $1 billion in deposits, it focuses on supporting minority-owned business in its region.

A fintech that may fundamentally change the system of banking is Good Money, a company that intends to practice what is calls "positive banking." This includes no minimum balance, no overdraft fees, no monthly fees, and no ATM fees at fifty-five thousand locations. Big banks in the United States charged $34 billion to their customers in 2017 for overdraft fees, in many cases because the banks delayed clearing a deposit the customer had made. With Good Money, every debit card purchase helps fund legal land titles for Indigenous communities in the Brazilian Amazon at no cost. The expense is paid for by a constant small donation of a portion of the interchange fee a merchant pays when you use the Good Money debit card.

Good Money will not be the last such innovator. Digital banking could completely change the relationship between individuals, families, communities, schools, and businesses and the flow and use of money worldwide. The tools are there. Too many money center banks are investing in destruction. Trillions of dollars in subsidies, loans, and equity are directed to the loss of the living world. The world is awash in money. That we get "richer" by destroying the wealth of the planet has long been known. But the opposite is equally possible. We have the funding to reverse global warming and restore the rich forms of life that reside on our lands and in our oceans. ●

Boreal forest trees clear felled to make way for a new tar sands mine north of Fort McMurray, Alberta, Canada.

War
Industry

The war industry was foretold by President Dwight D. Eisenhower in his Chance of Peace speech in 1953. "Every gun that is made, every warship launched, every rocket fired signifies, in the final sense, a theft from those who hunger and are not fed, those who are cold and are not clothed. . . . The cost of one modern heavy bomber is this: a modern brick school in more than thirty cities. It is two electric power plants, each serving a town of sixty thousand population. It is two fine, fully equipped hospitals. It is some fifty miles of concrete pavement. We pay for a single fighter with a half million bushels of wheat. We pay for a single destroyer with new homes that could have housed more than eight thousand people. . . . This is not a way of life at all, in any true sense. Under the cloud of threatening war, it is humanity hanging from a cross of iron."

If we reframe Eisenhower's speech for today, it looks like this: The cost of one B-2 bomber totals a new middle school in seventy-five cities, seventy-two solar power plants serving 4.15 million people, thirty-six fully equipped hospitals, and 281,000 electric vehicle charging stations. We pay for a single F-35 Lightning fighter jet with 22 million bushels of wheat. We pay for a single Zumwalt destroyer with new homes that could house more than 58,000 people.

As a former five-star general, Eisenhower was acutely familiar with the ins and outs of war. In his final speech upon leaving office in 1961, he used the term "military-industrial complex," calling it a self-justifying and self-perpetuating industry with "unwarranted influence." (A memo preceding that speech called it the "war-based industrial complex.") In 1939, before the onset of World War II, the United States had the nineteenth-largest army in the world, just behind Portugal. Today its military expenditures exceed the combined spending of China, Saudi Arabia, India, France, Russia, the UK, and Germany. There is currently no clear military threat to the United

States, yet it operates eight hundred military bases in eighty countries. To support the magnitude and complexity of this effort, more than one-half of military expenditures involve private contractors. The five largest arms-producing and military services companies in the world (excluding China, for lack of data) are American: Lockheed Martin, Boeing, Northrop Grumman, Raytheon, and General Dynamics. Their combined market value on the New York Stock Exchange is $424 billion. There is a threat to the United States. The threat is a rapidly changing climate.

Like all industrial systems, the war industry works to increase its own growth, income, security, and influence. But the full effect of the industry does not show up on the balance sheets of military contractors. The wounds, trauma, and death of innocent women, children, and civilians caused by warfare is called collateral damage. Soldiers who are permanently injured, physically and psychologically, are hailed as wounded warriors. In the United States, there are 4.7 million veterans with service-related disabilities, including impaired mobility, brain injuries, burn pit syndromes, cancers, third-degree burns, spinal cord damage, hearing loss, post-traumatic stress syndrome, alcoholism, homelessness, and loss of limbs.

Waging war and maintaining the capacity to wage war are degenerative activities, in that they harm life. Abetted by powerful lobbies in most every country, the world spends trillions of dollars annually on armaments, military bases, armies, air forces, navies, armored vehicles, fighter jets, aircraft carriers, nuclear bombs, and treating the aftermath of war—the trauma and physical injuries of millions of veterans around the world. The war industry is enabled by partners—politicians and lobbyists, often abetted by different forms of corruption. It is estimated 40 percent of all known government corruption originates in the arms trade. The war industry, especially weapons and arms manufacturers, grows and profits by looking the other way. Manufacturers insist that they are not responsible for machine guns, land mines, rocket-propelled grenades, and ammunition falling into "the wrong hands." There is only one government-sanctioned gun store in all of Mexico. The tens of thousands of guns used by the drug cartels are smuggled in, purchased at gun shops located within blocks of the border.

There are 164 countries with armed forces, 169 unsanctioned armies or militias, and 32 ongoing armed conflicts in the world as of this writing. Yet not a single country has a department of peace. There is a pressing need to make peace with the living systems of the earth. If we cannot come to peace with one another, it is unlikely we will do the latter. A department of peace is not about making enemies shake hands. Its function is to go upstream and determine why we became enemies in the first place. There are biological parallels here.

Darwinian is an adjective that has come to be associated with "survival of the fittest." That would easily describe the mindset of opposing armies. The phrase was coined by Herbert Spencer to defend his economic theories, not Charles Darwin's findings. Darwin meant that the fittest are "better designed for an immediate, local environment," which defines what humanity needs to do. All species, including humankind, come with a built-in characteristic known as mutualism. In life sciences, it is defined as ecological interaction between species that benefits both. For example, the mycelium (fungi) network in the soil is fed by root sugars and provides the plants with needed trace elements and compounds. Hummingbirds feed on nectar and deposit pollen from the male part of one plant onto the female part of another, to enable fertilization and the creation of seeds. The red-billed oxpecker perches upon impala eating ticks, blood-sucking flies, fleas, and lice. The Impala gets rid of its parasites and the oxpecker gets nutrient dense food.

Clearly, people can be warlike. But they also are mutualistic. Mutualism is interaction between two similar or different species in which both benefit. Marriage, family, clans, communities, well-managed companies, and sports teams all depend on mutualism. We have mutual insurance companies and mutual funds. Scientists believe that Homo sapiens dominated the larger, stronger, and now extinct Homo neanderthalensis because of their mutualistic relationship with their dogs and one another. It seems that human mutualism founders when authority is vested in larger and more powerful institutions and governments. It breaks down further when social media reinforces self-beliefs by feeding people an "individualized reality" of the world based on their online searches and behavior. When there is not a shared reality of news and current events, mutuality is impossible.

The climate crisis is unlike any crisis or problem ever faced by humanity. It is global, with no boundaries. It cannot be tackled, fought, controlled, mitigated, or curbed, as is often stated. Nor can it be weaponized: "Scientists Recommend These 4 Weapons in Our War Against Climate Change" reads one headline. If we understood what we face, we would not find war a useful metaphor to describe the solution. Global warming is a massive force beyond human comprehension, but it is not the enemy. How can a

There are twenty B-2 stealth bombers currently in service, and each can carry sixteen B83 nuclear bombs. A single B83 bomb is eighty times more powerful than the nuclear bomb dropped on Hiroshima. All sixteen bombs equal 1,280 Hiroshimas, and together are capable of destroying London, Paris, Berlin, Rome, Madrid, Zurich, Oslo, Stockholm, Copenhagen, Prague, Helsinki, Moscow, New York, Washington, D.C., Chicago, and Miami. There are approximately fifteen thousand nuclear bombs in the arsenals of nine countries.

balkanized, politicized, and weaponized world deal with a global atmospheric phenomenon that is dictated by the laws of physics? It cannot. The world will either change its capacity to respond or succumb to the effects. Could the war industry play a key role? At the moment it is a significant contributor to global warming. The total direct and indirect emissions created by armies and weapons is incalculable. And the damage to soldiers, women, children, lands, and oceans is immeasurable. And we will have no luck convincing armies to stand down or countries to go "defenseless."

However, there may be a middle way. Militaries could play a key role in securing our future, because the climate crisis threatens and undermines the security of everything—food, the economy, family, home, farm, land, fish, water, and health. It sounds far-fetched, but the conjoint armed forces of the world could cooperate to defend, secure, stabilize, surveil, and protect. It does all of these things now, but in a different context. As the impacts of floods, fires, droughts, and hurricanes increase in intensity and number, the world can either divide and devolve, or unite and evolve. It can recognize our common interest and realize that the tens of millions of individuals mobilized for

war can be deployed to make peace with the earth. Regeneration aligns human action with life's principles. Aligning society is about teaming, banding together, and joining forces toward a common end. If those phrases sound like a military recruiting advertisement, that's because these archetypal qualities bring us together. As we mentioned in Agency, the climate movement will become the largest movement in the world for one reason: weather will become more extreme, erratic, and punishing. The military is already deployed to aid in the aftermath of severe climate impacts, as we've seen in Puerto Rico, Honduras, Nicaragua, the Philippines, Australia, and California. It is not a big step to imagine the world's military getting ahead of the curve in order to educate and protect, to construct and build, to monitor and guide, and to cooperate and collaborate. ●

Chinese People's Liberation Army soldiers assembling during military training in the Pamir Mountains in Kashgar, northwestern China's Xinjiang region.

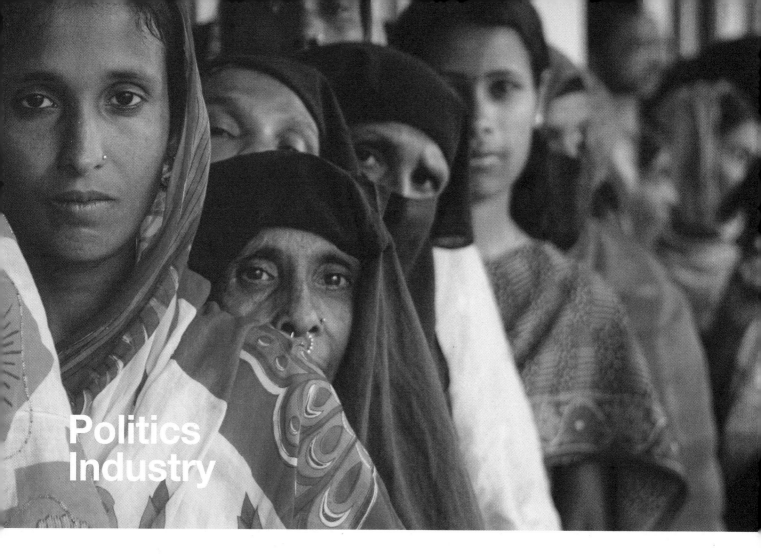

Politics Industry

Throughout the world, people are concerned about climate change and want action—Pacific Islanders who face rising seas, farmers who experience cyclonic storms, flood and heat victims, fisherfolk with empty nets, countries inundated by migrants, a generation that does not see a future. There is anxiety and it is increasing. In the United States, two in three Americans are worried about global warming, and one in four are "very worried" about it. In 2016, Pew Research Center surveyed 41,953 people in thirty-eight countries about eight possible threats to national security. In Africa and Latin America, most people said climate change was the greatest threat to their country. In African countries, 58 percent of people believe it's a major threat, while 74 percent said the same in Latin America. And 64 percent of people in European countries said global climate change is a major threat to their country. If it is that important to citizens, one might ask, why wasn't it important to politicians?

An industry is a system that produces goods or services in response to its consumers. Just as there is a pharma industry, automobile industry, and banking industry, there is a politics industry, invisible yet in plain sight. It may be one of the most destructive industries in the world, because it creates campaigns and ads that reject, downplay, and mock climate science, slowing adoption of policies and legislation that would benefit us all. It is paid to polarize, to broadcast disinformation, to burnish the image of oil and gas polluters, to create fear-based ads about renewable energy on behalf of incumbent industries, or to create advertisements for political candidates that contain outright lies. It is a global, multibillion-dollar industry that foments and thrives on discord. The politics industry requires conflict and opposition, and it accomplishes this by dehumanizing opponents. Dehumanization is another form of degeneration. We cannot regenerate our climate in a degenerative political climate.

Bangladeshi villagers line up to have their photographs and signatures taken and saved to an extensive database in Rajashi Division, some 200 kilometers northwest of Dhaka on March 16, 2008, as part of a United Nations Development Programme (UNDP) voting initiative. One of the main reasons for the deferment of the January 2007 elections was an inaccurate electoral roll that was not acceptable to opposition parties. The UNDP is supporting the Bangladeshi government's creation of a fresh voters list with photographs and fingerprints. It is the first time in Bangladesh that photographs are being included in the voters list. The completion of this list will eliminate fraudulent entries and build the nation's confidence in the credibility of parliamentary elections.

The outcomes of U.S. elections, and those in many other countries, do not reflect what voters want. They reflect what people fear. The politics industry is not designed or intended to serve voters. Like all industries, it serves itself. And like most industries, it stifles competition. It depends on maintaining highly opposing points of view. In the United States, it is a $20–$30 billion industry, made up of wealthy contributors, political action committees, law firms, advertising agencies, lobbyists, conferences, junkets, dark money, revolving-door appointments, and corporate contributions.

Prior to the November 2020 elections, the U.S. Congress had a 23 percent approval rating, yet House members are reelected 91 percent of the time, and senators with six-year terms 85 percent of the time. That is not a functional system. It is what happens when voters have only two choices. They vote against as much as for. There may be other candidates on the ballot with different political affiliations, but in a winner-takes-all system, splinter candidates become a "wasted" vote. The textbook example is the 2000 presidential election, which was decided by the Supreme Court. The justices overruled a recount effort in Florida where the difference between George W. Bush and Al Gore was 537 votes. Ralph Nader received 97,421 votes in the state and, given their similar politics, it was widely assumed that had he not been on the ballot, Gore would have become president.

The United States is a two-party system, a duopoly. We think of Republicans and Democrats as fighting one another, but they actually work together effectively to protect themselves. There is combat but not competition. There is a simple and effective way to create true competition that would give voters a wider choice without fear of wasting their vote; it is called ranked-choice voting.

Ranked-choice in a nonpartisan primary system would advance the top four or five candidates to the general election. This would allow candidates who are not aligned with the duopoly to be considered and gain attention from the media. As it stands now, they may still get on the ballot in the general election, but in most cases they enter as complete unknowns, given the primary process. The general election would employ a ranked-choice voting system to determine the winner. Ranked choice uses a preferential ballot on which the voter ranks the final candidates in terms of preference, first to last by number. The initial ballot count tallies each voter's top choice, and if someone wins more than half, they are declared the winner. If no one wins a majority, the candidate who has received the fewest votes is eliminated; voters who had chosen the eliminated candidate have their second choice counted and added to the totals. The process is repeated again until there is a majority winner.

Ranked-choice voting is used throughout the world,

in elections from Maine and New York City to India and Ireland, as well as in the Australian House of Representatives and the Academy Awards, in local elections, and within political parties for internal voting. The benefit of this system is that it encourages types of campaigns that are positive rather than unrelentingly negative. More smiles, less mud. A candidate would want to demonstrate the breadth of their appeal, as contrasted with the narrowness of fear-based politics. Being ranked high is the key to success.

In the current system, the candidate with a majority of votes wins. In a ranked-choice system, the candidate with the majority wins but gets there differently. In a multicandidate race, there may be no majority winner. Needing a high ranking from those who support another candidate leads to campaigns that build connection. It makes negative ads dangerous. It leads to what happened in Maine in 2018, when Mark Eves and Betsy Sweet, candidates for governor from the same party in the primary election, ran an advertisement in which they praised each other and requested that anyone choosing them first also rank the other one second.

If an election system promotes consensus rather than polarization, what people need and want becomes more relevant and doable. The influence of corrosive advertising paid for by unknown persons becomes strategically unwise. There are no wasted votes. Third parties can compete on a level playing field. Diverse political strategies that cut across doctrinaire party positions can be heard and may prove more appealing than party loyalty litmus tests.

If elected politicians emerge from a ranked-choice voting system, the likelihood of their continuing to respond to a broader base of their constituents is all but certain. Also likely are legislatures that benefit from building consensus rather than building political walls. They may see the opportunity to form nonpartisan commissions to supervise many activities, such as redistricting and gerrymandering, with open debates for senatorial, gubernatorial, and presidential candidates. A Congress elected by a consensus voting process might undo the morass of arcane rules that concentrate power in the hands of a few, rules developed by politicians themselves that are not based on the Constitution, rules that grant unaccountable and undemocratic powers to majority leaders in the Senate and the House. Such a Congress may see its way to understanding that public financing of political campaigns is the surest way to hear all of the voices.

A truly competitive process of choosing our representatives would, paradoxically, lead to more cooperation. It would moderate the influence of entrenched power, big money, and lobbyists, and reward those who more faithfully represent the needs of their electorate. The total number of people who vote in U.S. elections is dismally low.

Rather than insist that voting matters, the focus should be on changing the system so that votes do matter. In the United States in 2016, average voter turnout reached 67 percent in the eleven states where the presidential election was closest. In states where the margin of victory was greater than thirty points, average turnout was 56 percent. Voting based on fear leads to a government that is about triumph instead of collaboration.

More than $7 billion was spent on political ads during the U.S. 2020 election cycle. World over, most people realize that their political systems are corrupt, dysfunctional, and/or unworkable. There are notable exceptions in northern Europe, but they highlight how politics has become the single greatest obstacle to planetary renewal and regeneration. Government policies, laws, subsidies, taxes, and regulations can transform a country overnight. If 10 percent of the $12 trillion committed to COVID relief were allocated to climate relief each of the next five years, the world would have a greater chance to achieve the 2030 climate goal of a 50 percent reduction in greenhouse gas emissions.

Whether at the local or national level, exciting participatory experiments around the world are demonstrating that it is possible to regenerate democracies by giving ordinary voters a greater voice and influence with their representatives. Participatory budgeting is one example.

Pioneered in Brazil, it is now used in more than twelve thousand locations around the world. Citizens assemblies, where randomly selected representative samples of voters are given power in decision-making, are another. These have been used in Canada, the UK, France, and Belgium, where there is a permanent council of twenty-four citizens serving in parliament in its German-speaking region. The UK's national citizen assembly on climate change has shown that voters are willing to support climate policies more ambitious than their government.

As with many of the strategies offered in *Regeneration*, the path to truly competitive democratic elections begins locally, in cities, councils, provinces, and states, but it can be simultaneously sought on the national level. We can blame, shame, and bemoan our dysfunctional election system, but that does not effect change. To transform a system, it works best to start upstream, at the source and origin—in local, regional, and provincial elections. Either voters will appreciate it or they will not. If they do, and that is very likely, it can migrate to federal levels. ●

A protester holding a makeshift shield stands on Mohammed Mahmoud Street in downtown Cairo, as clashes erupt between protesters and riot police on the anniversary of the Mohammed Mahmoud protest and clashes during the "Arab Spring."

Clothing Industry

Clothing and fashion are two different things. Clothing is a need, fashion is a desire. Although the fashion industry is notorious for its lack of environmental transparency, the best data indicates that the apparel and footwear industry are responsible for 8 percent of global greenhouse gas emissions, about the same as the beef and pork industries. It is an industry powered by coal, oil, diesel, gas, jet A fuel, and bunker crude for ships. Clothing manufacturers use open-loop production cycles, which means that production waste streams go directly out of the factory onto land and into water. The industry consumes 21 trillion gallons of water annually and generates 203 trillion pounds of waste, which includes dyes, mordants, microfibers, heavy metals, flame retardants, formaldehyde, and phthalates.

It wasn't always this way. Human beings have been creating and weaving fibers for more than thirty thousand years. Dyed linen discovered in a cave in the Republic of Georgia was found to be thirty-four thousand years old. The *techno* in the word *technology* is derived from the Indo-European teks, "to weave." Technology applied to spinning and weaving gave birth to the Industrial Age. The cotton gin, spinning jenny, water frame, and power loom increased human productivity ten to twenty times over, which brought about a revolution in clothing affordability. Then as now, the textile industry was beset by inhumane working conditions and poverty wages. At the beginning of the Industrial Revolution people worked twelve- to

sixteen-hour shifts, six days a week, for ten shillings—one and a half cents per hour at that time. Women were paid half as much. Children who had to work because their families could not get by on their parents' wages were paid half as much as women. According to Fashion Checker, an NGO that is part of the Clean Clothes Campaign, 93 percent of the biggest clothing and sportswear brands surveyed do not pay a living wage, which means their workers cannot afford the basic necessities of life. Measured in today's dollars, mill workers in the early 1800s were making 34 cents per hour. That is the same wage that the majority of garment workers make in 2021, more than two hundred years later.

Fashion has grown from a $500 billion industry in 1990 to $2.5 trillion in 2019, making it the third-largest manufacturing industry in the world, behind automobiles and technology, employing one out of six people on the planet. More than 100 billion garments are produced annually, thirteen pieces for every person. The average American purchases sixty-eight apparel items per year, about one new garment every five days. Overconsumption is discussed by the environmental community as an individual responsibility, but the greatest overconsumer is the industry itself. Thirty percent of produced clothing never makes it to the customer. Garments are overproduced due to incorrect forecasts, fickle customer tastes, or unit-cost economics—ordering more makes the unit price less expensive. Clothing giant H&M had $4.3 billion worth of

unsold clothes in 2018. The combined heat-and-power station in Västerås, Sweden, replaced coal as a fuel and is burning municipal waste, including discarded clothing from H&M. Burberry admitted to burning $37 million of its product to prevent it from being discounted. In the United States, an estimated 35.4 billion pounds of garments were landfilled in 2020. More than 60 percent of clothing is synthetic and will remain in landfills for hundreds of years.

Because of fast fashion, the clothing industry boomed. Fast fashion is defined as cut-rate, substandard clothing that quickly samples and imitates the latest catwalk and celebrity fashions, employing industrial just-in-time production methods. Breathtakingly inexpensive clothing is produced around the clock, with lightning speed, in byzantine networks of hard-to-trace subcontracted factories, and shipped in jumbo jets to thousands of stores around the world, where it is sold at disposable prices. The industry delivers tens of thousands of down-market copies of a garment first seen two weeks earlier on social media. The cost: $2.40. Retail price: $9.99. Just as refrigerators are where food goes to die, closets from China to Germany are where clothes go to die. Perfectly good garments are not worn again for a variety of reasons, but primarily because they look dated. When thrown away, they are placed in bins to be shipped to developing countries, landfilled, or incinerated. One-third of garments in sub-Saharan Africa are discarded secondhand imports.

The retail clothing industry is aware of its impact. Criticism of the industry comes from the UN, the World Bank, NGOs, human rights organizations, fashion magazines, *The Washington Post*, and *The Wall Street Journal*. Numerous countering initiatives are being undertaken by the biggest fashion brands. The Ellen MacArthur Foundation has mounted an industrywide collaborative, Make Fashion Circular, that researches how the clothing industry can align with the Paris Agreement, the UN Sustainable Development Goals, clean water initiatives, and ocean plastic reduction. The New Textiles Economy addresses four key issues: (1) the phasing out of toxic chemicals and synthetic microfibers, both of which are found in the fish we eat and the water we drink; (2) changing the way clothing is made and marketed, in order to shift the perception of clothing as disposable to that of a durable good; (3) radically improving the materials used in clothing and promoting the collection and recycling of all clothing and fibers manufactured (only 1 percent of textiles produced for clothing are currently recycled); (4) moving to renewable energy and renewable feedstocks, such as biopolymer yarns instead of petroleum-based synthetics.

_____

(*Left*) A model presents the autumn/winter 2020–2021 creation of Dominnico during Mercedes Benz Fashion Week in Madrid, Spain, 2020.

(*Above*) Dump site for garment-factory waste found in the Export Processing Zone of Dhaka, Bangladesh.

The New Textiles collaborative is a company—an NGO-funded endeavor intended to create, in their words, a restorative and regenerative industry. As yet, it does not address human rights, worker safety, child labor, modern slavery, fair trade, living wages, labor conditions, codes of conduct, or animal welfare. Kara Kupe, a Māori clothing designer who produces in Bali, believes that exploitation and miserable working conditions for Black and Brown people "needs to be acknowledged as the most unsustainable practice in fashion." Carry Somers, the activist founder of the Fashion Revolution Foundation, which operates in a hundred countries, says, "We need transparency to be irreversibly entwined with every thread of every garment that will ever be made."

A key partner in the New Textiles economy is H&M, the pioneering $26 billion fast-fashion company, which is committed to being carbon positive in its supply chain by 2040 and utilizing only recycled yarns and sustainably produced materials by 2030. The retailer also has a detailed commitment concerning worker rights, living wages, and worker health and safety. What it has yet to detail is how it will continue the airborne, weekly deliveries to its 5,018 retail locations around the world, from Kazakhstan to Iceland. Nor do the commitments to reduce energy and material flow made by H&M and other large fashion companies address the level of consumption that they and social media influencers reward by making inexpensively constructed clothing "unfashionable" within weeks in the eyes of its youthful customers. In 2013, H&M launched the Garment Collective initiative, a take-back program that collects clothing for reuse and recycling. It launched another initiative called Recycling and Upcycling. Neither change the rate of production of new garments in an industry that constantly creates new trends. Fast fashion, which includes Zara and its twenty-four

brands, has resulted in more than twenty fashion seasons instead of the traditional two or four. Fast is the problem. Consumption is the issue. Growth is the cause.

One company epitomizes what could be called "slow" fashion. The Swedish company Asket has only one permanent season. That continues each year with tweaks and improvements to existing designs, but with no new designs. The idea is to break the cycle of overproduction and overconsumption by making the clothing you need, value, and keep for years. When you purchase an Asket garment, you get an Impact Receipt that details carbon dioxide, water, energy usage for materials, milling manufacturing, trims, and transport. Summary data at the bottom of the receipt details the minimum times a garment can be worn in its lifetime ("180 wears"), cost per wear, and impact per wear.

Asket is part of the worldwide response to the impact of the clothing industry: localization. Smaller companies are taking up the challenge to cut, sew, remake, restore, upcycle, and sell garments that are ethical, enduring, and kind to the earth. In a typical closet of the future, one-fourth of the garments might be purchased or rented from secondhand and consignment stores; one-fourth could be durable, long-lasting clothing that is sustainably sourced and produced with transparency, from seed to closet; another fourth would be clothing remade with patchwork textiles made from discarded clothing; while the final fourth is made of fibers that have been upcycled from plastic that does not shed microfibers. This new fashion industry is guided by a host of nonprofit organizations that are defining standards, insisting on transparency, and certifying the adopters of ethical principles. These include the Global Organic Textile Standard (GOTS), the Fair Wear Foundation, the Declaration on Fundamental Principles and Rights at Work of the UN International Labor Organization, the Fair Labor Association, the Sustainable Apparel Coalition, Responsible Wool Standard, Global Recycled Standard, Oeko-Tex Made in Green (no harmful substances), Organic Cotton Accelerator, Fair Trade USA, Social Accountability International, Fairtrade International's Small-scale Producer Organizations, the Transparency Pledge coalition, and the World Fair Trade Organization. It may be that no industry has as many initiatives, certification standards, and commitments to transform itself than the clothing industry.

The burgeoning slow-fashion industry employs organic cotton and hemp, ethical wool, and recycled fibers. Some are vegan and use no wool or leather. The companies publish their code of conduct, use compostable packaging, and monitor and trace their complete supply chain. Companies like Patagonia, Puma, Eileen Fisher, Levi's, Columbia, and H&M have taken the Transparency Pledge developed by civil society and freely share their entire list of suppliers, down to their address. But so too have hundreds of smaller companies such as Kuyichi and Kings of Indigo. Some slow-fashion companies ship by sea rather than use air freight for overseas shipments. And companies try to manufacture and design products that are made to last. Some companies do more. Outland Denim is a social enterprise that creates employment for Cambodian women who were sex-trafficked. Little Yellow Bird, in New Zealand, provides nutritious meals to migrant workers in Faridabad, India, who were displaced by the COVID pandemic and sources its cotton from a growers' co-op in the state of Odisha, in India.

A longtime leader in ethical clothing is Eileen Fisher, whose manifesto is to make "simple, well-crafted clothes for life." The designer's environmental and social initiatives are legion, but foremost is its Renew brand, which purchases garments back from customers and reworks them into beautiful outfits made of patches and scraps. Patagonia also leads in the effort to create a new clothing paradigm, with its Worn Wear program. It starts with making clothes that last. All garments can be returned to Patagonia to be repaired and resold. If there is too much wear and damage, the fibers are recycled. If "certified, pre-owned" stickers can apply to a used car, they can be applied to well-made clothing, since most of Patagonia's outerwear lasts longer than a car.

A pioneer in reimagining clothing is Lindsay Rose Medoff and her company Suay, which in her words "cleans up the mess created by larger corporations and disconnected consumers." Suay cuts and sews thrown-away garments from any maker into attractive, useful, affordable clothing. Remaking garments from discarded clothing and textiles requires 90 percent less energy and creates similar reductions in carbon emissions. Today, garment production is concentrated in four countries: China, Bangladesh, Vietnam, and India. Like local food, remaking creates local clothing and regional employment, regardless of the original fiber source.

The clothing resale market is thriving these days, in no small part because people need to pare down closets that got crammed. H&M's brand COS is launching a resale business for its customers. The global clothing secondhand market was $28 billion in 2019 and is projected to reach $64 billion by 2023. Reselling is sensible when clothing is long-lasting, durable, and made of sustainable materials. Resold disposable fast fashion makes very little difference to the footprint of the clothing industry.

While the transformation of the clothing industry is growing, ethically made clothing is unaffordable to the vast majority of people. Conversely, the word *ethical* is being used as a way to greenwash high-end, boutique clothing consumption. When people in lower-income countries were making their own clothing, it was affordable. Now that garment workers are producing clothing for affluent customers, it is not. Many countries have banned the importation of secondhand clothing altogether, because of its adverse impact. Now that clothing is being reused, resold, and recycled in wealthier countries, the best garments are selected and removed, and the rejects are sent overseas. Ghanaian clothing designer Samuel Oteng points out that if "you give someone something you don't want, it isn't helping; that is an insult."

Localization of clothing closes the most important loop of all: the connection between a maker and a customer. All sides of the current fashion dilemma take and share responsibility—the consumers who stoke the fads of fast fashion and the companies that create obsolescent products and exploitative marketing gambits. However, nothing changes a corporation faster than once-loyal customers who protest, walk away, or boycott. Corporations have changed our behavior by overwhelming consumers with marketing messages that prey upon insecurities and our innate need for social approval. That table needs to be turned so that clothing corporations are overwhelmed by revenue insecurity and a lack of social approval on the part of their customers. This is what makes big companies change, and many now do want to transform their supply chains and practices. They need to know those efforts are supported. And they need to understand that, as Patagonia demonstrates, more is not better, that the pathway to regeneration requires a drawdown of the number of garments they produce and a substantial increase in quality. Durable clothing maintains its value and the value of the earth. ●

Lindsay Rose Medoff, the founder of Suay, the multimillion-dollar clothing remanufacturing company located in Los Angeles, California.

Plastics Industry

You drink a Coke in thirty seconds and the bottle lasts for hundreds of years. The world produces approximately 407 million tons of plastic yearly, 30 percent more than the weight of all of humanity. It piles up on beaches, landfills, and roadsides. It floats in and upon the oceans in mammoth, spinning gyres of trash. Marine life is caught in abandoned plastic nets or eats broken bits of plastic and dies. Two hundred fifty marine species suffer from plastic entanglement, including turtles, penguins, whales, and dolphins. Plastic doesn't degrade naturally, it breaks down into smaller and smaller particles. Microscopic plastic accumulates in food chains, including in fish and plankton. Particles can be found on top of the highest peaks and at the bottom of the deepest sea trenches. Particles are found in apples, carrots, broccoli, and pears. If you see a dust cloud, it contains microplastics. If you see dust bunnies in the corner of your bedroom, they contain microplastics. Plastic contains a toxic mélange of chemicals that make them flexible, flame-resistant, and brightly colored. These carcinogens, neurotoxins, and chemicals disrupt our hormones, create infertility, birth defects, and cancer, and leach into streams and oceans and pollute groundwater sources. The UN believes that plastic is the "second most ominous threat to the global environment, after climate change."

Plastic trash is garish and ugly, and costly to haul away, because no one wants it. In 2019, American companies exported more than a billion pounds of plastic waste to over ninety-five countries in order to get rid of it. Developing nations struggle to manage rising amounts of plastic garbage and often have inadequate disposal systems, and imported plastic waste from the United States is either incinerated—which releases toxic chemicals into the air, causing death and disease—discarded along roadsides, or thrown into unregulated dumps.

The trouble with plastic starts with its origins: it isn't a natural substance. Less than 2 percent of all plastic is made from bio-based sources, such as cornstarch or sugarcane. The remainder originates as crude oil, natural gas, or other fossil sources. In a refinery, oil and gas are heated until they break down into different combinations of hydrocarbons, including naphtha, the feedstock for plastic. Two of naphtha's chemicals are ethane and propene, which are broken down further, often with an energy-intensive process called thermal cracking. These substances are then processed into different types of plastic by adding chemicals, including flame retardant, phthalates, or bisphenol A (BPA). The result is a synthetic material not found anywhere in nature, foreign to the planet's vast network of microbes. Decomposition is impossible. Plastic can break into smaller parts, but cannot break apart. Unless recycled or burned, plastic waste will linger for centuries.

The scale of the problem is staggering. In 1950, the plastics industry generated 2.2 million tons of product. In 2015, the total was nearly 407 million tons. Of the total amount of virgin plastic manufactured by companies since its invention in 1907, more than half has been produced in the past fifteen years, 60 percent of which ended up in the environment as waste. At least one trillion plastic bags are used globally every year. A million plastic bottles are purchased every minute. A huge amount of plastic waste is microscopic (less than five millimeters in size). A study found that an average-size load of laundry in a household releases more than seven hundred thousand microfibers into wastewater.

More than twenty-two tons of plastic enters the ocean every minute. By 2040, the plastic entering the oceans would blanket every yard of the world's coastline with one hundred pounds of plastic. More than half of all plastic trash in the oceans is less dense than water. This floating debris is often caught in slow-turning vortices located far from land. The most famous example is the Great Pacific Garbage Patch, a massive clockwise-spinning gyre in the Pacific Ocean that is roughly the size of Alaska and that contains every type of plastic product, including baby bottles, grocery bags, fishing nets, cups, and packaging.

Some chemicals used in the manufacture of plastic are toxic to humans. BPA can be absorbed through the skin. Microplastic particles containing these chemicals fall on our food as part of airborne dust. After ingestion, the particles can pass through the wall of our gut into our bloodstream, where they can lodge in organs, including the liver. People in high-income countries consume approximately five grams of microplastics each year, equivalent to one credit card.

The chemical factories that produce plastics are often located in economically distressed areas, especially near communities of color. Formosa Plastics, the giant petrochemical company based in Taiwan, which has a significant track record of pollution, targeted the small town of Welcome, in southern Louisiana, for a plastics manufacturing complex that included ten factories. They met with fierce resistance from the community, whose population is 98 percent Black. Community leaders noted that possible sites for the company's plastics complex in predominantly White communities had been omitted in an analysis by the U.S. Army Corps of Engineers. In 2019, Formosa agreed to pay $50 million to settle a lawsuit that ruled the company had been deliberately dumping plastic pellets and pollutants into Lavaca Bay, Texas, and nearby waterways for years. This was not the first time. The judge in the Formosa case called the company a "serial offender."

Companies like Formosa Plastics benefit from a banking system that provides billions of dollars of support without requiring limits to the pollution its clients generate. In 2020, the production and incineration of plastic contributed more than two billion tons of greenhouse gases to the atmosphere, equal to emissions from nearly five hundred large coal-fired power plants. At current rates of growth, greenhouse gas emissions from petroleum-derived plastics will grow to 6.5 billion tons by 2050.

In 2019, Break Free from Plastic, a global alliance of organizations, analyzed trash collected on beaches, city streets, and neighborhoods by seventy thousand volunteers across fifty nations. Coca-Cola was the largest plastic polluter (for the second year in a row). We need laws, policies, and other incentives that encourage or require companies to close the spigot of plastic waste. Responsibility for waste management and costs needs to shift from municipalities to manufacturers. To date, no plastic manufacturing company has taken responsibility for what it makes or the pollution caused by its products. Extended producer responsibility, also known as end-of-use accountability, is the only mechanism that will stop the worldwide explosion of plastic pollution and ecological damage.

Around the world, effective legislative, regulatory, financial, and legal methods are being used to prevent new coal-fired power plants from being built and shutting down existing ones. Plastic plants should be no different. We do not need single-use plastic. We need to rethink how we shop, sell, purchase, and live.

We need to rethink recycling. The vast majority of recycled plastic is mechanically ground to tiny bits and fibers, producing material of a lower quality—a process called downcycling. After one or two cycles, much of this plastic is discarded. Clothing fiber made from recycled single-use plastic bottles has become a popular alternative to polyester, a petroleum-based fabric, but all polyester clothes shed microplastics when they are washed, which end up in the ocean, where they do more harm than if they had remained a bottle. Alternatives are being developed

The Parola Biondo side of Manila harbor is home to twenty thousand squatters who live between port facilities and the mouth of the Pasig River. The river winds through downtown Manila and carries refuse that gets deposited on the shore beneath Parola's stilt houses. Here Rodello Coronel Jr., thirteen years old, the second of nine children in his family, spends the morning picking through the trash on shore, looking for recyclable plastic, which sells for 13 pesos (35 cents) per kilo. The next day he went to school in his uniform, with a small briefcase holding his homework papers. With a rapidly growing population, the slums of Manila have extended onto coastal mudflats and waterways that are very susceptible to flooding from storms and rising sea levels. The government is trying to move these people out of the hazard areas, but has agreed that they must be moved to new areas nearby from which they can reasonably commute to work (less than 25 kilometers).

that break down plastic into its original chemical feedstocks, which are then reconstituted into new products—this is called upcycling. One method involves heating plastic in an oxygenless reactor at high temperatures, which produces a liquid that resembles virgin plastic. A leader is Eastman Kodak, the famous camera company. Their plant in Tennessee can chemically break down plastic to the molecular level, enabling it to be reformulated into new products—and essentially recycled forever. A similar technology allows polyester to be recycled. Eastman plans to expand its technology to a wide variety of plastics, putting many carbon molecules back into use as part of a closed-loop system that has a lower carbon footprint than producing petroleum-based plastics. PerPETual recycles plastic bottles back into virgin-quality polyester, which ends up in clothing sold by companies like H&M, Adidas, and Zara, and which can be used to make plastic bottles.

The best solutions go beyond recycling: Replace single-use plastic with reusable and refillable containers. Ban plastic bottles for water. Make filtered water dispensers mandatory in stadiums, shopping centers, cities, businesses, and anywhere water is needed or water quality is lacking. Ban plastic single-use food containers. Companies like Searious Business and RePack in Europe and BarePack in Singapore provide reusable packaging for meals, coordinating the recovery of containers and working with online food delivery services to reduce waste. (With the tap of a finger you can refuse single-use plastic when ordering.) Loop is a delivery service that offers more than three hundred items, ranging from ice cream to hand soap to pet food. It brings the items to your doorstep in sturdy, refillable containers packed into a tote bag. When you're done, you place the empty containers back into the bag and call the company for pickup. In Chile, a company called Algramo ("by the gram") has created a vending machine system that dispenses liquid detergent and other cleaning solutions directly into refillable containers.

One of the premier solutions to plastic water bottles is Eau de Paris (Paris Water). In 2008, it took back its water rights from private companies Suez and Veolia. It

instituted the most sophisticated water filtration and puri-fication system in the world and placed more than a thou-sand free water-dispensing locations throughout Paris, some of which even offer sparkling water. Next to areas of concentrated use are vending machines selling reusable water bottles. Paris is determined to become the first plastic-waste-free water system in the world. Deputy Mayor Celia Blauel is in league with five hundred other cities, from Edinburgh to Milan, that want to take back water utilities from for-profit companies and replicate the quality and ubiquity of the Parisian system. After Paris took over from Veolia and Suez, water quality went up and water rates went down. Better water, more available, for less money.

To make recycling truly effective, close the loop and require a hefty deposit on every piece of plastic sold. In Germany, this is being done, and is called *Pfand*. The deposit fee is part of the price of the product. You get it back when you place the bottle in a "reverse vending machine," found in almost all German supermarkets. It scans and weighs the item and then issues you a voucher that you can redeem for cash. Pfand has been a big suc-cess. More than 95 percent of reusable bottles and cans are returned. An informal economy of collectors has devel-oped around the process, mostly composed of people on fixed or lower incomes.

In 2017, Kenya instituted an outright ban on making or selling plastic bags. Before the ban, an estimated 100 million plastic bags were being used per year, and bags clogged waterways and drainage systems and worsened flooding during the rainy season. In 2020, limits on single-use plastic were applied to parks and beaches. While locals have seen fewer plastic bags hanging from trees, and butchers have reported fewer plastic bags inside cows, there are signs that plastic is still being smuggled into the country from neighboring Somalia.

A similar scheme in Norway has reduced plastic bot-tle litter to nearly zero. And the quality of the returned bot-tles is so high that some have been reused more than fifty times. The deposit fee needs to be applied to all plastics, a sum high enough to incentivize people around the world to collect plastic trash, old or new. The recycling rate of plastic bottles in the United States is less than 30 percent. In Nor-way it's 97 percent. One half of the plastic washing up on Norway's shores still comes from Norway, yet half comes from other countries. We will know success when every seashore in the world has returned to normal. ●

(*Left*) SKM, a recycling company in Melbourne, Victoria, Australia, declared bankruptcy; its six major warehouses were full of recyclable materials awaiting processing. The Victoria government and the warehouse owners, Marwood Constructions, did not know how to deal with this material, which is largely unsorted and cannot be sold easily to other materials processors. With no one to process their household recycling, Victoria councils were forced to send thousands of tons of recyclable waste to landfills.

(*Above*) The chub (*Squalius cephalus*) is a freshwater carp found in Spain. Plastic in lakes and rivers can become deadly traps for aquatic life. Here, a broken piece of plastic tubing enwraps the chub, eventually causing deformation, deep wounds, and suffering.

Poverty Industry

"We, the patients of Zanmi Lasante (Partners in Health) in Cange, have a declaration we would like to put before all of you. It is we who are sick; it is therefore we who take the responsibility to declare our suffering, our misery, and our pain, as well as our hope. The right to health is the right to life. Everyone has a right to live. If we were not living in poverty, we would not be in this predicament today. As we scrape for life, we encounter death. We have a message for the big shots in organizations like the World Bank and USAID. We ask you to take consciousness of all that we continually endure. We too are human beings, we too are people. We entreat you to put aside your egotism and selfishness, and to stop wasting critical funds by buying big cars, constructing big buildings, and amassing huge salaries. Also, please stop lying about the poor. Stop accusing us unjustly and propagating erroneous assumptions about our right to health and our unconditional right to life. We are indeed poor, but just because we are poor does not mean we are stupid."

> —Excerpted from the Cange Declaration in 2001, given by Haitian Nerlande Lahens, one of the first impoverished persons in the world to receive antiretroviral therapies for HIV, an initiative instituted by Paul Farmer and Partners in Health that was denounced by the World Health Organization and World Bank as being too expensive and unsustainable.

"My suspicion is that bankers in general are not getting a lot of sex because they spend a lot of time screwing the poor."

> —Paul Farmer's response to Mead Over of the World Bank after he criticized HIV treatment for the poor as emotionally compelling but impractical and unaffordable.

Poverty is an extractive industry. It takes value from people, transfers it to others, and disvalues the producers. The impoverished struggle to gain fairness, whether it be in work, pay, health, education, or housing. They may dwell in makeshift shelters where there is pollution, insufficient sanitation, impure water, and marginal schools—if any. They suffer constant economic stress and lack of healthcare. In rural areas, poverty creates destructive forms of deforestation and desertification. Roads into formerly inaccessible areas built by mining and logging companies open land to farming companies deploying slash-and-burn agriculture, displacing traditional Indige-

nous stewards of the land. In the forests of Asia, Africa, and the Amazon, some species populations are crashing because of the hunting of wild bushmeat. Degeneration of land, water, forests, biodiversity, and human health is a cause of climate change. And climate change is yet another cause of poverty. Turning this vicious circle to a virtuous one is crucial to addressing the climate crisis.

The other harm caused by chronic poverty is the insensitivity of those who do not notice or have "no time to care" about such remote matters. Bryan Stevenson points out that how we treat the poor impoverishes us: "Our humanity depends on everyone's humanity. . . . Our survival is tied to the survival of everyone." This has never been truer than today. Reversing the climate crisis cannot be done by one country, one economic sector, one industry, one culture, or one demographic. There is not going to be a magic technology that will fix it. We cannot wait to see if experts, governments, or corporations figure out how to end the crisis, because they can't by themselves. The crisis, if it could speak, would tell us all that we have forgotten that we truly are a "we," and nothing less than our joint effort is sufficient to reverse decades and centuries of exploiting people and the earth. Climate change and poverty have the same root cause.

The idea of global poverty is a new concept, conceived in 1990, when the World Bank defined poverty as earning less than one dollar per day. It was called the International Poverty Line and has been used as a measure of scarcity and need ever since. Today, threshold income is pegged at $1.90 per day, an increase of ninety cents in thirty years. Going by that measure, the World Bank believes that poverty has decreased from 36 percent of world population to 10 percent by 2015. Living anywhere on $1.90 per day is called destitution, not poverty. To better describe the disparities that exist today, one month of poverty income is equal to one Lululemon bra, the hood ornament on a Mercedes, or two bags of Purina Dog Chow (with real chicken).

When the World Bank reported in 2015 that 700 million people lived in poverty, the UN Food and Agriculture Organization stated that 821 million people did not have enough calories to sustain minimal human activity that year, much less work and earn an income. In fact, 1.9 billion were food insecure, and 2.6 billion suffered from malnutrition. The International Poverty Line is monetary, and does take into account complementary needs, including the lack of public goods—education, nutrition, healthcare, and sanitation. The World Bank now refers to the $1.90 threshold as a measure of *extreme* poverty. If it were two dollars a day, would that be *normal* poverty? This points to the problem when lives are measured by income. Suffering cannot be monetized.

Different nonprofit aid organizations set the global poverty threshold at a more realistic seven to eight dollars

per day, an income level at which a family can achieve basic nutrition and reasonable life expectancy. Using that metric, the number of poor has increased from 3.2 billion in 1981 to 4.2 billion in 2015. In the United States, 62 million people work full-time but cannot live on their income. Nearly one in five Americans are poor, and 72 percent of them are women and children. Fifty-four percent of working African Americans do not receive a living wage.

Although per capita income doubled between 1980 and 2016, global poverty levels escalated by 31 percent. The cause is unambiguous: fewer people gained more income, and many people received less income. During those thirty-six years, only 12 percent of global income growth went to the poorest 50 percent of humanity. The remaining income growth, profits, and capital migrated to the top 40 percent of income earners, with the majority of income growth going to the top one-tenth of 1 percent of humanity. At the current rate of capital distribution, *The World Bank Economic Review* estimates that it will take more than two hundred years of current economic growth to end poverty, which means poverty won't end at all.

To better understand poverty, three questions are helpful. Question number one: Who benefits when someone suffers? This reveals root causes. Question number two: When was the last time you were in a room with an impoverished family or group? This demonstrates cultural gaps and lack of understanding. Question number three was asked by Pulitzer Prize–winning author Marilynne Robinson, and it may be the most important: Is poverty necessary? It is not necessary, but it is reinforced and exploited. Poverty is an industry. In the United States, there is a vast network of for-profit "human service" corporations, funded by state and federal agencies, that form a business empire worth hundreds of billions of dollars.

Financial support that is intended to go to children, the elderly, the disabled, and the underserved is skimmed

---

Solar rooftop panels and microgrids are being proposed for all of India. While Prime Minister Narendra Modi has ambitiously made these pledges, the question remains whether the 750 million Indians living on less than two dollars per day can afford or embrace green energy. Here, villagers tend to cows in Dharnai in Jehanabad, Bihar, India.

and siphoned off. As bizarre and untenable as this may sound, human suffering is a profit-making enterprise. For example, in the United States, companies reclassify foster children in order to claim extra benefits, which they direct back to the foster-care agencies—a practice known as revenue maximization but might better be described as kickbacks. Employees of agencies hunt for federal or state survivor benefits, which they will divert to their employer after earning a commission. The child knows nothing of this. She or he is a foster child with no parental protection or guidance. State governments are in on exploitation too. They redirect federal aid intended to serve the poor into state coffers. Another practice utilized by for-profit "care" enterprises is heavily sedating the young and elderly who reside in juvenile detention facilities and nursing homes in order to reduce staffing and labor costs.

Funding for prisons is an $80 billion industry in the United States. America is a country of mass incarceration—2.3 million are in prison, and 4.4 million are on probation or parole. Nearly half of the prison population is there for writing a bad check, petty theft, or possession of a substance. Private prison corporations lobby for stricter sentencing and longer confinement for minor crimes. In 1997, a thirty-nine-year-old Black man was sentenced to life imprisonment for stealing hedge clippers. Private prison companies have perverse incentives; rehabilitation and improved public safety are not in their best interest. To reify and enforce ongoing poverty, convicted drug use offenders cannot obtain food stamps or gain access to public housing, and they are unable to find employment due to their record.

Private companies have also found ways to turn a profit along borders by incarcerating migrants and refugees who have traveled thousands of miles for their survival. In Europe, companies like Swiss-owned ORS Service AG are contracted by governments to run for-profit detention centers. When migrants flooded into Europe in the past decade, ORS expanded its migrant "reception" services from Switzerland into Austria and Germany, replacing the nonprofits that usually assist migrants and refugees. The retired founder explained that when running refugee camps, "the margins are very low," so to maximize profit, "the key is volume." Some of ORS's camps were so overcrowded that thousands of women, children, and men were forced to sleep out in the open. In 2019, ORS had revenues of almost $150 million, whereas refugees were often not even allowed to work. Addressing European politicians, one Syrian refugee in the Moria camp, in Greece, proposed, "If you want to know the true meaning of fear, hunger, and cold, come and stay here in Moria camp for a month." In the eyes of for-profit service providers, refugees are assets to extract value from. Almost half of these "assets" are children.

Migration is the last resort when facing conflict or when crops fail from erratic weather patterns. The impoverished first cope by eating less, selling their possessions, and even removing their children from school. In a future where increased natural disasters and changing weather make parts of the planet uninhabitable, more will be forcibly displaced; people will have no choice but to leave their homes. Popular media narratives depict most refugees traveling halfway across the globe, but the vast majority of refugees come from poor countries and flee to other poor countries, and more than three-fourths of those who will be forcibly displaced stay within their country's borders.

In Somalia, floods, droughts, and conflicts drive people to leave their homes; more than 2.5 million people were displaced within the country as of 2019. The underlying condition of poverty forces families to see migration as the only option, since they do not have the resources to rebuild after houses are destroyed or enough savings to weather a bad harvest. More than half of the working population is unemployed in the country, so when natural disasters strike, like the 2016 drought that decimated livestock populations and croplands, economic conditions become especially dire. Families flocked to cities in hopes of finding opportunity, but instead, many found limited jobs, infrastructure, or public services. Water prices almost doubled, and families spent all their energy sourcing food, water, and other basic necessities. Many found it impossible to survive on their earnings and lived in constant fear of eviction or further displacement from natural disasters. Living for months or even years under these conditions, many become reliant on international aid, with very little say on the decisions that impact their lives. More than 96 percent of participants in one study said they did not feel they had been consulted about the aid they received and did not have a platform to voice their concerns. Under this model, when the next natural disaster strikes, families will be displaced, they will lose everything, and the cycle will repeat itself.

Globally, poverty alleviation is a multibillion-dollar industry involving governments, companies, celebrities, and charities that can create dependency rather than the means to create prosperity. Celebrity-graced distributions of shiploads or planeloads of surplus corn and wheat reinforce the idea that money and largesse are helping to solve the problem. It is kind, but it hasn't worked. The poor will remain poor until they are asked what they need—and someone listens. Invariably, mothers, daughters, fathers, and sons will point to the lack of fairness, justice, and opportunity in their country. The poor are like everyone else. They need care, time, energy, relationships—they need to be connected to resources in the form of people who do not go away after the photo shoot.

Desmond Tutu once said, "There comes a point where we need to stop pulling people out of the river. We need to go upstream and find out why they're falling in." Philanthropies and governments place significant money into programs to abate poverty. Activists are different. They go upstream to find out why people are being thrown into the river. *Regeneration* enlarges the conversation by making a simple point: The breadth of solutions, techniques, and practices that address the climate crisis absolutely address poverty. Poverty does not want to be "fixed." Poverty wants to fix itself. Those who are economically disadvantaged want to regenerate their well-being, villages, communities, schools, and cultures. They do so with tools, education, and collaboration. The most effective path to reversing warming is to turn to the people who are most impacted, with the greatest need, yet are the least heard—and then listen, support, and empower. The climate crisis will not be addressed unless the bulk of humanity is engaged. Statistically, that means those beset by poverty. *Regeneration* is about creating the conditions for self-organization. The poor know what to do. The more than four billion people who share some level of poverty will engage and act to counter climate change when social justice and climate justice are the same thing—more nutritious food, clean water, resilient and profitable agriculture, restored fisheries, accessible mobility, dignified housing, renewable electricity, free and safe education, and public health.

Reversing centuries of prejudice that have told us that some people are not good enough, smart enough, or deserving enough is a big task. Realizing that the climate crisis cannot be fixed with technology runs counter to what so many believe. Understanding that those with little or no financial resources impact the destiny of those who may have more seems illogical. At this time, all of humanity depends on all of humanity. In a globalized, digitized, hyperconnected world, we have become one system, just as the earth is one system. Humanity's common needs want to be synchronized, harmonized, and recognized. Regeneration creates abundance, not scarcity. It expands what is possible. It enlarges the human prospect. ●

A Rohingya Muslim refugee boy waits with others to receive food aid from a local NGO at the Kutupalong refugee camp in Cox's Bazar, Bangladesh. More than six hundred thousand Rohingya refugees flooded into Bangladesh to flee an offensive by Myanmar's military that the United Nations has called "a textbook example of ethnic cleansing." Refugee populations of Rohingya Muslims made the perilous journey on foot to the border, or paid smugglers to take them across by water in wooden boats. The Rohingya refugees fled to a different kind of suffering found in sprawling makeshift camps rife with fear of malnutrition, cholera, and other diseases. Aid organizations struggled to keep pace with the scale of need and the staggering number of children arriving alone—an estimated 60 percent. The "clearance operations" by Myanmar's army and Buddhist mobs occurred under the leadership of Aung San Suu Kyi, a Nobel Peace Prize laureate.

Offsets to Onsets

"It is not a gain to offset a loss." —Anonymous

If you travel, you are probably familiar with offsets. It is the money you pay to a company to offset the greenhouse gas emissions you are generating with your journey. A round-trip flight from Los Angeles to London might cost you fifty dollars. Seven days on a cruise liner might be eighty dollars. Neutralizing the carbon emissions from the twenty thousand miles you put on your car each year could set you back a hundred dollars. You can pay to offset the carbon footprint from other aspects of your life, such as the emissions from the electricity used in your home or business. But is the money making a difference to the climate crisis? Or is there a better way?

An offset is a promissory note. By paying, you receive a promise that the greenhouse gas emissions you're generating today will be offset by the elimination of an equal amount in the future. The location of the offset could be anywhere in the world, and the time frame could be short or long. Paying to replace an inefficient, carbon-spewing cookstove in a rural village with a clean stove could offset your emissions quickly. Planting trees may take years to achieve a similar reduction. The offset concept dates back to the Clean Air Act of 1970, when the U.S. Congress allowed big polluters to continue polluting in one place if they reduced emissions somewhere else. As concern about climate change rose during the 1990s, carbon polluters began to use renewable energy projects to offset their emissions. Today there are a variety of businesses and organizations that will calculate your carbon footprint and sell you an offset. That round-trip flight to London? It's equivalent to planting 110 seedlings. When will it offset your carbon? Difficult to say; ten to twenty years is a good guess.

Offsets have become popular recently with some of the largest corporations on the planet, including Amazon, Google, Nestlé, Disney, General Motors, Starbucks, and Delta Airlines. Procter & Gamble announced plans to spend $100 million on carbon offsets to neutralize a portion of its annual greenhouse gas emissions. Other purchasers include entertainers, sports organizations, food companies, universities, cities, even entire countries. Offsets have become a key part of overall plans to become carbon neutral—the means to achieve a net-zero balance between greenhouse gas emissions and reductions. Apple, for instance, promises to directly reduce its emissions by 75 percent before 2030 and use offsets to make up the remaining quarter. A steep reduction of emissions isn't a viable option for some businesses, such as airlines, which means they must use higher amounts of offsets to meet carbon-neutral goals. In 2021, an international aviation offset program called CORSIA went into effect, generating $40 billion in transactions over fourteen years, with the goal of offsetting 2.5 billion tons of carbon dioxide.

Most offsets are voluntary, but some can be used to meet mandatory emission reduction targets set by regulatory agencies. For example, a power plant can purchase offsetting "carbon credits" from brokers, allowing them to continue emitting greenhouse gases at levels that exceed state or federal limits. In order to sell a credit, a broker must verify a future reduction of greenhouse emissions. In order not to come up empty or short on promises, carbon trading schemes need to overcome a number of daunting challenges, the following three in particular. (1) Permanence: To be credible, the achieved emissions reduction must last indefinitely. For example, a newly planted forest must not later be logged or lost in a wildfire, releasing its stored carbon. (2) Additionality: The emissions reduction must be in addition to whatever was going to happen anyway. An offset doesn't count if a planned solar farm was going to be built regardless of the transaction. (3) Accounting: The emissions reduction must be carefully measured and monitored to ensure that the promissory note has been fully paid. Speculative deals, imprecise protocols, overpromises, and underdelivery of greenhouse gas reductions, as well as instances of outright fraud, have made meeting these challenges difficult over the years.

Today, certified standards and scientific methodologies for calculating offsets are well established and much more transparent, boosting confidence in carbon trading marketplaces. They help ensure that projects deliver emissions reductions while respecting social and environmental safeguards, especially any that may undermine the rights of marginalized and vulnerable communities. Still, as offsets become more popular, deceptive practices and hollow accomplishments continue to trouble the movement. One example is legacy credits, which are carbon credits purchased from a project, such as a wind farm, built years ago. While the renewable energy produced by a project may be displacing carbon-intensive fossil fuels, there's no carbon additionality. The net emissions reduction happened in the past. Selling legacy credits to a corporation to offset its carbon footprint, thereby allowing it to continue polluting, does nothing for climate change. Unfortunately, as many as 60 percent of credits being purchased involve dubious claims of additionality.

Another challenge involves money. Carbon credits can be lucrative to sellers and brokers, which creates a financial incentive that can override that larger goal of achieving a net reduction in greenhouse gas emissions, especially if the credits allow the buyer to maintain its carbon footprint. For example, carbon credits are sometimes sold from forests using a hypothetical threat of imminent logging. If cut down, the trees would release their stored carbon. However, if the trees are not cut down, then any "credit" sold to a company based on the threat of logging is essentially worthless, even if proper verification protocols were followed. While credits like these look good in the ledger sheets (and press releases) of a carbon-intensive industry, such as the cruise industry, from a climate perspective they are meaningless. The cruise ships sail on, and their emissions continue.

It's becoming a familiar story for offsets in general. Both the promised reductions and the achieved results of many projects are modest at best. The typical corporation offsets less than 2 percent of its total emissions. In some cases, the delivered reductions miss their goals because of natural setbacks, such as a forest fire, or the result of unexpected human interference. In other cases, the promised offset occurs too far in the future to have any meaningful impact on the climate crisis today. While offsets can buy time, they can also be a tactic for delaying deeper reductions. And there is a moral hazard when high-polluting companies transfer their obligations to reduce their emissions to less developed and more vulnerable parts of the world, propagating new injustices rather than resolving the current one. There is also a false equivalency between a pound of carbon dioxide emitted by a luxury activity in an industrialized country and one generated by an essential

Children of families who live in the Isangi Rainforest in the Democratic Republic of Congo. Offsets have stopped a former logging concession in the lowland tropical forests, an area that hosts 11 percent of the world's known bird species.

activity, such as feeding a family, especially in developing nations. While offsets have a role, the bottom line is clear: greenhouse gas emissions must be reduced now, not in the future, near or far. The reductions must be real, substantial, and immediate. We do not have time to waste.

Offsets do have benefits. As investments, they have acted as change agents in rural communities around the world. For instance, the Save80 project, in the small southern African nation of Lesotho, used credits to employ local women to create a program that has distributed ten thousand clean cookstoves to families, alleviating the need to cut trees for fuel and reducing the negative health consequences of inhaling toxic smoke. In Peru, offset money helps Indigenous peoples use drones and satellite data to spot early signs of illegal logging in their forests. In the United States, offset funds have paid for river and wetlands restoration projects that have improved riparian habitat for fish, beaver, and migrating birds. Offsets have supported soil-carbon-building projects in grasslands, including regenerative wool farms in Argentina, savannas in Kenya, and a cattle ranch in Australia employing regenerative grazing. Other projects include land conservation projects among Hadza communities in Tanzania; restoration of degraded forestland in Laos and Brazil; implementation of clean water projects, such as one in Honduras used by sustainable coffee growers; and forest protection in Canada.

The main problem with offsets is the word itself: *offset*. Neutralizing greenhouse gas emissions does almost nothing to chip away at the legacy carbon that has been accumulating in the atmosphere for decades. Virtually all climate scientists believe we have exceeded safe levels of carbon dioxide concentration, which requires significant reductions now. In 2019, total carbon dioxide emissions globally were forty-one gigatons, a rise of one-third since 2000. Offsets that promise to deliver their reductions ten or twenty years in the future are nearly useless in this regard.

We need onsets instead—activities by individuals, companies, and nations that remove more carbon from the atmosphere than they release and store this carbon for as long as possible in natural sinks, such as soils. Instead of simply neutralizing emissions, why not double or triple the reductions in order to work down the accumulation of carbon dioxide in the atmosphere instead? Traditional offset projects could be turned into onsets with additional carbon sequestration activity—measured, monitored, and verified by third parties. Benefits would include more jobs, more food security, and increased resilience to climate extremes.

The Sodo/Humbo Forestry Project is a good example. One of the poorest nations in the world, Ethiopia suffers from widespread land degradation that has crippled its agricultural sector, impacting 90 percent of the country's population. Exploitation has eliminated nearly all of Ethiopia's native forests and caused extensive erosion, reducing the land's capacity to handle increasingly severe cycles of floods and drought. The goal of the Sodo/Humbo project, located in southern Ethiopia, is reforesting the degraded slopes of mountains as part of a long-term restoration strategy. The work is implemented by local community members and uses a methodology developed in Niger called farmer-managed natural regeneration (FMNR), which rapidly regrows trees from existing stumps and rootstock at a fraction of the cost of planting trees grown in a nursery. The carbon sequestration and storage potential of FMNR has been demonstrated to be high. According to Gold Standard, an offset verification nonprofit founded by the World Wildlife Fund and other organizations in 2003, the Sodo/Humbo project will sequester an estimated 1.1 million tons of carbon dioxide. The cost to a buyer? Eighteen dollars per ton.

It's not just carbon. The Sodo/Humbo project has (1) generated two thousand local jobs, (2) restored eight thousand acres of land with indigenous tree species, including several threatened species, (3) created diverse habitats for numerous plants and animals, (4) reduced erosion, improved water infiltration, and increased soil fertility, (5) increased local sources of honey, fruit, and medicinal plants, and (6) improved the well-being of as many as fifty thousand people in the area who depend on the land for a sustainable source of food, fodder, and livelihoods. Additionally, a portion of the money generated is reinvested in local economic development as well as education and health programs.

The practice of offsets is not regulated so as to protect the rights of local Indigenous communities. Free-flowing rivers have been dammed for hydropower with "avoided emissions" carbon credits being sold to other countries or multinational corporations. Offsets credits for "sustainable forestry," a term that has no agreed-upon definition, do not include community rights. The UN wording in this area was weak; parties should "respect, promote and consider their respective obligations on human rights." *Should.* But in 2019, the UN abandoned that wording. In Article 6 of the Paris Agreement, which is supposed to address safeguards for people and the environment, the term *human rights* or *Indigenous* cannot be found. Today, countries that build dams or commodify nonnative, monoculture "forests" can claim carbon credits. At the same time, a company or country that purchases them can claim the same credits, a method of double counting that compounds the insensitivity to Indigenous people, cultures, and places. Offsets can easily become a transaction, by which the emissions of the global North can be "paid" for by appropriating traditional Indigenous lands in the global South to be used as carbon sinks and offsets.

Instead of paying off a promissory note for your carbon debt, an onset pays your debt forward. It makes a payment to another person or community, possibly disadvantaged, for a subsequent good carbon deed. Instead of simply neutralizing the emissions from the twenty thousand miles you put on your car for a hundred dollars, double the amount to two hundred dollars and pay forward the extra money to a verified project that draws down extra greenhouse gas emissions while restoring degraded land and improving the well-being of humans and nature. While it may take a while to see the benefits accrue, the course of action is proactive, not merely neutralizing. If two people paid their debt forward—or four people, or four hundred—measurable reductions in atmospheric carbon dioxide will occur. If a corporation doubled or tripled its purchase of a calculated amount of offsets—making them onsets—it would be paying forward a significant amount of goodness. It's the same principle we use with our children—we invest love and attention in them so they can go forward and do good work in their lives. ●

The Southern Cardamom Forest lies in southwest Cambodia and covers 1.24 million acres of relatively intact tropical forest. Offset payments fund rangers, who confiscate over fifteen hundred chain saws a year from illegal loggers. It is home to more than fifty endangered species, including the Asian forest elephant, clouded leopard, pileated gibbon, Siamese crocodile, and sun bear. Offsets prevent 110 million tons of carbon emissions and support the local communities in tenure registration, scholarship funding for higher education, and ecotourism projects.

Action + Connection

The final section of *Regeneration* is action amplified by connection. It contains offers, possibilities, ideas, some fun, and links to thousands of people and groups in the world who deeply care about the future. To continue the conversation beyond the book, you will find URLs that go to specific areas of the Regeneration website. It is an organized cornucopia of information, ideas, groups, videos, books, and people who are implementing regeneration worldwide and who welcome support and involvement.

The most-asked question about the climate crisis is what to do, where to start, and how to make a difference. When you see or read what climate change is doing to the earth, it is natural to feel overwhelmed, anxious, confused, or very small—I am just one person, or one small family. Kimberly Nicholas, the climate scientist and brilliant author of *Under the Sky We Make*, distills climate science into five facts. "It's warming, It's us. We're sure. It's bad." The fifth fact is that human beings have the ability to end the climate crisis. She recounts that until recently she and her friends didn't talk about climate even though they understood it well. While vociferous climate deniers have had no problem grabbing the nearest megaphone, the overwhelming majority of people, the 90 percent who understand that humans are creating a warmer planet, have remained largely silent. She wants to change that. So do we.

The early predictions of climate science are now being reported daily and experienced personally. Theory became reality, and yet as great as climate science is, it cannot save the day. Nor do we need more science to understand what to do. The majority of the people in the world understand there is a crisis. The bridge to ending the climate crisis is the awakening of that majority to take action.

What to Do

In his book *The Checklist Manifesto*, Atul Gawande outlines how to make decisions that generate effective action for highly complex problems. As a surgeon, Gawande created checklists similar to the ones pilots and copilots use before flying passenger jets. He wanted to reduce and eliminate medical error for doctors performing operations on one of the most complex systems in the world—the human body. No one, including medical doctors, fully understands the human body, but that does not prevent physicians from being effective surgeons. A checklist is developed from prior knowledge, experience, failures, and learning.

The climate crisis is similar. It is an extremely complex system, and there is no one who fully understands it. That can tend to make us believe only experts can solve the crisis. We unintentionally give our power over to technocrats, international leaders, or scientists, and hope they do something and get it right. Inspired by what he found in the building and construction industry, Gawande discovered a direct way to create a more effective system: *You push the power of decision-making out to the periphery and away from the center. You give people the room to adapt, based on their experience and expertise. All you ask is that [people] talk to one another and take responsibility. That is what works.* We are what works. Few of us are experts, but that does not prevent us from understanding what to do and how to do it. Climate checklists can guide our action.

Where to Start

A climate checklist is informed by straightforward principles. They help guide our endeavors, from farms to finance, cities to clothing, groceries to grasslands, and are applicable to every level of activity: people, homes, groups, companies, communities, cities—and countries too. The guidelines are yes or no questions. Every action either moves toward a desired outcome or heads away from it. The number one guideline is the fundamental principle of regeneration. The remaining are outcomes of that principle.

1. Does the action create more life or reduce it?
2. Does it heal the future or steal the future?
3. Does it enhance human well-being or diminish it?
4. Does it prevent disease or profit from it?
5. Does it create livelihoods or eliminate them?
6. Does it restore land or degrade it?
7. Does it increase global warming or decrease it?
8. Does it serve human needs or manufacture human wants?
9. Does it reduce poverty or expand it?
10. Does it promote fundamental human rights or deny them?
11. Does it provide workers with dignity or demean them?
12. In short, is the activity extractive or regenerative?

How you apply, score, or evaluate these principles is up to you. Most of what we do does not tick all the boxes. However, like a compass, it shows us the direction and where to go. By employing the guidelines, you pivot and begin, action by action, bit by bit, step by step to create regeneration in one's life. What am I eating? Why? How am I feeling? What is happening in my community? What am I wearing? What am I buying? What am I making? Etc.

Create a Punch List

A punch list is a personal, group, or institutional checklist. Because of the differences among people, cultures, incomes, and knowledge, there is no one common or correct checklist. The top "ten" solutions to reverse global warming are an abstraction. The true top solutions are what you can, want, and will do. The value of a punch list is that when you commit to something, things can happen. A punch list can be for an individual, family, community, company, or city. It is the list of the actions you or a group will undertake and accomplish over a predetermined span of time—one month, one year, five years, or more. You can make different lists for different time periods—this week and this year, for example. If you go to www.regeneration.org/punchlist you will find a kit, a worksheet, and more sample punch lists. In these two sample cases, the reduction emissions exceed 50 percent. You can compare yours with lists created by others, including some people you may know of. Our staff punch lists are there too. And if you would like to estimate the current carbon impact of your family, company, or building, you can go to www.regeneration.org/carbon.

A sample punch list, by a homeowner.

1. Install heat pumps and stop all fossil fuel use in my home—cooking, heating, and hot water.
2. Get an induction cooktop to replace a gas range.
3. Convert to an electricity source that is entirely renewable.
4. Set an annual clothing budget of seven durable garments per year.
5. Create a composting system in the backyard.
6. Reduce plane flights by 90 percent and purchase 5x carbon offsets for flights taken.
7. Gather and donate all unnecessary belongings and provide for the needs of others.

A punch list from a small food company.

1. Build supply chain transparency—research product sources and their environmental impact.
2. Locate regenerative/organic sources for vegetables, seeds, and grain.
3. Recruit and train workers from underserved communities.
4. Replace electricity for offices, warehouse and production with renewables.
5. Remove natural gas and install induction heating for tanks and vats.
6. Advance nutritional literacy for local schools and create health conscious cafeterias.
7. Specify recycled paperboard packaging and set timetable for eliminating plastic.

Climate Action Systems—Working Together

Human beings are social creatures. We like to work on problems, make commitments, and learn in groups. It is satisfying and more effective. Addressing the climate crisis involves our family, friends, community, workers, and others. One tool that helps facilitate collaboration is Climate Action Systems. It is a downloadable learning pod for solving climate issues. It can move around the world, endlessly self-propagate, go where it is wanted, and set up shop. You can invite however many people you wish and form a group learning pod that gets smarter the more it is used. It also:

1. Generates networks that can help solve climate issues.
2. Seeds, replicates, and propagates best actions and solutions, which become available to all.
3. Differentiates and modifies knowledge according to place, people, and culture.
4. Continuously analyzes outcomes and produces evolving insights.
5. Accelerates the flow of actions using conversational software.
6. Enables cross-sector collaboration between people, neighborhoods, or organizations.

Climate Action Systems was created by Rosamund Zander, Ilan Rozenblat and Harry Lasker and can be explored at our website www.regeneration.org/CAS.

Enlarging Our Focus—Nexus

Nexus are large, complex issues that intersect multiple institutions, geographies, cultures, and people, but which do not fall under a single category of action or impact. Plastic, global fishing fleets, and palm oil are three prime examples. Rather than feel powerless in the face of these enormous and/or distant threats, on the website we have an ever-growing "wiki" of climate actions for the most important Nexus, some of which are covered in *Regeneration* and some that are not, like Digital Consumption (the amount of energy we consume in our digital life). Here is a summary example of one, the Boreal Forest:

> The Boreal wraps around seven countries and comprises 4.2 billion acres. To save the largest intact forest system on earth requires education, activity, and initiatives coming from people and institutions around the world, especially Russia, Scandinavia, the United States, and Canada. This includes activism, political influence, economic pressure, consumer education, boycotts, and support for its native residents, primarily Indigenous and First Nation People. Boreal ecosystems are being eaten away and torn apart by mining companies, oil companies, and logging and paper companies. There are three companies that make plush toilet paper from boreal timber: Procter & Gamble, Kimberly-Clark, and Georgia-Pacific, corporate partners in the "tree-to-toilet" pipeline. Avoid purchasing these products. Buy recycled or bamboo toilet paper instead. Write or email the heads of those companies. The Natural Resources Defense Council report *The Issue with Tissue* is an excellent analysis. The Pew International Boreal Conservation Campaign, the Boreal Leadership Council, and the Boreal Songbird Initiative, which protects boreal habitat for billions of migratory birds, are very effective organizations. Put pressure on *The New York Times* and *The Washington Post* and inquire why they are buying their newsprint from Resolute Forest Products when they are doing such excellent climate reporting. Resolute filed intimidating lawsuits it could not win (known as SLAPP suits, strategic lawsuits against public participation) against Greenpeace and other activists who opposed their boreal forestry practices. Resolute lost in court and had to reimburse defendants $816,000 for intentionally filing an unwinnable lawsuit. We have asked our publisher, Penguin, not to purchase from Resolute. The paper in *Regeneration* is 100 percent postconsumer waste, but the matter stands. These are just some of the issues that impact the integrity of the boreal forest. There is also the tar sands, mining, mountaintop clearing for coal, habitat destruction, and illegal incursions into Native lands.

For each Nexus category, we include:
1. Clear descriptions of the issues, history, players, and impacts.
2. The specific parties actively causing degradation and damage.
3. The NGOs, activists, affected populations, and other institutions that are addressing the issue.
4. Addresses and emails of CEOs, politicians, or other people who are key decision-makers.
5. Products and companies to lobby, avoid, or support.
6. Links to videos, conferences, documentaries, articles, and papers.

All of this can be found at www.regeneration.org/nexus, and we welcome your help and participation to improve, add, and update.

Included on the website are:

Tropical Forests	Indigenous Rights/Culture	Banking and Finance
Palm Oil	Global Fishing Fleets	War Industry
Pollinator Extinction	Mangroves	Politics Industry
Wetlands	Tidal Salt Marshes	Clothing Industry
Beavers	Food Waste	Plastics Industry
Bioregions	Big Food	Digital Consumption
Grasslands	Regenerative Agriculture	Degraded Land Restoration
Marine Protected Areas	Microgrids	The Amazon Forest

The Goal

The solutions outlined here cut emissions, protect and restore ecosystems, address equity, and create life. One might call it a regeneration revolution. If implemented rapidly and globally, the initiatives outlined can avoid and sequester over 1,600 gigatons of carbon dioxide–equivalent emissions by 2050, which will meet both the 2030 and 2050 targets of the IPCC. Is this ambitious? Absolutely. Is it possible? Absolutely.

The solutions summarized in the table represent well-researched analytics by a team of scholars and researchers at Regeneration.org. Our research links to analyses from around the world. It shows we can cut energy emissions by half by 2028, and if we transform land use emissions, particularly agriculture and forests, it can become a net sink instead of net source of emissions by 2027. Together, these actions would meet what is considered necessary to stay below 1.5°C of warming. We focus on the two crucial areas, energy and nature. There are over four hundred different climate scenarios describing how to limit warming to 1.5°C, created by some of the top universities, institutions, and scientists in the world. This is brilliant and encouraging. It demonstrates how focused the world has become on the climate crisis. A significant number of the forecasts rely on nascent technologies that capture carbon dioxide from the air, liquefy it, and pump it deep underground. As much as one might wish there was such a third way, one in which 30 million carbon removal machines operating 24/7 until 2100 are installed across the world, we believe that hope is unrealistic, a type of thinking promoted primarily by fossil-fuel companies. Some treat the climate crisis as one would diagnose a patient with symptoms. We address the crisis as a system the needs healing. The way to heal a system is to connect more of it to itself. It could be said that everything in *Regeneration* is ultimately about recreating those connections and repairing the bonds that have been broken or sundered.

Our analysis focuses on what is possible today. The scenario emphasizes addressing current human needs. It makes serious amends to Indigenous People. It assumes that people will create food webs and "agrihoods" that diversify and localize food production. It means Big Food can no longer measure its success by the quantity of ultraprocessed foods it sells, but by how much renewable food it creates, food that restores people's health and regenerates the soil. It calls for protecting and setting aside 30 percent of our land and oceans within a decade. And it requires that the top 10 percent of income earners in the world (those making more than $38,000 per year), who are responsible for over half of the world's greenhouse gas emissions, appreciate that their ultimate well-being is inseparable from the well-being of all people and thus change their demands upon the planet.

There are some solutions that we chose not to calculate because we believe quantification inappropriately reduces complex topics to a deceptive or reductionist single metric. Ensuring universal access to education and healthcare for women has been tied to reducing population growth and thus to climate. Even though outcomes of greater access to education do lower birth rates, we believe education is a basic human right. In addition, people who have low carbon emissions due to poverty have the right to increase their carbon emissions in pursuit of a higher quality of life. Similarly, it is frequently noted that lands managed by Indigenous

Peoples hold disproportionately high levels of biodiversity, and that protecting Indigenous forest tenure is an effective way to protect terrestrial carbon stocks. That is true. However, returning stolen land to Indigenous Peoples is the only moral, effective, and correct action.

Avoided emissions in carbon dioxide equivalence (in gigatons)	2030	2040	2050
Agroforestry	5	16	26
Azolla Fern	1.9	3.6	5.4
Biochar	6	17	28
Boreal Forests	2	6.1	10
Buildings	28	89	167
Carbon Architecture	5.7	11	16
Clean Cookstoves	1.8	5.5	9.2
Compost	0.2	0.7	1.4
Eating Everything	9.9	40	93
Geothermal	4.2	16	35
Grasslands & Grazing	1.9	5.8	9.7
Industry	34	102	191
Mangroves	3.7	11	18
Mobility and EVs	38	122	226
Peatlands	7.8	24	39
Regenerative Agriculture	12	35	56
Seaforestation	3.2	14	31
Seagrasses	1.7	5.1	8.5
Solar	20	73	141
Temperate Forest Management	3	9.1	15
Temperate Forest Restoration	11	32	53
Tidal Salt Marshes	0.4	1.1	2
Tropical Forest Management	4.9	15	25
Tropical Forest Protection	18	54	90
Tropical Forest Restoration	40	120	201
Wasting Nothing	5.2	19	39
Wind	11	41	77
Total Emissions Avoided & Sequestered	**280.5**	**888**	**1613.2**

*Some solutions included in *Regeneration* are not included because they have no readily quantifiable impact on the future temperature of our planet, or because there is insufficient data, or because they are encompassed by other solutions. Read more about our methodology at www.regeneration.org/methodology.

Protect

Perhaps the most overlooked solution to global warming is safeguarding the carbon stocks on earth. We currently emit carbon by combusting ancient carbon stocks—coal, gas, and oil—but we also emit by destroying current carbon stocks in terrestrial ecosystems. Damage, degradation, or elimination of ecosystems result in the release of carbon dioxide and methane gases. The global estimates of organic carbon stocks, particularly soil carbon, are not precise. The calculations of total carbon stocks provided in the table—3,300 gigatons—is on the higher side of older estimates from the late 1990s and early 2000s. As of this writing, new maps of global soil carbon down to one meter depth are about to be released and we believe will present significant strides going forward in our understanding of terrestrial carbon.

Organic carbon stocks (in gigatons)	Soil	Biomass	Total
Boreal Forests	1,086	54	1,140
Deserts and Xeric Shrub	68	10	78
Grasslands	392	77	469
Mangroves	5	1	6
Mediterranean	26	6	32
Seagrasses	3	0	4
Temperate Forests	375	72	447
Tidal Salt Marshes	1	0	1
Tropical Forests	407	181	589
Tundra	527	8	535
Total Carbon Stocks	2,890	409	3,301

One More Thing

It's not your job to save the planet. The idea of saving the earth is a heavy burden and you can't do it anyway. Another belief that torques the mind is that carbon is bad. There is no such thing as carbon pollution. It is part and parcel of virtually everything we need, make, and touch, everything that is alive, delicious, astonishing, and sacred. We have placed extraordinary amounts of carbon into the atmosphere, and we know exactly how we did it. Today, we know how to bring it back home to bring the planet into balance. The earth is forgiving about what that balance should be. It is approximately the average level of atmospheric carbon dioxide seen for the past eight hundred thousand years. The carbon we bring home is the food needed to regenerate life on earth. When we feed the earth, we heal the climate. Regeneration is the default mode of life. You are able to read this sentence because the 30 trillion cells in your body are regenerating every nanosecond. We can kill, poison, burn, or quell life on earth, but when that ceases, regeneration begins. Now is the time to bring our life, practices, products, cities, agriculture, and all else into alignment with the living world and end the climate crisis. We cannot do this if we believe or assume others will do it for us. We have a common interest and that interest is served when we come together and join all forms and forces of life. Welcome to Regeneration.

—Paul Hawken

Afterword
Damon Gameau

My first glimpse at the cultural-defining power of a well told story took place in an Asian studies class in high school. During one particular afternoon, we were learning about the Austronesian expansion that took place around 3,000 BCE. These were the people who left the Asian mainland and headed for the vast island networks to the south. They set out on their adventures in the first ever catamarans and outrigger boats, covered in tattoos and carrying statues carved from jade. At the time, many of the islands they reached were thriving with dense forest cover and rich ecosystems that included giant turtles, birds, and fish. But the new arrivals brought with them a mindset fit for a large expansive mainland. By overfishing and over-hunting, plus felling excess trees for fuel and cropland, the explorers quickly knocked their new ecosystems out of balance. Many of these nascent societies collapsed, and some islands were completely abandoned. Many centuries later, as a new wave of catamarans and outriggers arrived, people came to understand that their survival depended on deep respect for the local ecology. They replaced domination with integration, treated nature as a gift that couldn't be abused, and saw themselves as a keystone species that would live in alignment with the land and sea to create ever more abundance.

What was imprinted on my adolescent brain was how these people understood the importance of embedding this wisdom into their culture. They intentionally created a way of life that would shape the behavior of future generations. And they did this by telling new stories and cultivating myths and metaphors. To this day, many of these islands have sustained their ecosystems.

For much of our existence, humans held animist beliefs in some form, a belief that a current of life runs through and connects all things, including trees, animals and rocks. Even today, the Achuar people living on the border of Peru and Ecuador don't even have a word for nature. They don't think it exists. They see no separation between themselves and their surroundings. With the spread of Christianity and the Scientific Revolution in the late sixteenth century, these animist beliefs were largely eradicated and a new story of nature was written, a story which saw humans as separate and superior to the living world. Francis Bacon, the father of modern science, said the researcher must "hound nature in her wanderings" in order to "find a way at length into her inner chambers." Sadly, this is a story that still pervades our culture today. The dominant narratives in our society, the myths and metaphors that are being cultivated are not being told by wise elders or experienced adventurers, but by advertising agencies on behalf of monolithic corporate entities. Our information ecology has become utterly polluted and this is having deleterious impacts on our health and the health of the planet. If we are to prevent the collapse of our own island, our beautiful island planet in the galaxy, then we need to tell better stories, to embed wisdom into those stories, and to cultivate respect and reverence for nature once again.

Our storytelling today falls predominantly into two categories. The first is analogous to a magician's trick. Our emotions are hijacked by mainstream media. Narratives arrive at the click of a finger, stories that seduce, blinding us to the fact that we inhabit a world that is dying.

The second category of storytelling is created with the best of intentions. For the past few decades, we have used alarming narratives, countless films and books, many with precise poetic detail, that purvey the destruction of the natural world. But at what cost? Neurological research shows that constant viewing of information laced with fear and anxiety can cause paralysis in people, shutting down the part of our brain that is crucial for problem solving and creative thinking.

What world could we build together if we incentivized and funded better stories, meaningful stories about our interconnected relationship with the living world? Or if we shared empowering narratives about individuals and communities that are restoring entire ecosystems, accounts, and tales of regeneration. For too long now, we've assaulted ourselves with a barrage of graphs, data, jargon, and lifeless statistics about life on earth. A new approach is required, a time-honored approach that involves telling better stories aimed directly at the heart. The role of the storyteller has never been more important. The real purpose of the artist, the poet, the songwriter, the author, or the filmmaker is to create and shape culture. That culture then determines what blooms or withers, what thrives or dies. In this moment, we need a revived culture with a radical empathy for the living world. And if our storytellers can't find a way, then the way cannot be found. Please tell those stories.

Damon Gameau is an artist, activist, and storyteller and the creator and director of the acclaimed film *2040: Join the Regeneration.*

ACKNOWLEDGMENTS

Damon Gameau, John Elkington, Giselle Bundchen, Laurene Powell Jobs, Bill and Lynne Twist, Cristina Mittermeier, Cyril Kormos, Rosamund Zander, Kasey Crown, Lou Buglioli, Natalie Orfalea, Tatiana Tilley, Mort Meyerson, Michael Dell, Brian von Herzen, Durita Holm, James Bullock, Julie Mills, Stefano Boeri, Stephen Roberts, Harry Lasker, Ilan Rozenblat, Rola Khoury, Janet Scotland, Lucrezia De Marco, Visra Vichit-Vadakan, Daniel Uyemura, Gillian Gutierrez, Raine Manley, Carla Yuen, Irene Polnyi, Elsie Iwase, Andrew Kessler, David Perry, Jennifer Betka, Michelle Best, Amy Low, Sarah Ezzy, Megan Dino, Patrick D'Arcy, Marybeth Carty, Danielle Nierenberg, Ryland Engelhart, AY Young, Susan Olesek, Margaret Atwood, Saul Griffith, Mary Reynolds, Anne Marie Burgoyne, Catherine Chien, David Festa, Suzanne Burrows, Michl Binderbauer, Leisl Copland, Mark Hyman, Rich Roll, John Cumming, David Cumming, Bryan Meehan, Charlie Burrell, Dave Chapman, Charles Massy, Sophie Pinchetti, Natalie Steinhauer, Mimi Casteel, Mitch Anderson, Kristina Fazzalaro, Chip Conley, Alejandro Foung, Jane Cavolina, Stephen Mitchell, Byron Katie, Emily Mansfield, Josephine Greywoode, Geoff von Maltzahn, Pedro Diniz, Katherine Mills, George Steinmetz, Ami Vitale, Chris Jordan, Jeff Jungsten, Karen Bearman, Kathryn Marshall, Sven Jense, Roy Straver, Tomislav Hengl, Linadria Porter, Bren Smith, Julianne Skai Arbor, Ann Chesterman, Anna Kaplan, Monica Noon, Brad Ack, Allie Goldstein, Spencer Scott, Alison Nill, Morgan Kelly, Meighan Visco, Bailey Farren, Elina Bell, Dominic Molinari, Noorie Rajvanshi, and Rebecca Adamson. With kudos to Anthony James, Daniel Christian Wahl, Philipp Kauffmann, Jonathan Rose, John Fullerton, Oren Lyons, Xiye Bastida, Yvon Chouinard, Pamela Mang, Bob Rodale, Ben Haggard, Bill Reed, Tim Murphy, Vandana Shiva, John D. Liu, Precious Phiri, Larry Kopald, Tom Newmark, Nathan Phillips, Rebecca and Josh Tickell, Thekla Teunis, Colin Seis, Geoff Bastyan, Jeff Pow, Tom Goldtooth, Xiuhtezcatl Martinez, Kris Nichols, Michelle McManus, Madonna Thunder Hawk, Nicole Masters, Dianne and Ian Haggerty, Terry McCosker, Tony Rinaudo, and Anne Poelina.

PHOTOGRAPHY CREDITS

GEORGE STEINMETZ: p. 22 (George Steinmetz), p. 48 (George Steinmetz), pp. 72–73 (George Steinmetz), p. 180 (George Steinmetz), p. 186 (George Steinmetz), pp. 194–195 (George Steinmetz), p. 201 (George Steinmetz), p. 216 (George Steinmetz), p. 218 (George Steinmetz), p. 236 (George Steinmetz)

AMI VITALE: p. 11 (Ami Vitale), p. 90 (Ami Vitale), p. 120 (Ami Vitale), p. 136 (Ami Vitale)

NATURE PICTURE LIBRARY: p. 5 (Guy Edwardes), p. 9 (Sven Zacek), p. 13 (Felis Images), pp. 18–19 (Doug Perrine), p. 25 (Alex Mustard), pp. 26–27 (Tim Laman), p. 33 (Claudio Contreras), p. 38 (Danny Green), p. 40 (David Allemand), p. 41 (Ashley Cooper), p. 46 (Nick Garbutt), p. 49 (Tim Laman), p. 60 (Jack Dykinga), p. 64 (Sumio Harada), p. 66 (Alfo), p. 67 (Klein & Hubert), p. 79 (Klein and Hubert), pp. 80–81 (Heather Angel), p. 82 (Wim van den Heever), p. 83 (Robert Thompson), p. 84 (Mark Hamblin), p. 108 (Gerrit Vyn), p. 114 (Bence Mate), p. 116 (Pete Oxford), p. 116 (Eric Baccega), p. 116 (Eric Baccega), p. 116 (Pete Oxford), p. 116 (Eric Baccega), p. 116 (Enrique López-Tapia), p. 117 (Bernard Castelein), p. 117 (Eric Baccega), p. 117 (Pete Oxford), p. 117 (Pete Oxford), p. 117 (Laurent Geslin), p. 117 (Pete Oxford), p. 147 (Tony Heald), p. 184 (Oliver Wright), p. 204 (Paul D. Stewart), p. 205 (Guy Edwardes), p. 223 (Ashley Cooper), p. 224 (Ashley Cooper)

GETTY IMAGES: p. 110 (Ricky Carioti/The Washington Post), p. 112 (Jeff Hutchins), p. 146 (Carl de Souza), p. 148 (Hufton+Crow/View Pictures/Universal Images Group), p. 162 (Stefano Montesi), p. 172 (CgWink), p. 199 (Roslan Rahman), p. 214 (Zhang Zhaojiu), p. 220 (Kate Geraghty/The Sydney Morning Herald), p. 226 (Stocktrek), p. 228 (STR/AFP), p. 229 (Lalage Snow), p. 231 (Ester Meerman), p. 232 (Burak Akbulut/Anadolu Agency), p. 233 (Storyplus), p. 238 (Jason South/The Age), pp. 240–241 (Prashanth Vishwanathan/Bloomberg), p. 242 (Kevin Frayer/Stringer)

ALAMY STOCK PHOTO: p. 44 (Jacob Lund), p. 104 (Derek Yamashita), pp. 106–107 (Farmlore Films), p. 119 (Romie Miller), p. 144 (Joanna B. Pinneo), p. 150 (Robert Harding), p. 160 (Mauritius Images), p. 164 (Greg Balfour Evans), p. 167 (DPA Picture Alliance), p. 169 (dpa Picture Alliance), p. 174 (Westend61), p. 174 (Ian Shaw), p. 174 (Buiten-Beeld), p. 174 (Phloen), p. 209 (Radu Sebastian), p. 239 (Paulo Oliveira)

NATIONAL GEOGRAPHIC: pp. 36–37 (Mike Nichols), pp. 42–43 (Frans Lanting), p. 62 (Peter R. Houlihan), pp. 68–69 (Alex Saberi), p. 92 (Klaus Nigge), pp. 94–95 (Erlend Haarberg), p. 170 (Jim Richardson)

OTHER: p. 6 (Stuart Clarke), p. 7 (Fernando Tumo), pp. 14–15 (Chris Jordan), p. 17 (Ines Álverez Fdez), p. 21 (Chris Newbert), pp. 28–29 (Neils Kooyman), p. 30 (Jay Fleming), p. 34 (Chris Jordan), p. 45. (Greenfleet/E O'Connor), p. 47 (NASA), p. 50 (Ute EisenLohr), p. 52 (Ute EisenLohr), p. 54 (Kilili Yuyan), p. 56 (Nathaniel Merz), p. 58 (Julianne Skai Arbor), pp. 70–71 (Louise Johns), pp. 74–75 (Charlie Burrell), p. 77 (Charlie Burrell), pp. 86–87 (Jillian), pp. 88–89 (Chris), p. 96 (NCRS Photo), p. 98 (Catherine Ulitsky), p. 99 (Frances Benjamin Johnson), p. 100 (Kim Wade), p. 102 (Russell Ord), p. 113 (Theo Schoo), p. 118 (Kilili Yuyan), p. 122 (Jerónimo Zúñiga), p. 123 (Mitch Anderson), p. 124 (Jonathan Nguyen), p. 127 (Lubos Chlubny), p. 128 (Philipp Kauffmann), p. 131 (Soul Fire Farm), p. 133 (Soul Fire Farm), p. 134 (Relief International Gyapa™ Project), p. 139 (Mary Reynolds), p. 140 (Mary Reynolds), p. 141 (Mary Reynolds), p. 143 (Mimi Casteel), p. 152 (Jason McLennan), p. 155 (Ronald Tilleman), pp. 156–157 (Michelle and Chris Gerard), p. 159 (Stefano Boeri), p. 166 (Michael Baumgartner), p. 176 (Dave Chapman), p. 177 (The Ron Finley Project), p. 178 (Dave Chapman), p. 179 (Eugene Cash), p. 182 (Courtesy of Pixy), p. 188 (Olga Kravchuk), pp. 190–191 (Rasica), p. 192 (Rainee Colacurcio), p. 196 (Ted Finch), p. 198 (Jaime Stilling), p. 202 (Courtesy of Raoul Cooijmans/Lightyear), p. 207 (Tzu Chen Photography), p. 210 (Jamie Stilling), p. 212 (Photonworks), p. 222 (Ton Keone), p. 234 (Courtesy of Suay), p. 244 (Joseph Wasilewski), p. 247 (Andrea Pistoles)